Illustrations of the Logic of Science

ILLUSTRATIONS OF THE LOGIC OF SCIENCE

CHARLES S. PEIRCE

Edited by
Cornelis de Waal

OPEN COURT
Chicago, Illinois

To order books from Open Court, call toll-free 1-800-815-2280, or visit our website at www.opencourtbooks.com.

Front cover illustration: "Corner of the laboratory of Pierre and Marie Curie," photo by photos.com, ©Getty Images.

Open Court Publishing Company is a division of Carus Publishing Company, dba ePals Media.

First printing 2014

Printed and bound in the United States of America.

Library of Congress Cataloging-in-Publication Data

Peirce, Charles S. (Charles Sanders), 1839–1914.

Illustrations of the logic of science / Charles S. Peirce ; edited by Cornelis de Waal.

pages cm

Includes bibliographical references and index.

ISBN 978-0-8126-9849-7 (trade paper : alk. paper) 1. Pragmatism 2. Science—Philosophy. I. De Waal, Cornelis, editor of compilation. II. Title.

B945.P43I45 2014

191—dc23

2013048870

Table of Contents

Preface

Work on this Open Court edition of the *Illustrations of the Logic of Science* began well over a century ago, on January 23, 1907. On that day Paul Carus wrote Peirce of his desire to publish in book form the six articles that had appeared in the *Popular Science Monthly* almost a quarter century before. In response, Peirce spent much time and effort trying to revise the articles and align them with how his thought had developed over time, but in the end he simply had to give up—he needed, or wanted, to change too much and no longer felt up to the task. The current edition contains a selection of these revisions, made in 1909 and 1910, as well as earlier revisions, made mostly when Peirce was working on his 1894 logic book *How to Reason*, which remained unpublished. The result is quite far removed from what Peirce, or Carus, imagined the republication of the *Illustrations* to be. It is not a polished "revised edition" of the original articles, but rather a scholarly work that allows the reader to connect some of Peirce's canonical papers with his later thoughts about the matter, without the latter having reached a publication-ready form.

While working on this edition I greatly benefited from the abundant resources at the Institute for American Thought at IUPUI, which houses the Peirce Edition Project, and from my own involvement as Associate Editor of the *Writings of Charles S. Peirce: A Chronological Edition.* I especially owe thanks to the General Editor at the time, Nathan Houser, for his continuous support and inspiration, and to my wife Kelly (Tully-Needler) de Waal, who was Assistant Textual Editor for the *Writings*, for many long discussions about editing, transcribing the *Popular Science Monthly* articles, proofreading, and for so much more. In addition, I want to thank my graduate research assistant Stephanie Harris for some additional transcriptions and for proofreading as well. This edition would not have been possible without the great support I received from the staff of the Special Collections Research Center at the Morris Library, Southern Illinois University, Carbondale, which houses the Open Court papers and correspondence, and from Harvard's Houghton Library, which houses the Peirce papers and correspondence. I want to thank both institutions also for giving their permission to use transcriptions of their manuscript holdings for this edition. I'm further a most grateful recipient of an Alwin C. Carus Research Grant and an IUPUI School of Liberal Arts manuscript preparation grant. Last but not least, a special, posthumous thanks goes to Max H. Fisch, who left behind thousands of small notecards that together form an easily accessible, immensely detailed, and truly impressive account of the life and work of Charles Sanders Peirce. And like much of Peirce's best work, this work too remained unfinished and unpublished.

Proposed Publisher's Preface (1913)

Charles Sanders Peirce's six papers of Illustrations of the Logic of Science first appeared in the *Popular Science Monthly* for 1877 and 1878 (vols. 12 and 13). They stand outside his chief contributions to logic—which consist principally of papers on what he called "the logic of relatives"—and are not mentioned by Schröder in his bibliography of Peirce's logical works.[1]

For some, the chief interest of these *Illustrations* will lie in their heralding the beginnings of Pragmatism. The term "Pragmatism" was "first introduced into philosophy," says William James,[2] "by Mr. Charles Peirce in 1878. In an article entitled "How to Make Our Ideas Clear," in the *Popular Science Monthly* for January of that year, Mr. Peirce, after pointing out that our beliefs are really rules for action, said that, to develop a thought's meaning, we need only determine what conduct it is fitted to produce: that conduct is for us its sole significance. And the tangible fact at the root of all our thought-distinctions, however subtle, is that there is no one of them so fine as to consist in anything but a possible difference of practice. To attain perfect clearness in our thoughts of an object, then, we need only consider what conceivable effects of a practical kind the object may involve—what sensations we are to expect from it, and what reactions we must prepare. Our conception of these objects, whether immediately or remote, is then for us the whole of our conception of the object, so far as that concept has positive significance at all." However, Peirce's statement that our beliefs are really rules for action "is," as Dr. Carus[3] has remarked, and as we can see from the second chapter below, "an explanation, not a principle, and the explanation is made so that we may rightly understand the nature of belief. Beliefs are never held at random; they serve a purpose and the purpose of a belief is ultimately to ensure a definite line of conduct. It is not probable that anyone would take exception to this. Professor James, however, goes beyond the original meaning of the term by changing the statement of fact into a principle, and he applies it to his conception of truth." For a criticism of pragmatism, we may refer to Mr. Bertrand Russell's *Philosophical Essays*[4] and the work of Dr. Carus already cited. But a criticism of pragmatism is rather a criticism of the doctrine held by William James and, under the name of "humanism," by Dr. F.C.S. Schiller, rather than of that held by C.S. Peirce. And it will surprise many to see that the term "pragmatism" was not used in Peirce's paper of 1878 which is reprinted in the second chapter below. Yet Peirce[5] has himself informed us that in the paper of 1878 he had set in motion the subject although not the word of pragmatism. He had only used this expression in oral conversation until James, who was not acquainted with him when he wrote *The Will to Believe* of 1897, had appropriated it and put his stamp upon it as a philosophical term.[6]

Peirce's own logical work is hardly touched upon in these essays, and his enduring work—which ranks next to De Morgan—on the logic of relations is not touched upon at all. And yet anyone who is even slightly acquainted with Peirce's later logical work will recognize in these essays the very characteristic thought and mode of expressing it of our author. The sixth chapter, in particular, contains much that is purely logical—such as a discussion of syllogisms; but, in the main, it is probability and induction to which these chapters on the logic of science are devoted.

Paul Carus
1913

NOTES

1. *Vorlesungen* über *die Algebra der Logik (exakte Logik)*, Vol. I (Leipzig, 1980), 710–11; Vol. II, Part II (Leipzig, 1905), 603.

2. *Pragmatism: A New Name for Some Old Ways of Thinking* (New York, 1907), 46.

3. *Truth on Trial* (Chicago, 1911), 5.

4. London, 1910, articles "Pragmatism" and "William James's Conception of Truth," 87–149.

5. "What Pragmatism Is," *Monist* (April 1905); Carus, *Truth on Trial*, 116.

6. *[It is unclear what made Carus think that James did not know Peirce before 1897. For a more accurate account, see the introduction and chapter 7.]*

Abbreviations

CN volume:page. *Charles Sanders Peirce: Contributions to* The Nation. 4 vols. Edited by Kenneth L. Ketner and James E. Cook. Lubbock: Institute for the Studies of Pragmaticism, 1975–87.

CP volume.paragraph. *The Collected Papers of Charles Sanders Peirce.* 8 vols. Vols. 1–6, edited by Charles Hartshorne and Paul Weiss. Vols. 7–8, edited by Arthur W. Burks. Cambridge, Mass.: Harvard University Press, 1931–58.

EP volume:page. *The Essential Peirce: Selected Philosophical Writings.* 2 vols. Vol. 1, edited by Nathan Houser and Christian Kloesel. Vol. 2, edited by The Peirce Edition Project. Bloomington: Indiana University Press, 1992–98.

HPPLS volume:page. *Historical Perspectives on Peirce's Logic of Science.* 2 vols. Edited by Carolyn Eisele. The Hague: Mouton, 1985.

NEM volume:page. *The New Elements of Mathematics.* 4 vols. in 5. Edited by Carolyn Eisele. The Hague: Mouton, 1976.

P followed by a number assigned by Kenneth L. Ketner et al. refers to a publication by Peirce as listed in Kenneth L. Ketner et al., *A Comprehensive Bibliography of the Published Works of Charles Sanders Peirce: with a Bibliography of Secondary Studies.* 3rd rev. edition (Lubbock: Institute for the Studies of Pragmaticism, 2012).

PM *Philosophy of Mathematics: Selected Writings.* Edited by Matthew Moore. Bloomington: Indiana University Press, 2010.

R followed by Robin catalogue and sheet number. Manuscripts held in the Houghton Library of Harvard University, as identified by Richard Robin, *Annotated Catalogue of the Papers of Charles S. Peirce.* Amherst: University of Massachusetts Press, 1967, and in Richard Robin, "The Peirce papers: a supplementary catalogue," *Transactions of the C. S. Peirce Society* 7 (1971): 37–57. The sheet number, assigned by the Institute for Studies in Pragmaticism, provides each page with a unique identifier. With some exceptions, the numbering follows the order of the manuscript on the microfilm, including the identification sheet that precedes it (causing a one-page discrepancy where the manuscript pages themselves are numbered). The full Houghton Library call number is: MS Am 1632 (R), where R stands for the Robin catalogue number.

RLT *Reasoning and the Logic of Things: The Cambridge Conferences Lectures of 1898*. Edited by Kenneth L. Ketner. Cambridge, Mass.: Harvard University Press, 1992.

SS *Semiotic and Significs: The Correspondence between Charles S. Peirce and Victoria Lady Welby*. Edited by Charles S. Hardwick. Bloomington: Indiana University Press, 1977.

W volume:page. *The Writings of Charles S. Peirce*. Edited by The Peirce Edition Project. 7 vols. to date. Bloomington: Indiana University Press, 1982–2010.

WMS refers to reorganized Peirce manuscripts as listed in the chronological volume of the writings.

References to the Open Court Collection held at the Special Collections Research Center in the Morris Library of Southern Illinois University Carbondale are referred to by collection number, box number, and folder number; e.g., 27.91.17.

INTRODUCTION

Words strain,
Crack and sometimes break, under the burden,
Under the tension, slip, slide, perish,
Decay with imprecision, will not stay in place,
Will not stay still.
—T.S. Eliot

Editing text is a risky business. This is true even if all we have is printed material, as with the various altered and expanded editions of John Stuart Mill's *System of Logic*. Should spelling and punctuation be modernized? What changes are truly Mill's? Could Mill, while making changes for later editions, have misinterpreted his own work? Should mistakes made by the author (or by past editors or compositors) be corrected? Even when extracting a scholarly edition wholly from previously published work, the potential for problems is endless. This increases exponentially when we must also sift through troves of unpublished manuscripts that are often repetitive, often incomplete, and always abandoned. Such is the case with Peirce's Illustrations of the Logic of Science, or Illustrations for short. The Illustrations first appeared in print in the form of six articles that were published in the *Popular Science Monthly* in 1877–78. There are only a few surviving prepublication manuscripts, and they were written roughly half a decade earlier with a quite different aim in mind. However, many post-publication manuscripts are related to the Illustrations. In the early 1890s Peirce revisited them when seeking to have his work republished in book form. What eventually came out of this is one fairly complete book manuscript, the 1894 *How to Reason*, which contains a number of the Illustrations articles, sometimes in significantly emended and expanded form. Though Peirce sent *How to Reason* to various publishers, the manuscript was never published.[1] Peirce returned to the Illustrations at the close of the first decade of the twentieth century, aiming to publish an updated version as a monograph with Open Court Publishing Company. By this time pragmatism had gained ground as a school in philosophy, mostly thanks to William James and Ferdinand Scott Schiller. James, moreover, vocally identified Peirce as the founder of pragmatism and identified the second Illustrations article as its birthplace.[2] Peirce, grateful for the long-overdue recognition, responded by seeking to carve out his place in it. He did so in part by discussing his contributions in the Metaphysical Club gatherings in the early 1870s, and in part by distancing himself from both James and Schiller, whose interpretations he thought were not only wrong, but also dangerous. Peirce responded extensively in the *Monist*, famously rebranding his view as pragmaticism, a term he believed was "ugly enough to be safe from kidnappers" (EP2:335). Part of this response was a renewed attempt to republish the Illustrations in book form. A selection of this is included here as chapters 7 to 9, and in the form of footnotes to the original articles. Peirce failed, however, to get much beyond several aborted prefaces and miscellaneous revisions. In the end Peirce bowed to repeated requests from Open Court publisher Paul Carus (1852–1919) to have the Illustrations reissued in their original form. In 1913 Carus prepared such an edition, composing a brief "Publisher's Preface" (which is included in this edition as well) and a few notes, but this final effort notwithstanding, the book did not make it into print. Published close to a century later, still with Open Court, the present edition is the first stand-alone edition of the *Illustrations of the Logic of Science*. Though plans existed to republish the papers as a single volume in Appleton's International Scientific Series, those plans never came to fruition.

The Illustrations originally appeared as a series of articles in the *Popular Science Monthly* in 1877–78. For the current edition only a selection of Peirce's later revisions could be included. It is oft argued that when several versions of the same text survived, one should publish the one that is most mature, as it is closest to the author's final intention. I have chosen to deviate from this practice. In the present case, which is not atypical for Peirce, the various versions differ considerably with regard to content, which is due in part to Peirce's proclivity to digression. Since painful cuts had to be made, I have opted to include those versions that are most useful to a reader who is interested in the subject of the Illustrations—the logic of science—and how Peirce's views about what he wrote in the Illustrations changed over time. The texts included, however, reproduce versions as Peirce originally wrote them as completely and as faithfully as the surviving manuscript material allows for—they are not composites of multiple versions. As for editorial intervention, I tried to be as conservative as possible. For instance, Peirce's spelling, which is an amalgam of American and British English enriched by personal idiosyncrasies, is retained unless it is thought to be too jarring to the reader. Applying Peirce's own pragmatic maxim to the process of editing, I have further sought to ensure that any editorial intervention that could conceivably cause a practical difference in interpretation was made in such manner that the reader can easily determine what was done to the text. Interventions, however, are kept to a minimum. The American historian Henry Adams once cynically observed, "It is always the good men who do the most harm in the world," and this is particularly true for well-intending scholarly editors. The temptation to improve upon a text is simply too great.

In this introduction I discuss the publication history of the Illustrations, running from its earliest prepublication material to the most recent posthumous renditions.[3]

The *Popular Science Monthly* was founded in 1872 by Edward L. Youmans (1821–1887) with the aim of putting scientific knowledge in the hands of the educated layman. It was a scholarly journal that contained eight to ten articles in each issue of about a hundred pages. The monthly was published by D. Appleton & Company till 1901, when it was sold to James McKeen Cattell (1860–1944). Fourteen years later, Cattell sold the *title* of the journal to the Modern Publishing Company—which used it for a general-audience science magazine (today's *Popular Science*)—while continuing the periodical under the more daunting title *Scientific Monthly.* The latter was published until 1958, when it was absorbed into *Science.*

In the monthly each of Peirce's articles was introduced as follows:

ILLUSTRATIONS OF THE LOGIC OF SCIENCE.

By C. S. PEIRCE,

ASSISTANT, UNITED STATES COAST SURVEY

The header is followed by an indication where the paper fits within the series and its title—for instance,

THIRD PAPER.—THE DOCTRINE OF CHANCES.

Reserving a discussion of the series title for later, I focus first on the author and what it meant to be an Assistant at the United States Coast Survey.

Charles Sanders Peirce (pronounced "purse") was born in Cambridge, Massachusetts, on September 10, 1839, as the second son of the distinguished mathematician and astronomer Benjamin Peirce. Charles Peirce is one of a handful in the history of thought that can truly be called a universal intellect. Robert Crease called him "a prolific and perpetually overextended polymath," and that pretty well sums him up.[4] He was deeply involved in the main currents of thought (mathematics, logic, experimental science), most of which were at the time in a rapid transition, and he made significant contributions to a great variety of areas, scientific as well as philosophical. Some have called him the American Aristotle, others the American Leibniz,[5] and it would certainly be no less appropriate to call him the American Leonardo. He did pioneering work on the magnitude of stars and the form of the Milky Way. He worked extensively determining the exact shape of the earth—designing instruments and improving methodologies. He invented a new map projection that gave a world map with a minimum distortion of the distance between any two points. He was a pioneer in mathematical logic and mathematical economy, did important work on Shakespearean pronunciation, engaged in experimental psychology, wrote several books on logic and mathematics (none of which were published), gave lectures on the history of science, developed a bleaching process for wood pulp, wrote on spelling reform, made calculations for a suspension bridge over the Hudson river, and was the first to use a wavelength of light to determine the exact length of the meter, making the standard of length no longer dependent on some physical artifact. Almost as an aside, in a short letter to his former student Alan Marquand, Peirce invented the electronic switching-circuit computer. Till then computing machines had all been wholly mechanical. None of these accomplishments, however, were of much help to Peirce. He died in abject poverty and mostly forgotten on April 19, 1914 in a small town called Milford, Pennsylvania. He was survived only by his second wife (whose identity is still a mystery) and by a disarray of more than a hundred thousand manuscript pages.

Peirce grew up in and around Harvard University, which had just begun its transformation from a rather unsuccessful small Christian college into the world-renowned institution it is today. His father, a professor of mathematics and astronomy at Harvard, felt that the young Peirce was destined for greatness and he took an active role in his education. The elder Peirce was an unconventional teacher who taught his students by inspiring them rather than by carefully guiding them through proofs and to the solutions of problems. This instilled in the young Peirce a habit of thinking things out for himself, which never left him. In this milieu Peirce was early on exposed to logic and chemistry. At twelve, he fell in love with logic after reading Richard Whately's *Elements of Logic*, a work that revitalized the study of logic in the English-speaking world.[6] Peirce claimed repeatedly that from that moment on, logic was his strongest passion. As a consequence he always displayed a keen, sometimes even overriding, interest in methodology when working as a scientist, and he sought to penetrate, as he liked to put it, "into the logic of things." Around the same time he was introduced to logic, Peirce's uncle, Charles Henry Peirce, helped him set up a small chemistry laboratory that was designed to work through Liebig's method of chemical analysis. On this method the student is given a number of bottles each marked with a letter of the alphabet and the student is asked to analyze their content using only an introductory textbook in qualitative analysis. In Peirce's case this was his uncle's translation of Stöckhardt's *Principles of Chemistry*, most likely supplemented with the *Nouveau Manuel Complet de Chimie Analytique*, of which Peirce's stained copy survives.[7] At sixteen, Peirce went to Harvard, obtaining the A.B. in 1859. In 1861 he entered Harvard's Lawrence Scientific School, graduating summa cum laude in Chemistry in 1863. In the same year that he entered the Lawrence Scientific School, Peirce was appointed a regular aide in the United States Coast Survey, which brings us back to his father. Benjamin Peirce played a key role in the establishment of scientific institutions in the United States, including Harvard's Lawrence Scientific School, the National Academy of Sciences, and the Smithsonian. In 1867, he reluctantly became the chief administrative officer, or Superintendent, of a rather troubled Coast Survey—a position he held till 1874. During his short tenure as Superintendent, Benjamin transformed the Coast Survey into an internationally recognized scientific institution. A central pillar of the new Coast Survey was gravitation research, and in line with this Benjamin changed its name to U.S. Coast and Geodetic Survey. On November 30, 1872 Benjamin put his son in charge of gravitation research, which required giving him the proper administrative clout. Consequently, on December 1, 1872, Charles Peirce was promoted to Assistant to the Superintendent of the Coast Survey. This meant that Peirce held the second highest rank in what was then America's premier scientific institution. In other words, for the *Popular Science Monthly* to have the Assistant of the Coast Survey write a series of papers was quite a catch. Peirce continued to work for the Survey for nineteen more years, devoting much of his energy to gravitation research.

More than any other decade in Peirce's life, the 1870s were dominated by the actual practice of scientific research. The two main areas Peirce was working in were astronomy and geodesy. Peirce's work in astronomy began in 1867 with his involvement with the Harvard Observatory, where he was appointed Assistant to the Director in 1869. Peirce's major research interest was to determine the relative brightness of stars, the purpose of which was to determine the shape of the Milky Way and the orientation of its axis. To do this Peirce used a modified Zöllner's astrophotometer that enabled him to directly compare the incoming light from a star with light from a kerosene lamp. Three polarizing prisms and a quartz plate then enabled him to modify the brightness and the color of the lamplight to have it match that of the star.[8] Peirce did most of his observations on the stars in 1872–1873. By subsequently comparing his findings with the earlier findings of Ptolemy, Ulugh, Sûfi, and Tycho Brahe, he was further able to address the issue of the variability of the brightness of stars. Peirce published his findings around the same time he published the Illustrations, in what was to become the only book he published in his lifetime: *Photometric Researches* (Leipzig: Wilhelm Engelmann, 1878).

In November of 1872 Peirce was put in charge of gravitation research at the Coast and Geodetic Survey. The purpose of this research was to map out local gravity (abbreviated *g*), by which is meant the acceleration the earth imparts to objects on or near its surface. Because our planet is not a perfect sphere of uniform density, there are slight deviations in the magnitude of *g* across its surface. Such deviations can be measured by swinging a pendulum at different locations, as the period of swing of a pendulum is a function of *g* and other factors that are relatively easy to control, such as the length of the pendulum. Carefully mapping the values of *g* would thus allow us to precisely determine the shape of the earth. However, since the variations in *g* are extremely minute, pendulum research requires immense precision. Even the smallest deviation—say when a drop in temperature causes the pendulum to slightly contract—suffices to discredit the result. Peirce spent countless long days swinging pendulums at various locations within the United States, from Key West to Ann Arbor, and in 1882 the Greely expedition swung a pendulum for him as far north as Ellesmere Island, the most northerly point of the North American continent (W6, sel. 30). Much of Peirce's pendulum research takes place in 1873–74. A location of particular interest is the Hoosac railway tunnel in northwestern Massachusetts where Peirce compared the values of *g* at the top and the bottom of the 1028-foot central ventilation shaft. This allowed him to estimate the mass of the earth, which is a function of *g*, the earth's radius, and the gravitational constant G (WMS 256; Summer 1874). In April of 1875, Peirce sailed to Europe to receive a newly designed pendulum made to his specifications and to test it against established values of *g* at various locations in Europe. One of Peirce's fellow passengers on his voyage to

Liverpool was the publisher W.H. Appleton. It was during this voyage that Appleton invited Peirce to write a series of articles on the logic of science for the *Popular Science Monthly*. It would take Peirce two more years and another trip to Europe to begin working on it in earnest, as much of the writing seems to have taken place shortly before the publication dates (Peirce regularly missed his publishing deadlines). To sum it all up, Peirce began working on the Illustrations following a period of intense empirical research. Moreover, the areas of research that Peirce was engaged in put him fairly directly into contact with that external permanency upon which our thinking has no effect, which comes to play such a central role in his conception of the fourth method of fixing belief: the method of science (see chapter 1).

Before delving deeper into the compositional history of the Illustrations themselves, I want to pay some attention to earlier work of Peirce that is clearly precursor to it. Later in life—mostly when seeking to establish himself as the founder of pragmatism—Peirce referred several times to a paper he read at the Metaphysical Club that he said contained the key points of both "The Fixation of Belief" and "How to Make Our Ideas Clear." This paper is often considered the place where pragmatism was first conceived. Letters from William James and from Thomas Sergeant Perry reveal that Peirce most likely read this paper sometime in the second half of November 1872. Peirce himself first reminisces about it in a letter to his former student Christine Ladd-Franklin dated October 28, 1904.[9] Peirce wrote Franklin that his paper was well received at the time and that "the ms. went around to different members who wished to go over it more closely." No manuscript matching this description, however, has been identified. Peirce continued his letter by explaining that in 1873 he worked to put "that piece into literary form" (most likely his 1872–73 attempt to write a logic book [W3, sels. 4–39]) and that after being invited to write some articles for the *Popular Science Monthly* he "patched up the piece" for the first paper, "The Fixation of Belief."

Does this correspondence indicate a lost manuscript, one of great historical importance no less? Possibly. However, given Peirce's almost obsessive urge to preserve everything he wrote, such a hypothesis should not be raised lightly.[10] It is certainly not the only option. It is not inconceivable that Peirce read to the Club parts of the logic book that he was working on at the time. This alternative hypothesis is confirmed by one of the draft openings of "My Pragmatism" of 1909, where Peirce explains that sections III–V of "The Fixation of Belief" and section II of "How to Make Our Ideas Clear" "reproduce almost verbatim" what he read to the Club in 1872 (R 620:53). With the exception of section II of "How to Make Our Ideas Clear," for which no manuscript material has been identified, the sections Peirce mentions reproduce almost verbatim three chapters of the logic book material. Moreover, in a letter of November 24, 1872, William James, one of the Metaphysical Club members, wrote to his brother Henry that Peirce "read to us an admirable introductory chapter to his book on logic the other day." Assuming James refers to the same event Peirce does, this not only connects the Metaphysical Club reading with a logic book, but also dates it after the relevant parts of the logic book material were written. On November 25, Thomas Sergeant Perry also wrote Peirce regarding the Metaphysical Club reading, asking whether he could publish the paper Peirce had read in the *North American Review*: "I write to beg you to let me have that paper you read the other night at Cambridge for the N.A.R. It ought to be published. I'll pay you bountifully & I *must* have it."[11] Nothing came of this, possibly because only days later Peirce was put in charge of pendulum experiments at the Coast Survey and soon thereafter was promoted to the rank of Assistant to the Superintendent. If what Peirce read was not a completed paper but rather parts of his logic book material, the added work pressure at the Survey and his work on the brightness of stars most likely made it impossible for him to grant Perry's request.

The absence of manuscript material for section II of "How to Make Our Ideas Clear," referred to in the "My Pragmatism" draft, is troubling, as this is precisely where Peirce would have spoken of the principle of pragmatism, presumably even using the term pragmatism itself—something he does not do in the published article. As we can see from other writings, however, Peirce easily could have discussed the pragmatic maxim, or more likely a precursor thereof, at the Metaphysical Club. For instance, in his 1871 review of the works of Berkeley, Peirce criticized Berkeley for claiming that a word can only have meaning if we have the corresponding idea. On this view, Peirce observed, all abstract ideas must be dismissed as meaningless (W2:483). In response, Peirce introduced a new criterion of meaning, which presages the pragmatic maxim: "Do things fulfil the same function practically? Then let them be signified by the same word. Do they not? Then let them be distinguished" (W2:483). In the logic book Peirce connected this idea of practicality with his doubt-belief theory. If the difference between doubt and belief is merely that they are different sensations, Peirce wrote, the distinction would be "almost without significance." Instead, he continued, "the sensible distinguishability is attended with an important *practical* difference" (W3:21, 1872; emphasis added). Earlier, in "Some Consequences of Four Incapacities," Peirce discussed another aspect of the maxim, that of conceivability, arguing that because "the meaning of a word is the conception it conveys, the absolutely incognizable [like Locke's *substratum,* or Kant's things in themselves] has no meaning because no conception attaches to it" (W2:238; 1868). In brief, when Peirce made his presentation for the Metaphysical Club in November of 1872, many of the trappings of the pragmatic maxim were already in place. Moreover, since historically the maxim emerged within the context of Peirce's discussions of reality, a discussion that is very much at the fore of his mind in the fall of 1872 (W3, sel. 15–20), it is not unlikely that he discussed some form of his famous criterion of meaning in his Metaphysical Club presentation.

Whether Peirce actually coined the term pragmatism, as he later repeatedly said he did, is a separate issue and is somewhat doubtful. The term appears nowhere in Peirce's manuscripts, nor in his correspondence, before James publicly called Peirce's maxim "the principle of pragmatism," in 1898. In contrast, James himself did use the word in the 1870s. In manuscript notes for his 1879 "The Sentiment of Rationality" James wrote, "The general principle of pragmatism proves every thing by its result," and, "the principle of 'pragmatism' . . . allows all assumptions to be of identical value so long as they equally save the appearances."[12] Hence, when Peirce was writing the Illustrations, James too was seeking to express the significance of our ideas in terms of their practical consequences, and he explicitly referred to it as the principle of pragmatism.[13] By 1898, however, James strongly preferred the term practicalism instead, most likely because of the negative connotations of the word pragmatism. The *Century Dictionary*, for instance, having defined the pragmatist as someone "who is impertinently busy or meddling," defines pragmatism as "busy impertinence" and described the pragmatizer as "a stupid creature [for whom] nothing is too beautiful or too sacred to be made dull and vulgar by its touch" (CD:4667–8). Such definitions cast some suspicion on Peirce's later claims that he deliberately decided not to include a philosophical definition of pragmatism in the dictionary because "it did not seem to me its vogue was sufficient to warrant that step" (CP 5.13; see also chapter 7 below).

Before discussing the Illustrations in more detail, a little more should be said of the early logic book material. The logic book is far from completed and the surviving material suggests various attempts at writing such a book, all of which can be dated 1872–73. As for its content, and its proposed content, the material falls conceptually somewhere between, on the one hand, the American Academy series of 1867 and the *Journal of Speculative Philosophy* series, and on the other, the Illustrations. What distinguishes the Illustrations from Peirce's earlier work is that in the Illustrations, the science of logic, as all science, must start from an examination of existing beliefs, doubts, and practices. This notion, that it is important for logicians to learn from the actual *practice* of science, is wholly absent in the logic book.

As said, Peirce met the publisher W.H. Appleton when they were sailing to Europe in April of 1875, and it was during this voyage that Appleton invited Peirce to contribute a series of papers on the logic of science to the *Popular Science Monthly.* As is noted also, it took him almost two more years to begin working on the project in earnest. On March 3, 1877, Peirce wrote his mother: "I am writing a paper for the *Popular Science Monthly* but it is not complete yet. I think when I have done one, I can write others more rapidly." For this paper, which appeared in December 1877 under the title "The Fixation of Belief," Peirce relied quite heavily on material that he had written for the logic notebook half a decade earlier. Comparing the text of "The Fixation of Belief" with the logic book material confirms what he wrote Christine Ladd-Franklin in 1904. Sections III and IV copy almost verbatim chapter 1, "Of the Difference Between Doubt and Belief" (W3, sel. 9; written between May 11 and 14, 1872) and chapter 2, "Of Inquiry" (W3, sel. 10; dated May-June 1872). In addition, much of section V comes from chapter 3, "Four Methods of Settling Opinion" (W3, sel. 11; also dated May-June 1872). Most likely sections I and II were composed only after Peirce had decided to explain the validity of synthetic inference in terms of the doctrine of probabilities in the third and fourth articles. The *Ohio Medical Recorder* wrote in its review of the November issue of *Popular Science Monthly*, "The place of honor is ably filled by the first of a series of "Illustrations of the Logic of Science," By C. S. Pearce [sic], assistant of the United States Coast Survey; the title of the paper is the "Fixation of Belief;" it will be read with interest and pleasure by those who are accustomed to criticize the bigotry and intolerance of the clergy."[14]

The second paper in the series is "How to Make our Ideas Clear." On November 2, 1877, Peirce wrote his mother that he wrote the best part of "How to Make Our Ideas Clear" between September 13 and 24 while sailing from Hoboken, New Jersey, to Plymouth, England. In a letter to his brother James Mills Peirce, written in December or January, Peirce wrote that he is sending him the second paper, adding that he had written it on board of a ship. In this paper Peirce distinguishes three criteria for the clearness of ideas, the third of which is what James later called "the principle of pragmatism." This principle is followed by several applications, mostly to concepts used in physics. The first of these, the application of the principle to the concept "hardness," would cause Peirce quite a bit of trouble. Though the general import of the paper is realist, this particular application is clearly nominalist, inviting a nominalist reading of the principle itself. Later in life Peirce says several times that he wrote the paper first in French, and even that he preferred the French version to the English one. I will return to that claim below when discussing the French publication of the first two papers.

On February 4, 1878 Peirce wrote his mother that the March number of *Popular Science Monthly* "will contain my article on the Theory of Chances which I fear they will have to divide into two." This is indeed what happened. The article was split in two, with the first half, "The Doctrine of Chances," appearing in March and the second half in April under the heading "The Probability of Induction." Circumstantial evidence suggests it is most likely that Peirce wrote "The Doctrine of Chances" and "The Probability of Induction" between December 1877 and March 1878. By publishing "The Doctrine of Chances" directly after "How to Make Our Ideas Clear," Peirce began his discussion of logical inference by discussing the theory of probability, which in the fifth paper returns as his ground for justifying induction. Hence, in contrast to traditional logic books, which typically focus their attention on the syllogisms and deductive reasoning, Peirce, with a clearer sense of what is valuable to science, started with an extensive discussion of induction, relegating deduction to a fairly short account in the sixth paper. Shortly after the fourth paper was published, Peirce wrote his mother: "I hear a good many people—Sylvester, for instance—express the opinion that the third and fourth papers are better than the first and second, which seems to me the most melancholy incapacity for judging such things."[15]

The fifth paper, "The Order of Nature," was most likely written from February to April 1878, as Peirce missed the deadline for the May issue of *Popular Science Monthly,* because he had "so many interruptions last month it was impossible for me to be ready in time."[16] In this paper, Peirce takes aim at the prevailing view, held most prominently by John Stuart Mill, that induction is justified by the order of nature. Induction is valid, Mill argued, because it contains a suppressed major premise that pronounces the uniformity "which we know to exist in the course of nature"; and it is this uniformity that allows us to draw the inference.[17] In its stead Peirce argued that what justifies induction is the theory of probability that he developed in the previous two papers. Later, when reflecting in the third person upon this article in a summary of his work for Friedrich Überweg's *Grundriß der Geschichte der Philosophie*, Peirce said of the uniformity of nature that,

> while it afforded opportunities for inductive reasonings, it does not constitute the general ground of validity of such reasonings. He also argued that as a fact there appears to be as little orderliness in the universe as we can conceive that a universe should have, and further that the degree of orderliness of the universe is relative to the mind that contemplates it, consisting merely in the breadth (*Umfang*) of that mind's interests.[18]

The sixth and final paper, "Deduction, Induction, and Hypothesis," was most likely written from April to June 1878, as this time Peirce missed the deadline for the July issue. The paper appeared in the August issue. Peirce discussed the three kinds of reasoning that can be extracted from the general form of the syllogism as different orderings of rule, case, and result. The three modes of reasoning Peirce distinguished are deduction, induction, and hypothesis (which, at times, he called abduction or retroduction). Much of the paper is devoted to establishing hypothesis as a mode of inference, something he does in part by drawing upon examples from the history of science.

The six articles here described originally appeared under the general heading "Illustrations of the Logic of Science." Some, including Peirce himself, have found this heading troublesome. The word "illustrations" suggests at best a rather thin thread of continuity throughout the papers. Rather than a coherent set of articles with a clearly defined topic, it suggests a series of rather detached articles with no aim of being comprehensive. Later on Peirce explicitly voiced his dislike of the heading for the series, which seems to have been the product of a compromise between him and *Popular Science* editor Edward Youmans (1821–1887), a person whom Peirce greatly disliked. The French translation of the series—on which more later—appeared under the title "The Logic of Science," which may have been more to Peirce's liking. In the spring of 1909 Peirce wrote that the Illustrations "appeared under the general title,—an inappropriate one, taken to please either the publisher or the editor, I forget which,—'Illustrations of the Logic of Science'" (R 620:61). At least on one occasion he even pretended that the articles originally appeared "without any title for the whole" (R 619:2). That Peirce amply displayed his disapproval of the heading does not tell us who proposed it and why. If it was the editor, he was most likely inspired by the opening section of "The Fixation of Belief," were Peirce writes that "each chief step in science has been a lesson in logic" and proceeds by briefly reviewing three such lessons. If it was Peirce, it may suggest that at the time he envisioned relying more heavily on actual examples, but veered away from that during the actual writing. It is clear, however, that the Illustrations break with Peirce's earlier work, in that he now very deliberately argues that the science of logic, like all science, must start from an examination of existing beliefs, doubts, and practices, which implicitly gives prominence to historical examples that can count as illustrations of the logic of science. In an offprint of "The Fixation of Belief" which Peirce prepared for possible republication, Peirce deleted the entire phrase "Illustrations of the Logic of Science" and replaced it with "Essays on the Reasoning of Science" (R 334:40, 1910).

Besides the *Popular Science Monthly*, Youmans also inaugurated the "International Scientific Series." The aim of this series was to publish work by the greatest scientists of all nations, preferably simultaneously in the principal modern languages. Arrangements were made for the publication of books in New York, London, Leipzig, and Paris, and later also in Milan and St. Petersburg. By 1888, 64 volumes appeared in the series. During the years 1878–1879, Appleton regularly included the Illustrations among the forthcoming volumes it listed in its books to advertise for them. No such plans materialized, however.

The failure of the book project raises the question whether the series in the *Popular Science Monthly* is complete, or whether it was broken off prematurely. Based in part on a comparison with the earlier logic book material, and in part on an 1881 letter Peirce wrote to his mother, Max Fisch concluded that the Illustrations remained unfinished (W3, Introduction). As Fisch pointed out, Peirce never touched upon the theory of the categories, nor upon the doctrine of signs, two central themes in his earlier work, and two themes that return in the 1894 logic book *How to Reason*. Moreover, in the 1881 letter to his mother Peirce confessed, "I am thinking of undertaking some more papers for the Popular Science Monthly though I can hardly screw myself up to that point yet" (W3:xxxvi), possibly suggesting that he was thinking of continuing the series. Long after the fact (in "My Pragmatism" of 1909—included as chapter 7 below) Peirce wrote that he had been engaged by Appleton "to write half a dozen articles for the *Popular Science Monthly* on the Logic of Science." Whether this means that the series was complete after all (as six articles were published) is impossible to say. What can be said is that a regular subscriber to the monthly who had just finished the sixth article would not naturally conclude that no additional installments were to follow. The series does not naturally come to a close, but ends rather with the observation that it makes good sense to distinguish induction from hypothesis (or abduction). This is an observation that strictly confines itself to the topic of the sixth paper, without any reference to the series as a whole. Adding further that two of the essays Peirce wrote were so long that he had to split them in two—so that arguably he wrote only four papers, not the six he said he had promised—I tend to agree with Fisch that more likely than not the Illustrations were unfinished. What counts against this interpretation is that whereas Peirce voiced many complaints about the original publication, especially to Carus when preparing a new edition of the Illustrations at the end of the first decade, he never says that the original series was left unfinished.

Shortly after the publication of the original series, the first two papers, "The Fixation of Belief" and "How to Make Our Ideas Clear," also appeared in French in *Revue philosophique de la France et de l'étranger* under the general heading "La logique de la science." The first appeared in December of 1878 as "Comment se fixe la croyance"; the second appeared the following month under the title "Comment rendre nos idées claires."[19] The two French versions are included in volume 3 of the *Writings*. In a letter to Christine Ladd-Franklin of October 28, 1904 Peirce wrote that during his 1877 voyage to Europe to attend the International Geodetic Association conference in Stuttgart, he wrote "an article about pragmatism in French" to practice his French. Peirce wrote also that he later "translated the steamer article into English and in that dress it appeared in the *Popular Science Monthly*." He also translated his earlier article, "The Fixation of Belief," from English into French.[20] In 1905, in an unpublished and unfinished manuscript called "Consequences of Pragmatism," Peirce explained that he wrote "How to Make Our Ideas Clear" in French to avoid the criticism he received from Youmans on "The Fixation of Belief," that it was too metaphysical for a scientific monthly: "The second article was entirely written on a French steamer, and was written first in French, although only an English publication was contemplated, with the idea that the temptations to be too darkly philosophical would by that means be diminished, and the editor be in some measure appeased" (R 289:3). This also suggests that Peirce wrote "How to Make Our Ideas Clear" in French before any arrangements were made with the *Revue philosophique* to publish French versions of the papers.

On several occasions Peirce remarked that he preferred the French versions above the English ones. For instance, in a note clipped to a collection of offprints entitled *Papers in Logic*, which Peirce left to the Johns Hopkins University Library, Peirce wrote: "The two French versions, which I prefer to the English of the same papers, derive their merit from the skill of M Léo Seguin, who was killed in Tunis in 1881." And in his 1903 Harvard Lectures, Peirce even went so far as to quote the French version of the pragmatic maxim, even though he was addressing an English-speaking audience (CP 5.18). The reference to Seguin suggests that Peirce was not solely responsible for the French. Peirce confirms this in a February 3, 1879 letter of reference to George Davidson, in which he praises Seguin's translation of "The Fixation of Belief." Comparing this text with other texts Peirce wrote in French, Gérard Delledalle concluded that Peirce did translate "The Fixation of Belief" into French himself and that Seguin corrected the translation, sometimes for grammatical reasons, sometimes to render the text more clear to the reader.[21] The French version, however, contains some dubious translations that make it hard to believe that it is superior to the English, as Peirce at various times claimed it to be. For instance, Seguin, a libertarian anarchist who in 1871 had been banished from France for his role in the revolt of the Paris Commune, was an avowed individualist who was ill at ease with the communal aspect of Peirce's theory of inquiry. Consequently, he "corrected" the text for it (see the footnotes to chapter 1 below). According to Delledalle, the French of "How to Make Our Ideas Clear" is superior to the French translation of "The Fixation of Belief," and the sort of interventions that were found there are absent here. Nevertheless, Delledalle continues, the English version, which is more precise and explicit, is superior to the French and there is no evidence that Peirce originally wrote the paper in French and subsequently translated it into English.

In 1887, close to a decade after the Illustrations were published, the wealthy zinc manufacturer Edward Charles Hegeler (1835–1910) founded the Open Court Publishing Company with the aim of stimulating the discussion of religious issues from a scientific viewpoint, and he appointed his future son-in-law Paul Carus (1852–1919) as its managing editor. Besides books, the company published two periodicals: the *Open Court* and the *Monist.* In July of 1890, Carus, who had read the Illustrations and was impressed by them, recruited Peirce to contribute an article on logic for the *Monist*'s inaugural issue.[22] Though Peirce missed the deadline, it proved the beginning of a lifelong and sometimes stormy relationship with the Open Court Publishing Company. Peirce contributed to both periodicals, and although he never published a book with the Open Court, a number of book proposals were discussed over the years, several of which involved the Illustrations.

In the winter of 1893, while he was working on a lengthy reply to Carus's critique of one of his *Monist* articles, Peirce visited Hegeler and Carus in LaSalle, Illinois. The experience must have been a good one as Peirce instantly submitted a proposal for a two-volume edition of mostly previously published work, to be titled *Collected Papers.*[23] In the proposal Peirce included the Illustrations, as chapters 11–16 of a total 42 chapters. The entire work, Peirce estimated, would run somewhere between seven and eight hundred octavo pages. On March 7, 1893, not having heard back, Peirce mailed a second proposal in which he changed the title of the work to *Quest for a Method.* In this proposal he discussed the contents for the first volume only. The new proposal also included the Illustrations, complete and in their original order.[24] Peirce further explained that he chose to call it *A Quest for a Method* to distance himself from the views he ascribed to Hegeler and Carus on the reconciliation of religion and science, which in his opinion essentially meant that religion had to adopt a new creed, namely the creed of science. Peirce rejected this as too superficial; it merely meant exchanging one creed for another. Peirce instead sought to dismiss all creeds, whether religious or scientific. As he put it in the letter, "My book is to be entitled *A Quest.* Now the very idea of a quest implies that what is said is not in harmony with any fixed creed, like yours. Worse yet, it is *a* quest for *a Method*, and what that method is cannot be predetermined. The presumption must be that it will be hostile to any creed formulated in advance." In this way *A Quest for a Method* continued the path Peirce had taken with the Illustrations, where the method of science is not something externally imposed upon inquiry, but itself a product of that inquiry. In what seems a different version of the same project, Peirce used the title *A Search for a Method* (R 592:2). Peirce, however, could not make himself republish his papers unaltered, and he began making sometimes-extensive revisions to the existing papers. Both *A Quest for a Method* and *A Search for a Method* open with a revised version of Peirce's 1867 "On a Natural Classification of Arguments" (see R 594 and R 592 respectively), with the revisions for the former being the more far-reaching.[25] A table of contents for the entire work survives as R 1583:2, and it shows that Peirce intended to include nineteen of his former publications in chronological order. It lists the entire Illustrations series as essays 8–13. Though several of the revised essays survive, it is quite evident that Peirce did not bring this project to completion. What happened instead was that it inspired him to embark upon a more ambitious project: the 1894 logic book *How to Reason*, to which we turn next.[26]

The aim of *How to Reason* is to make "the nature of inquiry into real facts illuminate that of demonstration from fixed assumptions, and *vice versa*" (R 397:2), which is commensurate with the previous projects. After significant reshuffling,[27] *How to Reason* is finally divided into nineteen chapters over three books (R 339:4): Of Reasoning in General, Demonstrative Reasoning, and Quantitative Reasoning.[28] An introduction on the association of ideas precedes the three books and they are followed by a set of recreations and exercises. Only three, possibly four, of the Illustrations articles make it into *How to Reason*, and they are not grouped together.[29] *How to Reason* starts off with a discussion of the categories and of signs—two topics discussed in the 1872–73 logic book material, but notably absent in the Illustrations. The next three chapters are grouped together under the heading "Transcendental Logic."[30] The first two, "The Materialistic Aspect of Reasoning" and "What's the Use of Consciousness?" lay the groundwork for "The Fixation of Belief," which appears as chapter 5.

In the first of these two chapters, Peirce argues that we can already discern what we call reasoning at the mechanical level of protoplasm. The argument is similar to what he had recently done in "Man's Glassy Essence," which appeared in the *Monist* in October 1892. The chapter concludes with the following observation: "A decapitated frog almost reasons. The habit that is in his cerebellum serves as a major premise. The excitation of a drop of acid is his minor premise. And his conclusion is the act of wiping it away. All that is of any value in the operation of ratiocination is there, except only one thing. What he lacks is the power of preparatory meditation" (R 405:10). In "What is the Use of Consciousness," Peirce argues against the mechanistic interpretation that the preceding chapter might elicit—an interpretation on which the role of consciousness is at best that of a spectator. In its stead Peirce argues that the only viable option is to consider consciousness as being part and parcel of the process itself: "The spectator is no longer on one side of the footlights, and the world on the other" (R 406:4). Next follows "The Fixation of Belief," in which Peirce explains how his doubt-belief theory is a natural outcome of this. Hence, in contrast to the Illustrations, where Peirce felt that he could simply start with the scientist's actual experience of doubt, Peirce here takes the view that significant groundwork is first needed. This groundwork includes several themes of his work in the 1860s, such as his theory of the categories and his semeiotics—themes that were also present in the 1872–73 logic book material. In chapter 6, "The Essence of Reasoning," Peirce gives an outline of traditional logic with the aim of explaining "the excellent and well-established terminology" of logic (R 408:2). As we will see shortly, Peirce later recycles the title of this chapter for a new edition of the Illustrations.

Book II of *How to Reason*, called "Demonstrative Reasoning," contains much of Peirce's technical work on logic. The bulk of it consists of mathematical logic (nonrelative as well as relative), to which are added two chapters under the heading of methodology. The first is an annotated offprint of the 1867 "Upon Logical Comprehension and Extension"; the second is the second Illustrations paper "How to Make Our Ideas Clear." The separation of "The Fixation of Belief" (chapter 5) from "How to Make Our Ideas Clear" (chapter 16) is interesting, given Peirce's later insistence, discussed below, that the two papers were originally conceived as one and that they are truly inseparable.[31]

Book III of *How to Reason*, is called "Quantitative Logic." It consists of three chapters: "Logic of Quantity," "The Doctrine of Chances," and "Induction Etc." The first (chapter 17) is a longish holograph manuscript that was written for the occasion (R 423). All that survives for "The Doctrine of Chances" (chapter 18) is a barely corrected typescript made from the published paper; and the final chapter (chapter 19), which Peirce listed on the table of contents under the no-doubt-provisional title "Induction Etc.," is altogether missing. Peirce could have intended to use the fourth Illustrations paper, "The Probability of Induction," for the final chapter 19. However, given that "The Doctrine of Chances" and "The Probability of Induction," were originally conceived as a single paper, it is also possible that Peirce intended the latter to also be part of chapter 18. There are other options for chapter 19 besides the fourth Illustrations paper. It is possible, for instance, that Peirce intended to use his 1883 "Theory of Probable Inference" (W4, sel. 64). Earlier, when discussing his logic book in a long letter to Francis Russell, Peirce lamented that "The Doctrine of Chances" was "the worst chapter in the book. A mere revision of my *Popular Science Monthly* article. This chapter demands amplification."[32] The last two Illustrations articles, "The Order of Nature" and "Deduction, Induction, Hypothesis," were not included in *How to Reason.*

How to Reason, possibly in the state in which it survived (it was almost, but not quite completed) was sent to publishers, but it was not accepted for publication.

Though "How to Make Our Ideas Clear" is generally considered the birthplace of pragmatism, as already mentioned, the term itself did not appear in print until 1898, when William James published his "Philosophical Conceptions and Practical Results."[33] It became a household term within philosophic circles not long after, when James used the term again in his 1901–1902 Gifford Lectures to help us "decide, among the various attributes set down in the scholastic inventory of God's perfections, whether some be not far less significant than others."[34] On both occasions James gave full credit to Peirce, calling it "the principle of Peirce," and identified "How to Make Our Ideas Clear" as his source. These events mark the beginning of many years during which Peirce sought to claim ownership of pragmatism while at the same time distancing himself from what other self-proclaimed pragmatists, like James and F.C.S. Schiller, were doing. The first major outcome of this is Peirce's 1903 Harvard Lectures, entitled "Pragmatism as a Principle and Method of Right Thinking."[35] After plans of publishing the lectures had fallen through, Peirce reconnected with Open Court, and in 1905 he published the first of a series of papers on pragmatism in the *Monist.* This is when, in a paper titled "What Pragmatism Is," Peirce famously baptized his own view "pragmaticism," a term he believed would be "ugly enough to be safe from kidnappers" (EP2:335). Two more papers appeared, but that series was never completed.

In their rendition of the Illustrations for the *Collected Papers*, Charles Hartshorne and Paul Weiss, included several notes that they drew from a manuscript entitled "My Plea for Pragmatism," which they dated 1903, the same year as the Harvard lectures. When compiling the bibliographical data at the end of volume 8 of the *Collected Papers*, Arthur Burks changed the date to 1909. Burks correctly identified the passage as coming from R 619:2–3, which is unambiguously dated, in Peirce's own hand, March 26,

1909.[36] No mention is made in that document, however, of "My Plea for Pragmatism," which Hartshorne and Weiss insert in square brackets. Peirce called it "The Import of Thought: An Essay in Two Chapters," which is the first chapter of a volume to be titled *Studies in Meaning*. Peirce used the phrase "My Plea for Pragmatism" in a different, undated manuscript that was recovered in 1969 from an old desk at Harvard that appeared to be full of Peirce manuscripts (RS 77), and in a letter to Paul Carus dated January 7, 1909.[37] Both the letter and RS 77 identify the first two articles of the Illustrations as Peirce's "plea for pragmatism." The first sheet of RS 77 opens with the book or series title *Essays on the Reasoning of Science*, followed immediately by what appears to be the title of the first essay "My Plea for Pragmatism." The latter is followed by a simple "Part I" and the opening paragraph of "The Fixation of Belief," suggesting that this opening essay was to combine the first two Illustrations articles, which matches what Peirce wrote on R 619:2. Combining the first two Illustrations papers is not unique to R 619, but is a general strategy characteristic of Peirce's 1909–1910 rewritings, and is in line with the compositional history of the original series, as the first two articles seem to have found their origin in Peirce's 1872 Metaphysical Club presentation (see above).

The attention that was drawn to the Illustrations due to the discussions about pragmatism may have played a role in Carus's renewed interest in seeing the Illustrations republished. In January 1907, he wrote Peirce that he wished to see the Illustrations "republished either by Appleton & Co., or by anyone who would bring it out," and hinted that he would be interested in publishing the series himself.[38] As noted earlier, the Illustrations had been Carus's own introduction to Peirce's writings and the reason why he had solicited Peirce to write for the *Monist* in the early 1890s. Peirce responded swiftly, and Carus convinced Hegeler that they should take on the book. As Carus explained in his answer to Peirce, "I have strongly recommended the publication of your book which I wish to bring out because I think that your papers on logic in the *Popular Science Monthly* are the best introduction to the new movement in logic that have been written in the English tongue, perhaps in all tongues."[39] Carus's typewritten letter mentions an advance of $100 with an additional royalty of $50 to be paid at a later date from the net returns. Carus further expressed his hope that he can receive the copy "very soon," so he can send it to the compositor before leaving for Europe. The positive tone of the letter is somewhat dampened by a handwritten insertion, no doubt inspired by Peirce's past failures to keep his promises: "I am just told that Mr. Hegeler is opposed to making any advanced payments, but I will do what I can do."

Notwithstanding a supportive letter of William James,[40] the project failed to get off the ground. The issue again resurfaced in December when Carus responded to a different proposal by Peirce that Carus would rather republish the Illustrations. In his response Carus is particularly critical of Peirce's later work: "although your articles contain valuable propositions, you enwrap them unnecessarily in such difficulties, and sometimes switch off your ideas without explaining why you do so, that you tax the patience even of our most attentive readers." "That you can write well," Carus continued, "is plainly shown in your former articles on logic in the Pop. Sci. M. I wish you would continue in that same forceful style which is at once instructive and to the point." Carus then reminded Peirce that about a decade earlier, when discussing the *Collected Papers* project mentioned above, Peirce had given him permission to republish the Illustrations, but added, "Yet I have not been able to do so."[41]

Another year passes until, in December of 1908, Carus rekindled his plan of publishing the Illustrations: "in thinking over and planning the scope of the work of the Open Court Publishing Company for the coming year, I think again of an old and cherished plan of mine which I had proposed to you some time ago, and which consists in the reproduction of your 'Illustrations of the Logic of Science.'"[42] "If I am not very much mistaken you gave me once the permission," Carus continued, "but I hesitated to do so because I was afraid of complications. Now there have elapsed thirty years since then and I think there is not the slightest doubt that you can dispose of it freely," and he repeated, with slight modification, the financial terms that he had given Peirce two years before: "Would you accede to my wish to publish it in book form for a honorarium of $100, and in addition thereto $50 for your trouble in revising these five [sic] essays, altering, adding or changing as you may see fit, and the proofreading."[43] From Carus's perspective the main reason for having the original articles revised was, as he explained, "to justify our taking out a new copyright."[44]

Peirce did not reply immediately. First he wrote his long-term friend and confidant Francis Russell. Francis Calvin Russell (1838–1920), an attorney and counselor at law in Chicago, was also a long-term friend of Paul Carus and was employed by Carus to vet articles, translations, and proposals related to logic for the Open Court Publishing Company. Peirce wrote Russell that Carus had offered him $100 for the copyright and the revision of the Illustrations. Noting that it would require the better part of a year to make the revisions, he called it "less than a starvation prize" and announced that he planned to seek out other publishers. Finally, he added, "anything from you in the way of faultfinding with definite passages of those articles will be valued by me.'"[45] Six days later, on January 7, 1909, Peirce officially replied to Carus's letter, asking him to keep the offer on the able for another three months, as he could not presently accept it,[46] and he included the following plan for revision:

> My plan is to reprint them with alteration except mere clerical ones *but*, firstly, to delete passages which were due to my trying to mollify Youmans's wrath at the engagement which old Mr. Wm. Appleton forced upon him. He had not an atom of the scientific spirit. He was devoted to Spencerianism just as he might have been to a scheme to get up a revolution in some country; and the idea of allowing an individual to express views in the Pop. Sci. that at all conflicted with Spencer's was disgusting to him. That accounts for various features of the articles, and as I was paid $150 an article, I was bound to take great pains with the style and all that. I still think them sound in the main; but I did not then see the whole truth of the matter. Consequently, in revising them, I might enliven them considerably by a little sparring match with my self on special points. Such passages to be in brackets, and perhaps with a different font of type. Besides that, secondly at the end of each, I should add a supplement setting forth what I have learned since then. The first and second of the original articles would make one essay in two parts, the third and fourth would make another. The fifth would be one part of an essay in two or three parts. The sixth would be replaced altogether by an account of the Methodeutic of Scientific Research; and I should add two chapters at least in the way of Illustrations of Science, setting forth, for example, the nature of the actual reasoning in detail for today's views on the chemical constitution of compounds. This would follow a method essentially different from Mach's treatment of dynamics;[47] for in chemistry there have been too many refutations of theories that always had been bad, i.e. unwarranted, from the very start. I shall insert also my classification of the sciences (perhaps the entire scheme, which has never been printed yet)[48] my criticism of Pearson in the Pop. Sci. for Jan. 1901,[49] and my *Evening Post* article on The Great Men of Science of the XIXth century.[50] The book will be much stronger and a good deal more popular (if I can succeed in making it so, with handwork,) than the original articles.[51]

The first two Illustrations articles taken together were to be called "A Plea for Pragmatism," and it seems that Peirce began his revision of the Illustrations almost immediately (RS 77).

Carus's response to Peirce's letter is supportive, but also cautious:

> Your letter just comes to hand and I will say in reply as follows: I give you all the time you want to revise it but I did not expect you would have to rewrite so much but of course if you have had to adapt yourself to Mr. Youmans I can

only wish myself that you would make such changes as you deem necessary, but please do not tamper too much with your own thought of former years. I remember of Goethe that he treated the writings of his younger years with great respect and did not change unnecessarily. Of Cantor I can only say that he changed his article on Pythagoras in the later edition and took out the best, most ingenious and most interesting part of his thought, so I hope you will not touch your own old thoughts but limit your revision to what is foreign to yourself in the articles. I had offered the pecuniary amount of revision on purpose of $50 because I did not want you to make supererogatory changes. However, as you state the case it is different.

As to the payment I am now willing to add another $100, and so I will pay you for the whole $250. I think very much of this article of yours and shall be very glad to have the revised copy within about three months, payment to be made of receipt of copy.[52]

Carus also emphasized again that there was little money to be made with a book like this, but that he hoped it would "contribute a good deal to clear up the ideas concerning pragmatism and the difference that obtains between pragmatism and pragmaticism."[53] In Carus's view, republishing the Illustrations would also have the good effect of drawing attention to Peirce's "more recondite work on the Algebra of Logic."[54]

On January 22 Peirce began an essay that was meant to be an extensive revision of the third and fourth Illustrations papers, a project he continued to work on till the end of that month (the essay survives as R 706). It is not till March 9, however, that Peirce officially agreed to work on the Illustrations, and not without first asking whether he could be paid the entire sum of $250 in advance, citing his wife's poor health as the reason:

If I could persuade you to advance the whole $250 on my solemn promise to devote all my energies after drawing plans for my wife's work (which are nearly complete now) exclusively to those two jobs for you; namely the completion of the revision of the Pop. Sci. articles and the article on my method of analyzing concepts (i.e. of defining them) a theory completely worked out now—then I would close with you with the understanding that your holding these articles should not be a reason for your not accepting new articles, being pestions of my System of Logic, which would be written with a view to their being widely read, and decidedly more perspicuous than my article in the Hibbert Journal of October 1908.[55]

Carus agreed, writing Peirce on March 15 that he will mail him a check for $250.[56] In response, Peirce began his revision of the Illustrations, presumably with the undated manuscripts R 628 and R 629. He called the volume *Studies in the Meaning of Our Thoughts* and the first essay was to consist of "The Fixation of Belief" and "How to Make Our Ideas Clear." This first essay, called "What is the Aim of Thinking?," Peirce explained, "contains the earliest formulation of the doctrine of Pragmatism" (R 628:2). Except for some minimal corrections to enhance their clearness, and a "few bracketed clauses to reinforce the original arguments," Peirce believed that the papers could be republished in their original form, while "such statements in it as are more or less incorrect are reconsidered in notes appended to this reprint" (R 628:2). On March 22, Peirce starts anew, opting for the shorter volume title *Studies in Meaning* under the pen name Charles Santiago Sanders Peirce. At this time he also started the practice of dating each sheet separately (R 630:2). Peirce wrote ten pages between March 22 and March 29, divided over three drafts that survive in R 618 and R 630.

It seems that Peirce only began working in earnest on the Illustrations after receiving his $250 check from Carus. The check arrived in the afternoon on April 3, which was a Saturday. Peirce confirmed its receipt the following Monday and began working on *Essays Toward the Interpretation of our Thoughts* (chapter 7 below) that Monday as well.[57] The letter and the check, he explained, "fired me with the desire to pitch in and make my 'Essays on the Meaning of Thought' as useful to a relatively large circle of readers as I possibly can."[58] With only a few intermissions Peirce worked till the end of October almost daily on his revision of the Illustrations. Roughly five hundred manuscript pages survive of this effort (R 620–40). This material covers several attempts. Peirce began his first main attempt on April 5 under the volume title *Essays Toward the Interpretation of our Thoughts*. He worked on it pretty much continuously until June 24, creating in the process a number of abandoned branches.

After a hiatus of about a month, Peirce returned to the Illustrations at the end of August. Peirce first attempted to start anew on August 24 (R 631–32), labeling that document simply "Preface," and abandoning it on August 27. Peirce continued his work on August 30, only to abandon the project a few days later, on September 1. On September 4 Peirce once again began afresh, doing the same again on September 7, and on September 8, at which point the volume title *Efsays on Meaning* was introduced, which Peirce chose to spell with a long *s*, even though the practice had fallen out of use in the beginning of the nineteenth century. The September 8 version is included below as Version I of chapter 8. Peirce worked on it till October 13. A week later he started over again, which resulted in his final run, written in a two-day period (October 22–23). This shorter version is included below as Version II of chapter 8. From what Peirce was doing it becomes increasingly clear that he is no longer aiming to just reprint all six of the Illustrations papers. As Peirce explains in an eight-page discarded draft composed September

4–6: "The remainder of the volume sets forth some of the matured ideas of a man who from boyhood to old age has longed and laboured to learn all he could of the right methods of the conduct of thought in the research of truth" (R 633:2–3). Instead of reworking the four remaining papers, Peirce is now seeking to develop the groundwork for a system of logic, considered as semeiotic.

Despite his significant output Peirce made but little progress. All he ended up with were a number of unfinished drafts of a preface he was dissatisfied with. He had barely begun working on the actual papers themselves, making the promised corrections and inserting new material between square brackets as he had told Carus he would do. Instead, Peirce got pretty much stuck in the very first paragraph of "The Fixation of Belief," trying to explain to a general audience why a study of logic is worthwhile. Unsatisfied with his progress, Peirce did not write Carus to keep him informed of the situation. On November 26, Carus wrote Peirce a diplomatic letter in which he inquired why he had not received the promised "corrected article on logic," and he was clearly dissatisfied with Peirce's reply to his inquiry. On December 6 Carus wrote Francis Russell, whom he had made responsible for readying the Illustrations for publication at the press:[59]

> I shall go to Chicago (Tuesday) and will be in my office Wednesday forenoon. I wish to talk over with you the problem of Peirce's logical papers. I paid for a revision of them $250, and maybe the condition of his present state of mind is such that the condition would merely mean a deterioration. Please look at the papers once more. Perhaps it would be advisable to republish them as they stand. He says that his memory fails him as to the $250 and what he promised.[60]

Two days later, on December 8, Carus sent Peirce a second letter with a detailed summary of previous correspondence in an attempt to force Peirce to fulfill his obligation. Peirce wrote back on December 14, explaining in a shaky hand that he was suffering from a really bad cold, after which he continued:

> Whether I did originally or not, certainly I did not *later* appreciate that you cared particularly to have the account of my method of logical analysis i.e. of making definitions separate from my Pragmatism essay, this latter being obviously a part of the former.
>
> I have been at work upon that ever since, but my style has become atrocious. I shall not be satisfied until what I write can be followed by any intelligent person *who is willing to take the trouble to think*, and not only *can* be followed but will be read with interest and pleasure. My internal difficulties are

> great, but I shall be able to overcome them. But I can't say how soon. I can't write much *per diem.* After three hours I become fatigued and find it best to stop; and I write very slowly because it is needful to be extremely accurate in this subject.[61]
>
> On receipt of your last letter, I laid aside what I was writing and began to write a separate article on my method of logical analysis.[62] I hardly think I can compress it into one article; but it shall count for one. I expect to have the first part of it ready for your April number of the *Monist*, but I won't promise absolutely. For my health is bad. Some days I cannot handle a pen at all; and every week I have more than one attack of a nervous nature in which I can move hardly a muscle and am in danger of falling. If I had not had so many that I know how to manage, I should fall often.[63]

The letter seems to confirm the concern Carus expressed in his letter to Russell—Peirce is no longer up to the task—and Carus again admonished Peirce to have the Illustrations published unchanged: "If after a perusal you are satisfied with them, I might publish them as they stand or perhaps I might add in an appendix notes of yours in which you state either a change of view or a dissatisfaction with the form in which they have appeared."[64]

It is unclear whether Peirce ever responded. On April 30, 1910, Carus wrote him again urging him to submit a corrected copy of the Illustrations for publication:

> You have promised again and again to send me the revised copy of your logical papers. I wish you would bear your promise in mind and send them at your earliest convenience. I am anxious to publish them, not only because I think they are good, but also to keep before the public that kind of pragmaticism I deem to be sound. You wanted to make alterations and perhaps write a few words of introduction for the new edition.
>
> Perhaps it would not be advisable to change too much because the papers are good as they stand, and the little shortcomings which they may have according to your present conception, or which may have slipped in through the influence of Mr. Youmans, are perhaps not very important. Please let me hear from you, and if possible send the corrected logical papers.
>
> You ought to do so not only to make good your promise but in your own interest as an author and a thinker.[65]

It would take Peirce a long time to respond. On 19 July he drafted a letter to Carus taking up Carus's suggestion to publish the articles pretty much as they were originally written (RL 77:208–17). Peirce acknowledged that it would be best to reprint the Illustrations as they originally appeared, but with three exceptions (RL 77:209):

1st Minor changes to render the English etc. better.
2nd The Articles of 1877 Nov. and 1878 Jan. to have a common title, say "Pragmatic Clearness of Thought," and be divided into two parts, namely, the two articles, each with its original title.
3rd With a prefatial note setting forth what I conceive to be the main error of the Essay, as it originally appeared.... The error of the Essay lies in its Nominalism.

The draft letter, which is incomplete and most likely unfinished, derails in a detailed description of the nominalism-realism debate and its implications. Somewhat dramatically, Peirce folded his sheets and wrote only on half of them, "so that, if I would suddenly die in one of the fits to which I am almost daily subject, what I write could be set up and printed."

It seems that Peirce was unable to complete the letter, leaving Carus once again in the dark, as about two weeks later, on August 1, Carus mailed Peirce a letter in which he practically threatened Peirce that he will publish the Illustrations unaltered, and asked for a clear statement from Peirce that he owns the copyright while also making clear that he doesn't really need such a statement to proceed:[66]

> I have been waiting in vain for a review of your papers "Illustrations of the Logic of Science." You had even forgotten that I had paid you for them, and I am afraid you have again forgotten that after some reluctance you had acknowledged your promise. It was for sundry considerations including the right of re-publishing your Logical papers, revising and proofreading them, writing a new preface, if you were so minded, etc., that I paid you in April 1909 the sum of $250. You have not yet revised your articles and I think that I might publish your articles as they stand. They may contain some objectionable points, which you do not mean to endorse but they are good *as they stand* and contain many valuable ideas. Perhaps it would even be sufficient if you would point out to me the passages which you would have altered and where as you stated in a former letter you yielded to Mr. Youmans's wishes, but at any rate I should like to have your acknowledgment under your own signature that the Open Court Publishing Company has acquired the right of republishing these essays, in which statement you have also to insert that (as you informed me in a former letter) *you*, not Appleton, hold the copyright and are the sole owner of these said essays. *I wish to be* on the safe side in case Appleton or their successors or heirs would later on trouble me, and claim an infringement on their rights. Yet I may add that the copyright has either or will soon run out for the papers were published in 1878. I do not wish to seem to appropriate the article without just a title. I could prove my claims by submitting to any judge our reference to these papers scattered through sev-

> eral letters of correspondence, but I would prefer to have the statement made in *one* document; either in your own handwriting or under your signature. I enclose a form which I wish you would sign and return.[67]

The form which Carus enclosed was short and to the point. Peirce was asked to acknowledge that he owned the copyright and to "authorize the Open Court Publishing Company to publish [the Illustrations] in such a form as *they* may deem advisable" (emphasis added). Peirce responded by drawing up a contract of his own.[68] His version of the contract is dated August 26. Peirce gave the Open Court the right to publish the Illustrations, acknowledged that he had been given that right by Appleton, and added that he "should be glad if they would include with them the article that appeared in that same Monthly in the number for January 1901 either in whole or omitting the first four and a half or five pages."[69] Peirce further granted the Open Court the right to publish the papers, albeit alone or collectively, in the *Monist*. Next he stipulated two conditions: (1) The papers may only be published, whether alone or together, when they are accompanied with a "statement of a change of views on my part as I may furnish upon notice that the printing is to commence at once, my Appendix to be mailed by me within a month of my receiving such notice." (2) Peirce be given the opportunity to, as he put it, "make such corrections in the copy to be used by the printer as I can write clearly in the side margins." Peirce further added, in square brackets, "I entirely abandon my original wish to make further alterations, since I find it far more feasible to state how I have changed my mind in an appendix or appendices" (RL 77:248). The contract, which is signed and dated, survives among the Peirce papers, together with a signed cover letter. There is no indication that Carus ever received such a contract, and the presence of Carus's original contract among the Peirce papers, pristine and unsigned, suggests that Peirce never sent that contract either (RL 77:245).[70]

Contractual issues notwithstanding, Carus's strategy worked: Peirce mailed him a 52-page letter (reproduced below as chapter 9) in which he described his poor health and announced that he would limit himself to a preface in which he stated "in general terms how what I then say ought to be altered," adding that "I will here (if my strength holds out) try to indicate the points I should make." The letter, which Carus received on August 29, is undated, though its latter part is dated separately August 23.[71] The revisions suggested in this letter are far more extensive than what is implied in the contract—the urge to do more than what could be written clearly in the margins of the compositor's copy was clearly far too strong. It is difficult to assess when exactly Peirce worked on *Essays on the Reasoning of Science* (R 334), an undated, marked-up offprint of the Illustrations with a number of significant additions, however handwriting and context suggests that it might have been composed around this time as well. (The alterations are included as footnotes in chapters 1 and 2 below). From manuscripts R 703–4 it is clear that Peirce worked extensively on the third Illustrations article, "The Doctrine of Chances," that same month. These manuscripts are reproduced as notes and an appendix in chapter 3.

Carus responded the very same day he received Peirce's bulky letter, writing, "Your letter and enclosure came duly to hand this morning and I hasten to acknowledge their receipt and express to you my best thanks."[72] In the same letter he warned Peirce that he might not be able to publish the book in the near future: "I only *wish to be ready* to do so *whenever* the time may arrive. I have in preparation an English edition of Couturat's *Logic* and when I see the attempts of others to deal with the logical problem I wish you not to be forgotten, hence my anxiety to be ready to bring you before the public as soon as the time comes."[73] Financial considerations were an issue, and possibly Carus still worried about copyright.[74] That same day Carus also wrote Russell, telling him that he received a letter from Peirce that morning, including "an introduction to his papers on Logic, published some time ago in the *Popular Science Monthly*, which I bought from him." Peirce, Carus added, "feels the approach of his age and seems to think that it would be better not to touch the old papers because he is no longer capable of doing the work to his satisfaction."[75]

Two days later, on September 1, Carus wrote Russell again, offering him the job of editing the Illustrations:

> I'm in receipt of a long letter from Peirce which I think of handing over to your tender mercies. It would involve on you some work which I am sure you will like. It is an explanation of how his logical works should be treated, what other article should be inserted, and what alterations should be made. If you wish I will hand over to you the editing of the papers after you have finished with the Couturat book. Please let me know at once, and I will forward the original copy. Peirce wants it returned but it may be better to have it first copied. He wants to work it out into good shape so as to have it ready for publication for use in connection with the republication of these articles. I assume that you know to what articles he refers, and I also assume that they are in your hands. You might use your own copy and make ready the whole as soon as you can. I am almost inclined to publish the work now and have it set in Chicago provided the copy of the MS. will be in good shape so as not to cause unnecessary corrections in the text. I have a feeling that

> Peirce may die at any time, and I would like to have these articles in as good a shape as possible brought before the public. I have slight fear that he will not improve the articles, and I want you to watch over it. He is apt to insert his later aberrations and at least even though they may be valuable his more intricate notions. This may especially be true of the insertion of his theory of graphs.[76]

Russell replied the following day, stating optimistically "I have Peirce's consent to use all that I have learned from his non-published MSS and letters to me," adding that Peirce himself,

> will never round up his *extensive* projected work on Logic.... the *lure* of study and of anticipated discovery has such a dominance with him that he *cannot* buckle down to the drudgery of working out a report of his progress. He is essentially and constitutionally an explorer and discoverer and not a good expositor at all. He is diffuse on making an unimportant point and laconic and blind to the last degree on many a point of prime import.[77]

In response, Carus forwarded Peirce's letter to Russell, who had three copies typed up at the Open Court's Chicago office on September 17.[78] One copy went to Russell and another to Carus, while the original letter was returned to Peirce.[79] Carus, however, remained unsure whether Russell had actually accepted the job. Almost two months after Russell had the copies made, on November 7, Carus wrote to him, "I have not heard from you for a long time. I wish to know whether you accept my proposition to edit Peirce's logical papers including additions which he suggested to me to make. Please let me hear from you soon because I am anxious to publish the logical papers in good form."[80] Russell responded two days later, writing, "As to Peirce's logical papers, I am at work on them and am delighted to have the task of editing them. Hope to report some progress very soon."[81] Progress clearly did not come soon enough for Carus, as on November 22 he wrote Peirce:

> I am rather disappointed that nothing is doing concerning the republication of your logical papers. I handed it over to Mr. Russell, and he is anxious to do the work. In my opinion he is well prepared for it for he has read every line you have written, and is a most enthusiastic admirer of yours. ... He will accomplish the work by adding footnotes and references to the text and when this is done the result of his labors shall be submitted to you for your approval. I believe that he is better fitted than even you yourself to revise the work for the simple reason that he will stick to the plan more as an editor than you would do. You seem to be anxious to introduce new material which would be foreign to the subject of these logical papers. Your existential graphs should

> not be introduced here nor your theory of probability. They are a subject by themselves and would only burden this splendidly popular exposition of logical papers. Accordingly I advise you to leave out what is not absolutely necessary for the present purpose.
>
> I hope to be able to have this book of yours edited by Russell ready by the end of this year, and would like to be able to present you with a copy during the spring of 1911.[82]

Russell too was disappointed with the lack of progress. Also on the 22nd he wrote to Carus:

> I wrote Peirce a long letter nearly a month ago and expected an answer very soon in reply but he has not replied yet and I have written him again. In my first letter I told him what I understood to be the plan and asked him if I was well informed.... What I personally have been at work on is going over the papers studiously and getting together and collating data for *footnotes*. His extensive though scattered writings afford a wealth of resources to one who like me has kept up with his output, for the elucidation and amplification of every point he makes in the papers in question. My own persuasion is that he has said in some form or other all he has got to say that he *can* say without completely overlaying the papers with a mass of recondite text that will completely extinguish the popular quality of the papers as they are originally made up.
>
> For instance he wants to put in a paper on the Existential Graphs and then he wants to write a new essay on probability depending largely for its justification on these Graphs. If you will look over your copy of his last letter you will see.[83]
>
> In other words I think we could make up a better book without him and still not under-represent him nor mis-represent him. There is not a point touched upon in the papers that he has not within ten years past expressed his mature ideas upon. Now these when collected and referred to in footnotes and perhaps in some instances quoted at length would give a better *survey* of his views in the [work,] as he put them forth in the papers for popular consumption than anything he will be likely to furnish.
>
> Another thing, if he was simply going to furnish a Preface that would be all right enough but he proposes to alter the order of the papers put a chapter on Graphs &c &c. Now would that break up the unity of the original schema? I fear so."[84]

In his response, Carus gave Russell the green light to begin editing the papers, and he agreed with Russell that the Illustrations should not be burdened with the existential graphs and Peirce's views on probability, as doing so would "spoil the logical papers."[85] Afraid that Peirce might die before the project could be completed, Carus was eager to see the work commence, and hoped that the edited version could be sent to the compositor before the end of the year.

Carus also planned to write Peirce about it, but it seems that a bad fall may have prevented him from doing so. He had broken tendons from both kneecaps and, as he explained to Marie de Souza Canavarro, he would be "laid up for some time at the Presbyterian Hospital in Chicago, and … keep both legs in plaster casts for a long time."[86] Russell visited Carus in the hospital on December 19. The Illustrations must have been a topic of conversation, as later that day Russell asked the Chicago office of the Open Court to return to him his personal bound copy of the Illustrations, explaining that he needed it to edit the papers, and he also asked for Carus's copy of Baldwin's *Dictionary*, which he needed to write his notes.[87]

On Christmas Eve, Peirce wrote Carus voicing his strong reservations about Russell adding anything to the Illustrations:

> A postal card from Russell informs me that you do not want the book (i.e. my papers in the Pop. Sci. of 1877–8) added to or subtracted from or altered over very much. Only to have noted my present dissents.
>
> I feel that I owe you my thanks for this decision. I hope no notes will be added, in any form, by anybody else; for I should regard it as very unfair to exclude my own explanations and print in the volume any other person's.[88]

Peirce continued by mentioning the last tortures of his wife's pulmonary tuberculosis, problems with carpenters, masons, and plumbers, the early frosts, and the critical condition of his own health, all in an attempt to show Carus "why I have been unable to get much done on this book, though I have written a good deal which your kind decision will enable me to publish separately, after a reasonable interval." At the end of the letter, Peirce returned to Francis Russell and his role in republishing the Illustrations:

> Our friend F.C.R. [Francis Russell] has I suppose seen his best days. I wrote a few words of thanks to him for his attempt to say in *The Monist* of July 1908 what I was driving at in the last part of my article therein. I did it simply because he so hopelessly missed the point as to convince me that it would be foolish to attempt to set him right, and I wanted to spare the poor fellow's feelings. His wonderings about the doctrine of parallel's (which might better be called the principle of similar triangles) *fixed* my estimate, and I hope he won't do much on the volume to be printed. But, above all, I don't want to hurt his feelings; and I think you had better not show him this letter.[89]

To further ensure that Carus would not share the letter with Russell, Peirce added at the top, firmly underlining each word: "Don't read this to F.C.R."

In a technical sense, Carus conformed to Peirce's wish. He did not show Russell the letter, but instead he paraphrased it:

> I have a letter from Mr. Peirce, which he does not want me to send to you, although I have no objection to telling you what he thinks. He complains about being short of cash, speaks again of his dying wife, and thinks that you have become feebleminded, a thought which it se[ems] you heartily reciprocate. He is afraid of your notes, and wants me to incorporate only his own dissent from his former position. I would propose to have your comments on him published as a comment in the MONIST, and we can use that article as a kind of introduction in a second edition of his logical papers. What do you think of this plan?[90]

The letter clearly quenched Russell's enthusiasm, as he altogether stopped working on the Illustrations. Nothing happened until April, when Carus pretty much threatened to fire him:

> I wish you had sent me some of the work which is in your hands. I have been waiting for it and I ought to have something to show that you are engaged with the Open Court Publishing Company.
>
> Today I send you word that I would like to see you next week so as to explain to you how it is impossible to continue our engagement. It may be disappointing to you, but I do not see how under the circumstances, I can avoid it. I hope to be in Chicago during the next week.[91]

Russell responded defensively, writing, "Last winter you wrote me that Peirce *kicked* against my editing the book and you either said or implied that I was to suspend my work on that behalf, and write an article on the book and the philosophy of Peirce after the publication of the book," and he added that he had been working diligently on the latter.[92] To this Carus responded with a scathing letter, denying that he ever suggested that Russell should suspend his work and accusing him of not doing the work for which he was being paid. "I never told you to stop the work," Carus wrote, "I told you not to insert your critical views in the book itself but to make the book ready by going over the Peirce articles, make your explanations so far as Peirce would not object and bring out your criticism and your own views in a special article to be published in either *The Monist* or *The Open Court*." In contrast to both Peirce and Russell, Carus steadfastly maintained that the original papers were good as they stood, and that they should be interfered with as little as possible. The differences were ironed out and on April 13, Carus wrote an optimistic letter urging Russell to finish up editing the Illustrations:

> I have not heard from Peirce for a longtime, but I think it would be easy for you to let the Peirce logical papers be set in Chicago. The changes ought to be few, if any, only where you are positive that Peirce has changed his views. You can send him proofs and he will naturally reply and write what changes he wants to make while reading the proofs.[93]

Not much happened after that. More than two years later, in June of 1913, having heard nothing from either Peirce or Russell, Carus sent the Illustrations to the British logician Philippe Jourdain for a consult.[94] Since 1910, Jourdain had been in charge of the new London office of the Open Court Publishing Company, and he had become deeply involved not only with the journals—Jourdain took over the editorship of *The Monist*—but also with book publications. Jourdain received the Illustrations on the 7th of July,[95] and after writing Carus several times that he was working on them, finally explained to Carus on August 8, "I did very little in the way of alterations of Peirce's papers on logic, as I do not think that old papers, when republished, should be cut about more than necessary. They should have a good sale, and I hope they will be published soon."[96]

Jourdain's response gave Carus the confidence he needed to edit the Illustrations himself. Fairly certain that the articles would be free of copyright, he added several notes and wrote the "Publisher's Preface" that is included in the current edition as well.[97] On August 23, 1913, he sent a copy of his preface to Peirce, adding the following in an accompanying letter:

> I have gone over your Logical Papers and have written a preface of which I enclose a copy. I believe you will be quite pleased at having these lectures on Logic published. You will remember that I paid you for the republication, further that you have assured me that you possess the right to republish it. Moreover the copyright of the *Popular Science Monthly* either has expired or will expire very soon. In case your copyright should be doubtful in any way, I would rather wait until the copyright expires.
>
> You wish to have it revised because you think that you could improve several passages, but I believe that the article is very good as it stands. Moreover I have submitted this series of articles to one of the most critical modern logisticians, Mr. Philip E. B. Jourdain. He has returned the copy to me, and approves its publication as it stands.[98]

Peirce quickly responded, explaining his long silence by telling Carus that on December 13, 1911, he slipped on the well-polished floor in his study breaking a few ribs, "the broken ends sticking into the pleura and almost stopping my breath." This, Peirce continued, long prevented him from doing any strenuous intellectual work. He further reiterated his wish to have his review of Pearson's *Grammar of Science*[99] included, or some of his earlier papers, such as his "old paper on Logical Extension and Comprehension" (W2, sel. 6), for which, he added, "I have a lot of notes ... which I will send you as they are pasted into stiff covers and can be easily looked over."[100] Carus was open to the suggestion of including the Pearson review, but suggested in lieu of republishing earlier papers Peirce write a resume of his theories that "will whet the appetite of readers to study your system more carefully in your former publications."[101] Carus further warned Peirce that he did not think the Open Court would make any money of the papers ("Books on logic are not much bought."), but that it was his main intention to keep Peirce's thought before the public.

Two months before Peirce's death Russell wrote a long letter to Carus, complaining that for a long time Peirce was not answering his letters. Russell continued his letter with a portrait of Peirce that is markedly different from his often quoted obituary of Peirce. "Peirce," he writes, "is a case of a remarkably penetrating mind, well-furnished with appropriate information but just as remarkably lacking in synthetic faculty. I don't believe he has ever been able to master and deploy the several *very, very* fertile insights and atsights he has had so as to organize the same into a sound system with what was already known." Russell cites the Illustrations as an example:

> No better example of the 'scrappiness' of his work can be cited than his six papers in the *Popular Science Monthly.* The domain of logic as traditionally conceived, that is to say, formal logic, embraces 1st a doctrine of Terms, 2nd a doctrine of the Proposition, 3rd a doctrine of the Syllogism, and 4th a doctrine of Fallacies. To this many logicians have annexed a doctrine of Induction. To me it is evident that Peirce would also add a doctrine of Hypothesis. But in the *Popular Science Monthly* Papers Formal Logic or the doctrine of necessary reasoning is referred to only in an almost incidental way while most of the text is not properly logical at all unless Methodology is to be taken as properly logical while his doctrine of Induction and his doctrine of Hypothesis is very cursorily treated. Nothing whatever therein leads up to any branch of Calculative (Boolean) Logic or to the extension of the same to the exigencies presented by the use of relative terms, that is to say, the Logic of Relations. In fact the papers are "*scrappy,*" "*scrappy,*" "*scrappy.*"
>
> Even with respect to these scraps Peirce seems not to be satisfied with them... He wants to alter paper No.2 in respect to the illustrations of the applications of his test of *meaning.* Also he wants to enunciate a new doctrine of mathematical probability by holding that Laplace and Mill have a vicious influence in this regard. This is as much as to say that Mill's doctrine of induction has no

> sound basis, for materialistic probability as opposed to subjective probability must be founded upon a perfectly sound mathematical theory of probability and any doctrine of induction founded upon subjective probability is fallacious as has been shown by Boole and others.[102]

Peirce died on April 19, 1914. Three days later, Carus asked Russell to write Peirce's obituary for the *Monist*, adding that he is still planning to publish the Illustrations, but that Russell better not mention it as he is does not yet know how and when to publish them.[103]

The subject of republishing the Illustrations resurfaces three years later, in December 1917. Carus had received a letter from Irving C. Smith from Harvard, who Carus explained to Russell was working on a posthumous edition of Peirce's work, which presumably was to include the Illustrations, though Smith called them "Peirce's stories in logic."[104] In his response to Smith, Carus wrote that several years earlier he had purchased the rights to those papers from Peirce "for a couple of hundred dollars," and had not yet "come to a conclusion concerning the republication of these papers on logic." Smith, as it turns out, was assisting Morris Cohen, whose 1923 *Chance, Love, and Logic,* was to be the first posthumous edition of Peirce's papers. Carus's claim that he possessed the rights to the Illustrations did little to stop Cohen. *Chance, Love, and Logic* consists of two parts. The first, "Chance and Logic," is nothing other than the Illustrations; the second, "Love and Chance," reprints five other articles Carus had the copyright to: the five cosmological papers that Peirce published in the *Monist* in the early 1890s. Carus himself never ended up publishing the Illustrations, nor did he live to see *Chance, Love, and Logic* appear in print. He died in 1919, four years before the book came out.

Chance, Love, and Logic is not the only posthumous work to include the Illustrations. Possibly the most influential rendition appeared in the *Collected Papers* of the 1930s. In the *Collected Papers*, however, the Illustrations is divided over several volumes, and even where three of the articles are included back-to-back, they are out of sequence.[105] The *Collected Papers* editors did include some of Peirce's later reworking, but they did so in a way that makes it difficult to see where their printed text was truly drawn from, leading to significant confusion, as Peirce's views shifted over time. In 1986 a far more responsible version appeared in volume 3 of the 30 volume *Writings of Charles S. Peirce: A Chronological Edition.* This edition, however, includes only the papers as they originally appeared in the *Popular Science Monthly* and in *Revue Philosophique de la France et de L'Étranger*, reserving subsequent revisions by Peirce for later volumes, yet to be published. The current edition, which began with an email to Paul

Carus's grandson, André Carus, seeks to combine Peirce's original *Popular Science* articles and Peirce's subsequent attempts at reworking them in a single volume, and it does so in a way that makes it easier to see when Peirce altered what, and for what purpose. This volume too, however, is unlikely to contain the final word on the Illustrations: words do not stay still.

Cornelis de Waal
2013

NOTES

1. Some details can be found in Cornelis de Waal, "The History of Peirce's 1894 Logic Book," *Peirce Project Newsletter* 3.2 (2000): 4–5. *How to Reason* is forthcoming as volume 11 of the *Writings of Charles S. Peirce.*

2. The most influential attribution probably occurs in William James's *The Varieties of Religious Experience: A Study in Human Nature* (New York: The Modern Library, 1902), 434–35. James's first attribution in print appeared four years earlier in "Philosophical Conceptions and Practical Results," *The University Chronicle* (Berkeley, California). The latter was reprinted, with some modifications, as "The Pragmatic Method" in *Journal of Philosophy* 1 (1904): 673–87. It is found also in *Pragmatism: A New Name for Some Old Ways of Thinking* (New York: Longmans, 1907), 46–47.

3. An earlier account of the publication history of the Illustrations is found in Donald R. Koehn, *Charles S. Peirce's "Illustrations of the Logic of Science,"* MA Thesis, University of Illinois at Urbana-Champaign, 1966.

4. Robert Crease, "Charles Sanders Peirce and the first absolute measurement standard," *Physics Today* (December 2009): 29–44, 39.

5. Whitehead called Peirce the American Aristotle in a letter to Charles Hartshorne. Paul Weiss in his *Dictionary of American Biography* article on Peirce called him the American Leibniz.

6. Raymie E. McKerrow, "Richard Whately and the revival of logic in nineteenth-century England," *Rhetorica* 5.2 (1987): 163–85.

7. Julius Adolph Stöckhardt, *The Principles of Chemistry, Illustrated by Simple Experiments* (Philadelphia: E. H. Butler, 1861); Heinrich Will, F. Voehler, Justus Liebig, and Francois Malepeyre, *Nouveau Manuel Complet de Chimie Analytique* (Paris: Librairie Encyclopédique de Roret, 1855).

8. For a detailed discussion of the instrument, see Klaus Staubermann, "The Trouble with the Instrument: Zöllner's Photometer," *Journal for the History of Astronomy* (2000): 323–38.

9. The letter is included in Christine Ladd-Franklin, "Charles S. Peirce at the Johns Hopkins," *Journal of Philosophy, Psychology and Scientific Methods* 13.26 (21 December 1916): 715–22; see esp. 718–20.

10. This being said, with regard to the preservation of manuscript material we should distinguish between the period before Peirce moved to Milford and the period thereafter. Whereas Peirce moved around quite a bit during the early part of his life, in Milford he remained in a single location that was spacious enough to not having to throw anything away.

11. A little over a year earlier, Peirce published in the *North American Review* a long essay inspired by Frazer's edition of Berkeley's *Works*. Peirce's review received in turn a positive review by Chauncey Wright in the *Nation*. See W2, sels. 48–50.

12. William James, *Essays in Philosophy* (Cambridge: Harvard University Press, 1978), 352, 364.

13. See e.g., ibid., 335.

14. *Ohio Medical Recorder* 2.5 (October 1877): 236.

15. Charles Peirce to Sarah Mills Peirce, 8 April 1878.

16. Ibid.

17. John Stuart Mill, *A System of Logic: Ratiocinative and Inductive* (8th edition; New York: Harper & Brothers, 1882): Bk. III, Ch. iii, § 1, 225.

18. Written by Peirce in response to an October 26, 1904 request by Matoon Monroe Curtis, who was responsible for the section "Geschichte der Philosophie in Nord America," for Friedrich Überweg and Max Heinze, *Grundriß der Geschichte der Philosophie.*

19. "Comment se fixe la croyance," *Revue philosophique* (December 1878): 553–69; "Comment rendre nos idées claires," *Revue philosophique* (January 1879): 39–57.

20. Ladd-Franklin, "Charles S. Peirce at the Johns Hopkins," 719.

21. See Gérard Deledalle, "Peirce's First Pragmatic Papers (1877–1878)," in *Charles S. Peirce's Philosophy of Science: Essays in Comparative Semiotics* (Bloomington: Indiana University Press, 2000), 23–36.

22. That the Illustrations inspired Carus to recruit Peirce can be surmised from a letter Carus sent to Peirce on January 30, 1907 (27.91.23; reference to sources held in the Open Court Archives at SIU's Morris Library are by collection, box, and folder number).

23. Charles Peirce to Edward Hegeler, undated, but most likely mid-February 1893 (27.91.7).

24. The table of contents for *A Quest of a Method* lists in historical order three papers of the 1867 American Academy series (W2, sel. 3, 4, 6), all three 1868 *Journal of Speculative Philosophy* papers (W2, sel. 21–23), the 1871 Berkeley review (W2, sel. 48), the Illustrations, the 1883 "Theory of Probable Inference" (W4, sel. 64), and the *Monist* papers of the early 1890s (W8, sel. 23, 24, 27, 29, 30).

25. Both versions of this first essay are scheduled to be published in an appendix in W11.

26. This work is posthumously also referred to as the *Grand Logic* (see e.g. CP 8:278), possibly to contrast it with Peirce's 1902–3 *Minute Logic*. There is no evidence that Peirce ever referred to this work (or any other work) by the name *Grand Logic.*

27. See Peirce's September 8, 1894 letter to Francis Russell for two earlier tables of contents, the first describing "its present condition," the second "how I would like to improve it." The first table of contents includes the first four Illustrations papers; the second matches the description of the second volume of the announced twelve-volume *Principles of Philosophy* entitled *Theory of Demonstrative Reasoning*, which Peirce states is "substantially ready" (CP 8:284). Judging the letter to Russell, this second volume might have included "The Fixation of Belief" and "How to Make Our Ideas Clear," though Peirce's description of the fifth volume, *Scientific Metaphysics*, suggests he might have planned to use them there.

28. A table of contents for *How to Reason* can be found in CP8:278–80, which also indicates which parts were published in the *Collected Papers*. A new edition of *How to Reason* is forthcoming as volume 11 of the *Writings*.

29. The same is true for the first table of contents given in the September 8, 1894 letter to Francis Russell, where "The Fixation of Belief" is listed as chapter 4, "The Doctrine of Chances" as chapter 13, "The Theory of Probable Inference" as chapter 14, and "How to Make Our Ideas Clear" as chapter 15.

30. In an September 8, 1894 letter to Francis Russell devoted primarily to *How to Reason*, Peirce explains: "By transcendental I mean in Kant's sense that is the epistemological aspect."

31. In the first table of contents in his September 8, 1894 letter to Russell, Peirce even went so far as to move the third and fourth Illustrations papers between the first two.

32. Charles Peirce to Francis Russell, September 8, 1894.

33. The paper is included in *The Writings of William James: A Comprehensive Edition*. Edited by John J. McDermott (Chicago, 1977), see esp. 347–48.

34. William James, *The Varieties of Religious Experience: A Study of Human Nature* (London: Longmans, Green, and Co.), 444f.

35. The most complete edition of the lectures, edited by Patricia Ann Turrisi, is found in *Charles S. Peirce, Pragmatism as a Principle and Method of Right Thinking: The 1903 Harvard Lectures on Pragmatism* (Albany: SUNY Press, 1997).

36. *Collected Papers*, vol. 8, 300, entry G 1909–1; see also Robin's catalogue, R 229).

37. The contents of this desk are not included in Richard Robin's catalogue (which had been published two years earlier). They were catalogued separately by Robin in "The Peirce Papers: A Supplementary Catalogue," *Transactions of the Charles S. Peirce Society* 7.1 (1971): 37–57. This discovery was also made after volume 8 of the *Collected Papers* appeared. The letter to Carus is preserved in the Open Court Archives at the Morris Library, Southern Illinois University. This January 7, 1909 letter of Peirce to Carus is found at 27.91.25; two drafts of this letter are preserved at Harvard, RL 77:198–203.

38. Paul Carus to Charles Peirce, January 23, 1907, 27.91.23; RL77:190.

39. Paul Carus to Charles Peirce, January 30, 1907, 27.91.23; RL77:191.

40. William James to Paul Carus, February 18, 1907, 27.91.7.

41. Paul Carus to Charles Peirce, December 2, 1907, 27.91.23.

42. Paul Carus to Charles Peirce, December 9, 1908, 27.91.24.

43. Ibid.

44. Ibid. A letter from M.A. Sacksteder, the Open Court's office manager in Chicago, to Paul Carus, dated 9 February 1914, and unrelated to Peirce or the Illustrations, sheds some light on the copyright situation: "As regards the copyright law, the first period is 28 years and then by making a proper application, a renewal may be made for 14 years thus making the total length of the copyright 42 years" (27.12.14).

45. Charles Peirce to Francis Russell, January 1, 1909 (RL 387:445).

46. A draft of the letter, dated January 6, reveals that part of the reason why Peirce could not accept it was the work he was still engaged in for the *Monist*; Peirce did not mention this in the letter he sent (RL 77:199).

47. The reference is to Ernst Mach's *Science of Mechanics*, the English translation of which was published by the Open Court in 1893. For this translation Peirce had been an advisor.

48. For a general overview of Peirce's classification of the sciences see Beverly Kent, *Charles S. Peirce: Logic and the Classification of the Sciences* (Montreal: McGill-Queen's University Press, 1987).

49. Charles S. Peirce, "Pearson's *Grammar of Science*. Annotations to the First Three Chapters," *Popular Science Monthly* 58 (January 1901): 296–306 (P802).

50. Charles S. Peirce, "The Century's Great Men in Science," *Evening Post*, New York City (Saturday, January 12, 1901), section 3, page 1, columns 1–3 (P779).

51. Charles Peirce to Paul Carus, January 7, 1909, 27.91.25; a draft is preserved at Harvard (RL 77:202–3).

52. Paul Carus to Charles Peirce, January 11, 1909, 27.91.25.

53. Ibid. On the difference between pragmatism and pragmaticism, see especially Peirce's 1905 "What Pragmatism Is," which Carus published in the *Monist* (EP2, sel. 24, esp. 335).

54. Paul Carus to Charles Peirce, February 6, 1909, 27.91.25.

55. Charles Peirce to Paul Carus, March 9, 1909, 27.91.25.

56. Paul Carus to Charles Peirce, March 15, 1909, 27.91.25. Inter-office correspondence leaves the impression that Carus pushed for Peirce to get paid, even though the company's publishing funds were low at the time.

57. Charles Peirce to Paul Carus, 27.91.25.

58. Ibid.

59. Paul Carus to Francis Russell, October 20, 1909, 27.91.25.

60. Paul Carus to Francis Russell, December 6, 1909, 27.91.25.

61. Looking at the manuscripts written for the Illustrations during 1909, we can estimate how long Peirce worked each day and the speed of his writing, as he gives each sheet a date and time stamp.

62. This work is preserved as R 643–50.

63. Charles Peirce to Paul Carus, December 14, 1909, 27.91.25.

64. Paul Carus to Charles Peirce, December 17, 1909, 27.91.25. Note, though, that on December 25, 1909 Peirce wrote William James, "my Essay on Pragmatism, as it will appear in the revised edition, is more important than the Evolution of Laws of Nature series"—the latter is a reference to the papers Peirce wrote for the *Monist* in the early 1890s.

65. Paul Carus to Charles Peirce, April 30, 1910, 27.91.26.

66. This letter is practically identical to one Carus wrote on July 27, 1910, but apparently did not send (27.91.26).

67. Paul Carus to Charles Peirce, August 1, 1910, 27.91.26. On Peirce's copy of the letter, but absent on the carbon copies, Carus played with the possibility of publishing the Illustrations without any mention of their former publication by Appleton.

68. The contract Carus sent Peirce is preserved at Harvard as RL 77:245; Peirce's (draft) versions of the contract are preserved as RL 77:246–51.

69. RL 77:247. Peirce did not make any reference to copyright issues related to the 1901 paper—a paper that was written for the Monthly after Appleton had sold the magazine.

70. In the margin of the cover letter, which Peirce also held back, Peirce wrote, "Of course I will sign that paper & forward it. I can't at the moment lay my hands on it" (RL 77:253). There are no signs that he ever did send Carus a contract.

71. Further down Peirce entered a third date, this time only stating the month without identifying the day, suggesting he did not complete the letter on the 23rd. Furthermore, in the unsent cover letter of August 26 Peirce mentions "a long letter that should have gone two days ago" (RL 77:252–53).

72. Paul Carus to Charles Peirce, August 29, 1910, 27.91.26.

73. Ibid.

74. On the copy of the letter Peirce received (RL 77:254) Carus had written by hand "At present our funds are low and I have to economize"—a comment not found on the carbon copies of the typed letter.

75. Paul Carus to Francis Russell, August 29, 1910, 27.91.26.

76. Paul Carus to Francis Russell, September 1, 1910, 27.91.7.

77. Francis Russell to Paul Carus, September 2, 1910, 27.91.7.

78. See Paul Carus to Francis Russell, September 3, 1910, 27.91.7, and Catherine Cook to Paul Carus, September 19, 1910, 27.14.3. The typescript is preserved at 27.91.27.

79. Most likely the third typescript was mailed to Peirce together with the original holograph, as one sheet of the typescript survives among the Peirce papers at Harvard (RL 77:242). This sheet, which is in a very poor shape, contains corrections that appear to be in Peirce's hand.

80. Paul Carus to Francis Russell, November 7, 1910, 27.91.26. Though Carus forwarded Peirce's letter to Russell on September 3, he did not interpret Russell's September 2 letter as a sufficiently clear statement that Russell would take on the project. When forwarding the letter, Carus wrote: "You can understand that

I myself would not have the time to edit Peirce's papers. If I would wait for him he would never do it, and so I would be very glad if you would take charge of it" (Paul Carus to Francis Russell, September 3, 1910, 27.91.7).

81. Francis Russell to Paul Carus, November 9, 1910, 27.14.5.

82. Paul Carus to Charles Peirce, November 22, 1910, 27.91.26.

83. This letter is reproduced below as chapter 9.

84. Francis Russell to Paul Carus, November 22, 1910, 27.91.26.

85. Paul Carus to Francis Russell, November 28, 1910, 27.91.26.

86. Paul Carus to Marie de Souza Canavarro, December 6, 1910, 27.40.6.

87. Francis Russell to Lydia Robinson, December 19, 1910, 27.91.26. James Mark Baldwin, *Dictionary of Philosophy and Psychology*, 3 vols. (New York: Macmillan, 1901–05). Peirce's 181 contributions to the dictionary, all signed C.S.P, are catalogued as P 761–78, 806–970.

88. Charles Peirce to Paul Carus, December 24, 1910, 27.91.26.

89. Ibid.

90. Paul Carus to Francis Russell, December 30, 1910, 27.91.26.

91. Paul Carus to Francis Russell, April 1, 1911, 27.91.27.

92. Francis Russell to Paul Carus, April 5, 1911, 27.91.27.

93. Paul Carus to Francis Russell, April 13, 1911, 27.91.27. In the Peirce papers at Harvard the first two sheets are preserved of an undated holograph editor's preface to the Illustrations that is clearly in Russell's hand. In rather bland prose Russell wrote that the papers "are issued in their original text," thereby preserving the format "upon which the voluminous literature of Pragmatism has arisen" (RL 387b:471–2).

94. Philippe Jourdain to Paul Carus, July 5, 1911, 27.43.77; earlier Jourdain had listed "Peirce's papers" among several prospected book publications (Jourdain to Carus, November 28, 1911, 27.43.77).

95. Philippe Jourdain to Paul Carus, July 7, 1911, 27.43.77.

96. Philippe Jourdain to Paul Carus, August 8, 1913, 27.43.77; see also Jourdain to Carus July 9 and 11 in same folder. In the latter, Jourdain writes: "Peirce is getting on quickly."

97. The Publisher's Preface and a printers' copy of the Illustrations articles, lightly annotated by Carus, are preserved in 27.91.40.

98. Paul Carus to Charles Peirce, 23 August 1913, 27.91.28.

99. Charles S. Peirce, "Pearson's *Grammar of Science.* Annotations on the First Three Chapters," *Popular Science Monthly* 58 (January 1901): 296–306; P802. Draft material related to this is preserved as R 1434.

100. Charles S. Peirce to Paul Carus, August 28,1913, 27.91.28. Most likely this is a reference to an undated notebook preserved in R 725 (WMS 170), which has the original publication pasted in it with ample handwritten annotations. Peirce used these annotations when updating the article for *How to Reason* (R 421).

101. Paul Carus to Charles Peirce, December 10, 1913, 27.91.28.

102. Francis Russell to Paul Carus, February 20, 1914, 27.91; Russell's Obituary appeared as "In Memoriam Charles S. Peirce," *Monist* 24 (July 1914): 469–72.

103. Paul Carus to Francis Russell, April 22, 1914, 27.91.7.

104. Irving Cranford Smith, a student of physics and mathematics at Harvard, published anonymously "Further Bibliography of the Writings of Charles S. Peirce," *Journal of Philosophy* 15.21 (October 10, 1918): 578–84, which is a supplement to Morris Cohen's "Charles S. Peirce and a Tentative Bibliography of His Published Writings," *Journal of Philosophy* 13.26 (December 21, 1916): 726–37. Smith was also involved in Cohen's *Chance, Love, and Logic: Philosophical Essays* (New York: Harcourt, Brace, and World, 1923), the first posthumous edition of Peirce's papers.

105. In the *Collected Papers* the Illustrations papers are found at the following locations: "The Fixation of Belief" (CP 5.358–87); "How to Make Our Ideas Clear" (CP 5.388–410); "The Doctrine of Chances" (CP 2.645–68); "The Probability of Induction" (CP 2.669–93); "The Order of Nature" (CP 6.395–427); "Deduction, Induction, and Hypothesis" (CP 2.619–44).

Chapter 1

THE FIXATION OF BELIEF

Peirce opens his "Illustrations of the Logic of Science" by developing a general theory of inquiry while seeking to avoid introducing metaphysical preconceptions. Peirce characterizes inquiry as the process through which we seek to alleviate doubt and (re)gain belief, and he argues against a notion of truth that is independent of belief. After spending considerable time defining doubt and belief, Peirce distinguishes four methods for attaining (or fixating) belief, adding that the aim of the series is to further explicate the fourth method, the method of science. In the process Peirce advocates for a social epistemology in which the fixation of belief is a decidedly communal affair. Over time Peirce expresses various misgivings about the article, many of which are made explicit in the various attempts to rewrite the paper for republication.

"The Fixation of Belief" first appeared in The Popular Science Monthly *12 (November 1877): 1–15 as the first of six articles published under the general heading "Illustrations of the Logic of Science." The text below is as printed in the* Popular Science Monthly. *It is only minimally edited and italic brackets surround text added by the editor. A French translation appeared as "Comment se fixe la croyance" in* Revue Philosophique de la France at de L'Étranger *6 (December 1878): 553–69. The footnotes contain discrepancies with the French version and alterations made by Peirce when preparing the text for inclusion as the fifth chapter in his 1894 logic book*

How to Reason *(R 407, with drafts in R 1002, 1009—hereafter referred to as HTR) and as part of the opening essay in* Essays on Reasoning *of circa 1910 (R 334, hereafter referred to as ER). Sections III and IV, and most of section V Peirce copied almost verbatim from chapters 1–3 of his unpublished and unfinished 1872–73 logic book. Where known the date of composition is given. A discussion of the history of the article, its French translation, and the various attempts at rewriting it, is found in the introduction.*

I.*

Few persons care to study logic, because everybody conceives himself to be proficient enough in the art of reasoning already.† But I observe that this

*In HTR (1894) the first three sections are completely rewritten. Rather than including it as a footnote, the three sections are reproduced as an appendix to this chapter. In ER (ca.1910) section I. is preceded by the following brief introduction (R 334:2¬4):

> A title to any writing ought to inform whomsoever may be fitted to read it, as well as can conveniently be done in a few words, what the nature of it will be found to be. Since my title *[Essays on Reasoning]* perhaps fails to do this I will at once explain it. Now Reasoning is generally regarded as the subject of Logic; but to have entitled this essay a discourse upon Logic would have conveyed a singularly false notion of it. At least fifty different definitions of Logic have been given. But the overwhelming majority of them follow those by which Aristotle inaugurated the systematic analysis of argumentations, in confining that subject, some exclusively, all in the main, to the study of the distinction between such as lead only to conclusions which absolutely must be true, so long as their premises, or assumed initial facts are so, and those "fallacious" reasonings that aim at being evident but fail to attain that aim. As long as such reasoning as that was considered to be the highest function of the human intellect (excepting for perhaps very rare operations supposed to be inspired by some divinity, on unique occasions when nothing at all like a reasoning process could be discerned,) it seemed quite needless to confirm the conclusions of reasonings by observation, since they were either evidently beyond cavil or else worthless. The result, as anybody who knows what the real nature of reasoning is might anticipate, was that the whole business of intellectual men came to consist in endless disputations, of which the premises for the rank and file were the idlest fictions, while for the really eminent minds they were sentences of ecclesiastical authorities who might or might not have dreamed of their words ever receiving such interpretations as were put upon them by the disputant who quoted any of them for the purpose of confounding his opponent. As for the system of theology that was so built up, if it had really no better foundation than it professed to have, and no instinctive good sense had furtively leaked in, in spite of all the apparatus to exclude it, and keep the doctrine stanch, it would have been even more unwholesome than it was.

The *Collected Papers* include at this point a note (CP 5.358n) wherein the editors quote what they identify as the introduction to an essay entitled "My Plea for Pragmatism" which was to combine the first two Illustrations articles into one. Though the editors date the manuscript 1903, the text they reproduce proves identical with

satisfaction is limited to one's own ratiocination, and does not extend to that of other men.[‡]

We come to the full possession of our power of drawing inferences the last of all our faculties, for it is not so much a natural gift as a long and difficult art.[§1] The history of its practice would make a grand subject for

---

R 619:2–3 which is dated, by Peirce, March 26, 1909. For a more detailed description see the introduction. The *Collected Papers* reproduces only the first paragraph, which is reproduced here as well. The remainder is biographical, resembling what is found in chapter 7 below. From R 619:2–3:

> The two chapters composing this Essay [titled "The Import of Thought: An Essay in Two Chapters"] were first published, without any title for the whole, in the *Popular Science Monthly* for November 1877 and January 1878. A French version by the author (the second having, in fact, been first written in French on board a steamer in September 1877,) appeared in the *Revue Philosophique*, Vols. VI and VII. They received as little attention as they laid claim to; but some years later the potent pen of Professor James brought their chief thesis to the attention of the philosophic world (pressing it, indeed, further than the tether of their author would reach, who continues to acknowledge, not indeed the *Existence*, but yet the *Reality*, of the *Absolute*, nearly as it has been set forth, for example, by Royce in his *The World and the Individual*, a work not free from faults of logic, yet valid in the main). The doctrine of this pair of chapters had already for some years been known among friends of the writer by the name he had proposed for it, which was "Pragmatism."

[†]In what appears to be a draft of ER, Peirce writes instead: "Few care to study logic, because almost everyone conceives himself to be an adept in drawing conclusions already;" (RS 77:2).

[‡]"My Pragmatism" of 1909, reproduced below as chapter 7, comes down to an aborted elaboration upon this first paragraph.

[§]In the quickly aborted *Studies in the Meanings of Our Thoughts* (1909) Peirce continues as follows, replacing the period with a comma (R 629:2–3):

> and though the pure eyes of reason of many children have a penetration that they are apt to lose during their teens, yet this is the one sense that, with assiduous practice, will retain its edge beyond the military years, and even be keener at seventy than it was at sixty.

A more cynical version is found in the draft of ER quoted from above (RS 77:2–3):

> We certainly come to the full possession of our power of reasoning the last of all our faculties, and only after all the others are on the road of decay. For though young children are often exceedingly penetrating in certain limited fields of thought, in mathematics for example, yet it usually happens that as maturity approaches a dense ~~veil of stupidity~~ cloud passes over the azure of their intellect and clears away so slowly that it is only in the fifties or sixties that ~~most minds thoroughly ripen~~ *[the]* reasoning power of those who reason best attains its ~~perfect~~ best ripening; how fully this best ripening is depends on how much pains they have taken, not merely to shatter Bacon's four idols but to perfect themselves in all the positive parts of this long and difficult art.

a book. The mediæval schoolmen, following the Romans, made logic the earliest of a boy's studies after grammar, as being very easy. So it was, as they understood it.* Its fundamental principle, according to them, was, that all knowledge rests on either authority or reason;† but that whatever is deduced by reason depends ultimately on a premise derived from authority. Accordingly, as soon as a boy was perfect in the syllogistic procedure, his intellectual kit of tools was held to be complete.

To Roger Bacon, that remarkable mind who in the middle of the thirteenth century was almost a scientific man, the schoolmen's conception of reasoning appeared only an obstacle to truth. He saw that experience alone teaches anything—a proposition which to us seems easy to understand, because a distinct‡ conception of experience has been handed down to us from former generations; which to him also seemed perfectly clear, because its difficulties had not yet unfolded themselves. Of all kinds of experience, the best, he thought, was interior illumination, which teaches many things

---

*An incomplete square-bracketed insertion survives on R 630:6, 1909:

> *[*Speaking*]* classical Latin correctly required considerably more training in ancient times than speaking English does now; and to speak and write it with elegance was a much greater accomplishment than it is to use English as well; and therefore to rank logic with grammar and rhetoric did not imply that it was a *trivial* matter, in the modern sense. Moreover, logic had, during the XIIth and XIIIth centuries made considerable progress beyond its condition when Appuleius wrote in the IInd century of our era. In fact it well represented the real state of thought in the middle ages.

†In R 630:6–8, 1909, Peirce revised the remainder of this paragraph as follows and added a new paragraph in square brackets:

> but in practice every formula of reason was deferred to the authority either of the infallible church or else to that of Aristotle; and although the audacious Abelard insisted that Aristotle might have erred, he was constrained to admit that "doubtless he never did."
>
> [The training of boys was almost exclusively in disputations, which were hedged with rules, called *Obligationes,* for keeping them to the point at issue. They entered into these exercises with all their might and disputed incessantly from morning till night. When a syllogism struck a disputant straight in the head, his only possible reply would be "distingue," with an explanation that in admitting the premisses, he had in one of them taken the middle term in one sense, and in the other premiss in another sense. Since this situation was continually recurring, they became extremely acute in drawing distinctions, acquired great exactitude of thought and language, without often penetrating to the real substance of the questions in dispute (when real substance they had), and learned to command great concentration of mind in keeping in view all the points of disputations whose intricacy often became bewildering to an ordinary intelligence; while they were apt to lose all practical good sense.]

‡In 1909 (R 630:8) Peirce uses the word "adequate" instead.

about Nature which the external senses could never discover, such as the transubstantiation of bread.*[2]

Four centuries later, the more celebrated Bacon, in the first book of his *Novum Organum*, gave his clear account of experience as something which must be open to verification and reëxamination. But, superior as Lord Bacon's conception is to earlier notions, a modern reader who is not in awe of his grandiloquence is chiefly struck by the inadequacy of his view of scientific procedure. That we have only to make some crude experiments, to draw up briefs of the results in certain blank forms, to go through these by rule, checking off everything disproved and setting down the alternatives, and that thus in a few years physical science would be finished up—what an idea! "He wrote on science like a Lord Chancellor,"† indeed.‡

The early scientists, Copernicus, Tycho Brahe, Kepler, Galileo, and Gilbert,§ had methods more like those of their modern brethren. Kepler undertook to draw a curve through the places of Mars;** and his greatest service to science was in impressing on men's minds that this was the thing to be done if they wished to improve astronomy; that they were not

---

*In 1909 (R 630:8) Peirce makes small modifications to the entire paragraph, without making any substantial changes, and then inserts the following paragraph in square brackets:

> [So Galileo, three centuries later, relied much upon *il lume naturale;* but the objects this light enabled him to see were very different from transubstantiation; and it is very important to distinguish any light of nature or of grace from *experience.* Experience, in the proper sense of the term, is all that one has *gone through.* It consists in the *events* of one's life. But a "light" is a faculty enabling its subject to recognize the characters of what future experience may put before him. Neither the one nor the other, nor any combination of these two alone can *teach* him anything, if we understand by "teaching" the communication of the skill and power to conduct oneself so as to attain a desired result; although both are indispensible to such teaching. What else is requisite it is the object of these studies as far as may be to ascertain.]

†The *Collected Papers* inserts the following footnote, marked as being added by Peirce at a later date but without disclosing where they found it: "Cf. J. Aubrey's *Brief Lives* (Oxford ed. 1898), I, 299."

‡In ER (R 334:41, ca.1910) Peirce adds, replacing the period with a comma: "as Harvey, a genuine man of science said."

§In ER (R 334:41) Peirce adds Harvey to the list.

**[Original Peirce footnote:] Not quite so, but as nearly so as can be told in a few words. In R 333:12 Peirce writes "This is not an exact statement, but it is near enough for the purpose." In ER (R 334:41, ca.1910) Peirce deletes the footnote, removes the semicolon after "Mars," and inserts: "and to state the times occupied by the planet in describing the different parts of that curve; but perhaps"

to content themselves with inquiring whether one system of epicycles was better than another, but that they were to sit down to the figures and find out what the curve, in truth, was. He accomplished this by his incomparable energy and courage, blundering along in the most inconceivable way (to us),* from one irrational hypothesis to another, until, after trying twenty-two† of these, he fell, by the mere exhaustion of his invention, upon the orbit which a mind well furnished with the weapons of modern logic would have tried almost at the outset.[3]

In the same way, every work of science great enough to be remembered for a few generations affords some exemplification of the defective state of the art of reasoning of the time when it was written; and each chief step in science has been a lesson in logic. It was so when Lavoisier and his contemporaries took up the study of chemistry. The old chemist's maxim had been, *"Lege, lege, lege, labora, ora, et relege."*[4] Lavoisier's method was not to read and pray, not to dream that some long and complicated chemical process would have a certain effect, to put it into practice with dull patience, after its inevitable failure to dream that with some modification it would have another result, and to end by publishing the last dream as a fact: his way was to carry his mind into his laboratory, and to make of his alembics and cucurbits instruments of thought,‡ giving a new conception of reasoning, as something which was to be done with one's eyes open, by manipulating real things instead of words and fancies.§[5]

The Darwinian controversy is, in large part, a question of logic. Mr. Darwin proposed to apply the statistical method to biology.**[6] The same thing had been done in a widely different branch of science, the theory of gases.†† Though unable to say what the movements of any particular molecule of

---

*In ER (R 334:41) Peirce deletes "in the most inconceivable way (to us)".

†In the 1878 French translation (W3:340) this number is reduced to twenty-one (*vingt et une*).

‡In ER (R 334:41) Peirce calls them "veritable instruments of thought".

§In an undated fragment, most likely written in the 1870s, Peirce adds (R 333:13): "Since his time the result of the internal operation of the mind, has no longer been the conclusion reached, but has only been to ask the question to be put to nature." This fragment is recatalogued as WMS 190 and dated summer 1872 (W3:548).

**CP 5.364n quotes from an unidentified manuscript dated 1903 (however see earlier note and introduction): "What he did, a most instructive illustration of the logic of science, will be described in another chapter *[not identified]*; and we now know what was authoritatively denied when I first suggested it, that he took a hint from Malthus' book on Population."

††In ER (R 334:42) Peirce, changing the period into a semicolon, adds "but he had been more particularly impressed by Malthus' book on Population."

a gas would be on a certain hypothesis regarding the constitution of this class of bodies, Clausius and Maxwell were yet able,[*] by the application of the doctrine of probabilities, to predict that in the long run such and such a proportion of the molecules would, under given circumstances, acquire such and such velocities; that there would take place, every second, such and such a number of collisions,[†] etc.; and from these propositions were able to deduce certain properties of gases, especially in regard to their heat-relations.[7] In like manner, Darwin, while unable to say what the operation of variation and natural selection in any individual case will be, demonstrates that in the long run they will[‡] adapt animals to their circumstances.[8] Whether or not existing animal forms are due to such action, or what position the theory ought to take, forms the subject of a discussion in which questions of fact and questions of logic are curiously interlaced.[§]

II.

The object of reasoning is to find out, from the consideration of what we already know, something else which we do not know. Consequently,

---

[*]In ER (R 334:42) Peirce replaces "were yet able" with "had yet been able, eight years before the publication of Darwin's immortal work".

[†]In ER (R 334:42) Peirce changes "number of collisions" into "relative number of collisions".

[‡]The editors of the *Collected Papers,* presumably on their own initiative, changed "will" into " will, or would" (CP 5.365), adding a subjunctive reading which is in line with Peirce's later view, but is not present in the original.

[§]In ER Peirce adds the following paragraph (R 334:42,12–13, ca.1910, marked B):

> Malthus, who thought out his celebrated book at an early age, and during the inundation of French emigrés (and his father, by the way, had been a personal friend of Rousseau,) was able to discern certain truths apt to be veiled from gentlemen of leisure, —the class to which the exceptional man, the Darwin who is to unearth and give to the world a prepotent idea, must necessarily belong. Among those truths was the doctrine that a population can maintain its numbers and continued existence only at the price of an incessant "struggle for existence." This was the phrase of Malthus; it was adopted by Darwin, whose whole theory is the result of considering biology from the point of view of an economist. It is an obvious corollary from this principle that if pain, anxiety, and fear are the greatest stimuli to exertion, far from regarding them as evils, we should think of them as the main safeguards of our race, and persons infected with anything like hedonism and its offspring, sentimentalism, as disseminators of the seeds of racial extinction.

In a draft of this note (R 334:14–15), Peirce adds the following:

> It was chiefly by reflecting upon species, as populations struggling to sustain themselves, a point of view so foreign to his own class of society, that Darwin was led to that line of study which has already revolutionized biology, and which must in the future deeply modify our ideas of morality and perhaps even of metaphysics.

reasoning is good if it be such as to[*] give a true conclusion from true premises, and not otherwise. Thus, the question of its validity is purely one of fact and not of thinking.[†] A being the premises and B the conclusion,[‡] the question is, whether these facts are really so related that if A is B is. If so, the inference is valid; if not, not.[§] It is not in the least the question whether, when the premises are accepted by the mind, we feel an impulse to accept the conclusion also. It is true that we do generally reason correctly by nature. But that is an accident; the true conclusion would remain true if we had no impulse to accept it; and the false one would remain false, though we could not resist the tendency to believe in it.[**]

We are, doubtless, in the main logical animals, but we are not perfectly so. Most of us, for example, are naturally more sanguine and hopeful than logic would justify. We seem to be so constituted that in the absence of any facts to go upon we are happy and self-satisfied; so that the effect of experience is continually to contract our hopes and aspirations. Yet a lifetime of the application of this corrective does not usually eradicate our sanguine disposition. Where hope is unchecked by any experience, it is likely that

---

[*]CP5.365n, from an unidentified MS dated 1903 by the editors: "I.e., be dominated by such a habit as generally to give."

[†]In *Studies in the Meanings of Our Thought,* Peirce phrases the previous two sentences thus (R 628:4):

> Consequently, reasoning is good if it be calculated to give a true conclusion from true premises; but otherwise it is bad. Thus, the question is purely one of fact and not of thinking or of any disposition to think in one way rather than another.

[‡]In ER, Peirce writes (R 334:42): A being the fact stated in the premises and B that concluded,

[§]In *Studies in the Meanings of Our Thought,* Peirce phrases the previous two sentences thus (R 628:4):

> Let A stand for the antecedently assumed state of things asserted in the collective premiss, and C for the state of things asserted in the conclusion to be consequent upon A. Then the whole question is whether or not, to the extent the reasoning supposes, [whether it professes to be necessary reasoning, or probable, or advantageous in the long run,] it be true that if A is realized in fact, C will be realized in fact, too. If so, the inference is valid; if not, it is fallacious.

[**]In a draft sheet for *Studies in the Meanings of Our Thoughts*, Peirce phrases the previous three sentences as follows (R628:3):

> It is not, in the least, the question whether the acceptance of the premises inclines us to accept the conclusion, too, or not. True, our brains or our minds function in such a way that, like the logical machines that men have invented and constructed, they usually do turn out the right conclusion. But the true conclusion would remain true if we had no impulse to accept it, and the false conclusion remains false in those cases in which all men for two thousand years have accepted it.

our optimism is extravagant. Logicality in regard to practical matters* is the most useful quality an animal can possess, and might, therefore, result from the action of natural selection; but outside of these it is probably of more advantage to the animal to have his mind filled with pleasing and encouraging visions, independently of their truth; and thus, upon unpractical subjects, natural selection might occasion a fallacious tendency of thought.†

That which determines us, from given premises, to draw one inference rather than another, is some habit of mind, whether it be constitutional or acquired. The habit is good or otherwise, according as it produces true conclusions from true premises or not; and an inference is regarded as valid or not, without reference to the truth or falsity of its conclusion specially, but according as the habit which determines it is such as to produce true conclusions in general or not. The particular habit of mind which governs this or that inference may be formulated in a proposition whose truth depends on the validity of the inferences which the habit determines; and such a formula is called a *guiding principle* of inference.[9] Suppose, for example, that we observe that a rotating disk of copper quickly comes to rest when placed between the poles of a magnet, and we infer that this will happen with every disk of copper. The guiding principle is, that what is true of one piece of copper is true of another. Such a guiding principle with regard to copper would be much safer than with regard to many other substances—brass, for example.

A book might be written to signalize all the most important of these guiding principles of reasoning. It would probably be, we must confess, of no service to a person whose thought is directed wholly to practical subjects, and whose activity moves along thoroughly-beaten paths. The problems which present themselves to such a mind are matters of routine which he has learned once for all to handle in learning his business. But let a man venture into an unfamiliar field, or where his results are not continually checked by experience, and all history shows that the most masculine intellect will ofttimes lose his orientation and waste his efforts in directions which bring him no nearer to his goal, or even carry him entirely astray. He is like a ship in the open sea, with no one on board who understands the rules of navigation. And in such a case some general study of the guiding principles of reasoning would be sure to be found useful.

---

*In ER Peirce adds the following parenthetical remark (R334:42): (if this be understood, not in the old sense, but as consisting in a wise union of security with fruitfulness of reasoning,).

†CP 5.366n, from an unidentified MS dated 1903 by the editors: "Let us not, however, be cocksure that natural selection is the only factor of evolution; and until this momentous proposition has been much better proved than as yet it has been, let it not blind us to the force *[of]* very sound reasoning."

The subject could hardly be treated, however, without being first limited; since almost any fact may serve as a guiding principle. But it so happens that there exists a division among facts, such that in one class are all those which are absolutely essential as guiding principles, while in the others are all which have any other interest as objects of research. This division is between those which are necessarily taken for granted in asking whether a certain conclusion follows* from certain premises, and those which are not implied in that question.† A moment's thought will show that a variety of facts are already assumed when the logical question is first asked. It is implied, for instance, that there are such states of mind as doubt and belief—that a passage from one to the other is possible, the object of thought remaining the same, and that this transition is subject to some rules which all minds are alike bound by. As these are facts which we must already know before we can have any clear conception of reasoning at all, it cannot be supposed to be any longer of much interest to inquire into their truth or falsity. On the other hand, it is easy to believe that those rules of reasoning which are deduced from the very idea of the process are the ones which are most essential; and, indeed, that so long as it conforms to these it will, at least, not lead to false conclusions from true premises. In point of fact, the importance of what may be deduced from the assumptions involved in the logical question turns out to be greater than might be supposed, and this for reasons which it is difficult to exhibit at the outset. The only one which I shall here mention is, that conceptions which are really products of logical reflection, without being readily seen to be so, mingle with our ordinary thoughts, and are frequently the causes of great confusion. This is the case, for example, with the conception of quality. A quality as such is‡ never an object of observation. We can see that a thing is blue or green, but the quality of being blue and the quality of being green are not things which we see; they are products of logical reflection. The truth is, that common-sense, or thought as it first emerges above the level of the narrowly practical, is deeply imbued with that bad logical quality to which the epithet *metaphysical* is commonly applied; and nothing can clear it up but a severe course of logic.§

---

*In ER (R 334:43) Peirce replaces "whether a certain conclusion follows" with "why a certain conclusion is thought to follow".

†In ER (R 334:43) Peirce replaces "in that question." with "in such a question."

‡In ER (R 334:44) Peirce replaces "a quality as such is" with "a quality, as such, is".

§ER concludes this section (there labeled Chapter II) with the following four paragraphs, marked A (R 334:44, 7–11):

III.

We generally know when we wish to ask a question and when we wish to pronounce a judgment, for there is a dissimilarity between the sensation of doubting and that of believing.[10]

But this is not all which distinguishes doubt from belief. There is a practical difference. Our beliefs guide our desires and shape our actions. The Assassins, or followers of the Old Man of the Mountain, used to rush into death at his least command, because they believed that obedience to

---

This entire Chapter II springs from a definition of "good" reasoning that makes that "goodness" to be purely a question "of fact, and not of thinking"; and the reason for so defining good reasoning is that it is fact, and not opinion, that we aim at. But we have no means of ascertaining any fact whatever except from appearances. In order even to know what is present to my mind at a given instant, I must put the question to myself; and by the time I have done that the "given instant" is passed, and I must trust to fallible memory. It is just the same if I ask what will be present to my mind a second later. As long as this question is occupying me, *that* is what is in my mind; and if anything besides that is present, a process of thought, to disengage it is needful, and when this is done, the instant in question is past. So there is no way of getting at any matter of fact, even if it be a fact of thinking, except through thinking. Now that thinking is fallible was doubtless my motive for defining "good" reasoning objectively. At any rate, everybody *thinks* that thinking is fallible; and if there be any kind of thinking that is not so, how can it be certainly known not to be so?

The only science that professes to go upon *certain* reasoning is mathematics; and none of the reasonings of mathematics *appear* to the ablest and closest students of reasoning to be more certain than those which have been adduced to show that none of the reasonings that support the *principal* theorems of geometry have to their apprehensions at all this character of absolute necessity.

It seems to me that that power of the mind by which we are led to pronounce (though not infallibly,) one reasoning (usually trivial) to be *necessary*, another to be *trustworthy*, a third to make its conclusion *worth consideration*, with the denials of these judgments, is a true "instinct," in the sense in which we speak of the wonderful instincts of pigeons, wasps, and salmon,—meaning a faculty which guides them right in the main, but which cannot reasonably be explained as consisting or due to reasoning. That is to say, though our reasonings may be supported by other reasonings, yet their ultimate support is not to be so explained, and is therefore to be called an *Instinct*. We know that the Instincts of dogs, canaries, gold-fish, etc. are capable of wonderful modification by wisely conducted exercise and training; and consequently the fact that human reasoning is susceptible of improvement (and [it] is sufficient to read Aristotle's *Organon*, to say nothing of his *Metaphysics* and *Physics*, to be convinced that it is so) is no objection to the opinion that it ultimately rests upon an Instinct analogous to those of the so-called "brutes" and insects.

This chapter must not be closed without again pointing out that a reasoning is much like any other business in that to consider safety as its sole possible recommendation without reckoning upon the enrichment of knowledge that it will bring if it is successful would be the most arid way of shaping one's course of thought and lead only into the shifting Sahara of scholasticism. But that fuller discussion cannot be taken up in the present essay: sundry simpler considerations must first be dealt with.

him would insure everlasting felicity. Had they doubted this, they would not have acted as they did. So it is with every belief, according to its degree. The feeling of believing is a more or less sure indication of there being established in our nature some habit which will determine our actions.[*] Doubt never has such an effect.[11]

Nor must we overlook a third point of difference. Doubt is an uneasy and dissatisfied state from which we struggle to free ourselves and pass into the state of belief;[†] while the latter is a calm and satisfactory state which we do not wish to avoid, or to change to a belief in anything else.[‡] On the contrary, we cling tenaciously, not merely to believing, but to believing just what we do believe.

Thus, both doubt and belief have positive effects upon us, though very different ones. Belief does not make us act at once, but puts us into such a condition that we shall behave in a certain way, when the occasion arises. Doubt has not the least effect of this sort, but stimulates us to action until it is destroyed.[§] This reminds us of the irritation of a nerve and the reflex action produced thereby; while for the analogue of belief, in the nervous system, we must look to what are called nervous associations—for example, to that habit of the nerves in consequence of which the smell of a peach will make the mouth water.[**]

---

[*]The *Collected Papers* editors insert a longish footnote drawn from the text included here as an appendix (at the end of this chapter).

[†]The *Collected Papers* editors insert a second footnote drawn from the same text.

[‡]*[Original Peirce footnote:]* "I am not speaking of secondary effects occasionally produced by the interference of other impulses." In ER Peirce modifies the note as follows (R 334:44): "I am not speaking of accidental effects occasionally superinduced by reflexion or the interference of other impulses."

[§]In the 1872 version the second half of the sentence runs thus (W3:22): "but stimulates us to try to destroy it." In ER this sentence runs as follows (R 334:45): "Doubt has not in the least such active effect, but stimulates us to inquiry until it is destroyed."

[**]The *Collected Papers* editors again insert a longish footnote drawn from the text included here as an appendix (see below). In ER Peirce adds the following two paragraphs, marked C, while deleting the section number and break, thereby combining sections III and IV (R 334:45,16):

> It is to be observed that there are two ~~grades~~ varieties of doubt, a more acute *definite* doubt whether a certain proposition be true or not, and a commonly milder *indefinite* doubt as to what is the character of a given subject of speculation.
>
> Besides belief and doubt, there is a third attitude of mind, that of calm ignorance, whether consciousness or unconscious,—the latter the birth-stage of thought, the former its death.

IV.[*]

The irritation of doubt causes a struggle to attain a state of belief. I shall term this struggle *inquiry,* though it must be admitted that this is sometimes not a very apt designation.

The irritation of doubt is the only immediate motive for the struggle to attain belief. It is certainly best for us that our beliefs should be such as may truly guide our actions so as to satisfy our desires; and this reflection will make us reject any belief[†] which does not seem to have been so formed as to insure this result. But it will only do so by creating a doubt in the place of that belief.[‡] With the doubt, therefore, the struggle begins, and with the cessation of doubt it ends. Hence, the sole object of inquiry[§] is the settlement of opinion. We may fancy that this is not enough for us, and that we seek, not merely an opinion, but a true opinion. But put this fancy to the test, and it proves groundless; for as soon as a firm belief is reached we are entirely satisfied, whether the belief be true or false. And it is clear that nothing out of the sphere of our knowledge can be our object, for nothing which does not affect the mind can be the motive for[**] a mental effort. The most that can be maintained is, that we seek for a belief that we shall *think* to be true.[††] But we think each one of our beliefs to be true, and, indeed, it is mere tautology to say so.[‡‡]

---

[*]In 1893 Peirce began reworking "The Fixation of Belief" for inclusion in a larger work, which resulted in a book on logic entitled *How to Reason* (for a further discussion, see the introduction). In the process Peirce removed the first three sections, replacing them with a much shorter text that is added here as an appendix at the end of this chapter.

[†]In HTR "reject any belief" is replaced with "reject every belief".

[‡]This sentence is omitted in the French version of 1878. CP 5.375n adds the following from an unidentified manuscript that the editors date 1903 (see also the introduction): "Unless, indeed, it leads us to modify our ideas."

[§]In HTR (R 407:5) "of inquiry" is replaced with "of inquiring".

[**]In HTR (R 407:6) "motive for" is replaced with "motive to".

[††]In the 1878 French version, the translation of "that we shall *think* to be true"—*que nous pensons vraie* (the entire phrase is set in italics)—lacks the modifier "shall" (W3:345).

[‡‡]The CP editors add another note drawn from a non-identified manuscript that they date 1903 (CP 5.375n): "For truth is neither more nor less than that character of a proposition which consists in this, that belief in the proposition would, with sufficient experience and reflection, lead us to such conduct as would tend to satisfy the desires we should then have. To say that truth means more than this is to say that it has no meaning at all."

That the settlement of opinion is the sole end of inquiry is a very important proposition. It sweeps away, at once, various vague and erroneous conceptions of proof. A few of these may be noticed here.

1. Some philosophers have imagined that to start an inquiry it was only necessary to utter a question or set it down upon paper, and have even recommended us to begin our studies with questioning everything!* But the mere putting of a proposition into the interrogative form does not stimulate the mind to any struggle after belief. There must be a real and living doubt, and without this all discussion is idle.[12]

2. It is a very common idea that a demonstration must rest on some ultimate and absolutely indubitable propositions.† These, according to one school, are first principles of a general nature; according to another, are first sensations. But, in point of fact, an inquiry, to have that completely satisfactory result called demonstration, has only to start with propositions perfectly free from all actual doubt. If the premises are not in fact doubted at all, they cannot be more satisfactory than they are.‡

*In HTR the remainder of this paragraph is replaced with the following (R 407:6–7):

> Now we may well think it inevitable that many of our opinions have some falsity mixed up with them; how can it be otherwise? And this reflection will make us modest and eager to learn better of those men who disagree from us. Still, so long as we cannot put our finger on our erroneous opinions, they remain our opinions, still. It will be wholesome enough for us to make a general review of the causes of our beliefs; and the result will be hat most of them have been taken upon trust and have been held since we were too young to discriminate the credible from the incredible. Such reflections may awaken real doubts about some of our positions. But in cases where no real doubt exists in our minds, inquiry will be an idle force, a mere whitewashing commission, which were better let alone. This fault in philosophy was very widespread in those ages in which Disputations were the principal exercises in the universities, that is, from their rise in the thirteenth century down into the eighteenth, and even so to this day in some catholic institutions. But since these disputations went out of vogue, this philosophic disease is less virulent.

†In HTR the remainder of this paragraph is replaced with the following (R 407:7–8):

> These, according to one school, are first principles, broad axioms; according to another, they are first data of sense. But this is pursuing a chimera. We cannot do better than begin with propositions of which we have no doubt. We have to acknowledge that doubts about them may spring up later; but we can find no propositions which are not subject to this contingency. We ought to construct our theories so as to provide for such discoveries, first, by making them rest on as great a variety of different considerations as possible, and second, by leaving room for the modifications which cannot be foreseen but which are pretty sure to prove needful. Some systems are much more open to this criticism than others. All those which repose heavily upon an "inconceivability of the opposite" have proved particularly fragile and short-lived. Those, however, which rest upon positive evidences and which avoid insisting upon the absolute precision of their dogmas are hard to destroy.

‡In his 1872 Logic Book, Peirce puts it thus (W3:14):

3. Some people seem to love to argue a point after all the world is fully convinced of it. But no further advance can be made. When doubt ceases, mental action on the subject comes to an end; and, if it did go on, it would be without a purpose.*

[I]t is false to say that reasoning must rest either on first principles or on ultimate facts. For we cannot go behind what we are unable to doubt, but it would be unphilosophical to suppose that any particular fact will never be brought into doubt.

In ER (ca.1910), Peirce replaces this paragraph by the following (R334:45, 20–21; marked E):

2. Aristotle, the first to study thought, unto the accomplishment of any purpose,[1] says not a word about how to impregnate reasoning with new life, nor even of how to increase its security. His main aim (as long as he has science and not mere persuasion in view,) is to render it absolutely certain, without apparent suspicion of any impossibility's lurking there. Hence the idea of *Demonstration*, an argumentation whose premisses are entirely believed and are quite true, while the operation of reasoning is absolutely trusted and is absolutely infallible; so that its conclusion becomes perfectly known and free from error.†† In our days Pure Mathematics and Logic are the only sciences (outside of the Catholic and Orthodox Churches, and within Christendom, at any rate,) that continue to make any pretention to being demonstrative; and Geometry being acknowledged to be, if not fictitious as to its absolute exactitude, at least subject to serious doubt in that respect, (owing to the study of electricity and its sister sciences), unless we regard Time to be an external reality, Pure Mathematics is reduced to Arithmetic, and becomes a science of mere ideas, as logic undoubtedly is, valid only in so far as our Rational Instinct happens to be so. But all this with its accompaniments, so far from affording any countenance to distrust of science or to either Philosophical or Religious Scepticism, ought to be regarded as, on the whole, a most tremendous confirmation of our confidence in rational endeavour and of our trust in God.

1. Though the mathematicians had taken some steps on that route.

†† In a draft of this insertion Peirce continues thus (R 334:22–24):

But we shall find that the greater the Security of a way of reasoning the less its power of impregnating any germ of knowledge with a principle of growth; or say, the greater its Security the less its Prepotency. The only science that still continues, in our day, to profess itself *demonstrative* throughout, is Mathematics. Now I shall not suppose my Reader to be very proficient in mathematics, but if he has studied the old fashioned elementary Geometry, with its Axioms and its Theorems, he must, I think, have remarked that the proofs fell into two classes, those that were of childish simplicity, and those that only such students as had specially clear geometrical intuitions were able at once to see the force of. Now I will add a modest essay, as a supplement to the present one, in which, running through the first Book of Euclid's *Elements*, I will show that every theorem of any particular prepotency,—every one not almost nauseously childish,—is really not demonstrative; so that the gifted pupil, though he imagines himself to have been rationally convinced by the reasoning in the book, certainly must be mistaken, since that reasoning was not rational: he has discerned the truth, just as the writer of the book did, by his perhaps correct, but unanalytical and instinctive idea of Space.

*The *Collected Papers* editors add a note drawn from an unidentified manuscript that they date 1903 (CP 5.376n): "Except that of self-criticism. Insert here a section upon self-control and the analogy between Moral and Rational self-control." (See the introduction on the 1903 date.)

V.

If the settlement of opinion is the sole object of inquiry, and if belief is of the nature of a habit, why should we not attain the desired end, by taking any answer to a question which we may fancy, and* constantly reiterating it to ourselves, dwelling on all which may conduce to that belief, and learning to turn with contempt and hatred from anything which might disturb it? This simple and direct method is really pursued by many men. I remember once being entreated not to read a certain newspaper lest it might change my opinion upon free-trade.[13] "Lest I might be entrapped by its fallacies and misstatements," was the form of expression. "You are not," my friend said, "a special student of political economy. You might, therefore, easily be deceived by fallacious arguments upon the subject. You might, then, if you read this paper, be led to believe in protection. But you admit that free-trade is the true doctrine; and you do not wish to believe what is not true." I have often known this system to be deliberately adopted. Still oftener, the instinctive dislike of an undecided state of mind, exaggerated into a vague dread of doubt, makes men cling spasmodically to the views they already take. The man feels that, if he only holds to his belief without wavering, it will be entirely satisfactory. Nor can it be denied that a steady and immovable faith yields great peace of mind. It may, indeed, give rise to inconveniences, as if a man should resolutely continue to believe that fire would not burn him, or that he would be eternally damned if he received his *ingesta* otherwise than through a stomach-pump. But then the man who adopts this method will not allow that its inconveniences are greater than its advantages. He will say, "I hold steadfastly to the truth, and the truth is always wholesome." And in many cases it may very well be that the pleasure he derives from his calm faith overbalances any inconveniences resulting from its deceptive character. Thus, if it be true that death is annihilation, then the man who believes that he will certainly go straight to heaven when he dies, provided he have fulfilled† certain simple observances in this life,‡ has a cheap pleasure which will not be followed by the least disappointment.§ A

*In ER (R 334:46) Peirce changes "any answer … and constantly" into "as answer to a question any we may fancy, and then constantly".

†In HTR (R 407:11) "he shall have fulfilled".

‡In 1872 Peirce writes, following the comma (W3:24): "however he may have behaved in this life".

§The CP editors add here a note drawn from an unidentified manuscript that they date 1903 (CP 5.375n): "Although it certainly may be that it will cause a line of conduct leading to pains that deeper reflection would have avoided."

similar consideration seems to have weight with many persons in religious topics, for we frequently hear it said, "Oh, I could not believe so-an-so, because I should be wretched if I did." When an ostrich buries its head in the sand as danger approaches, it very likely takes the happiest course. It hides the danger, and then calmly says there is no danger; and, if it feels perfectly sure there is none, why should it raise its head to see? A man may go through life, systematically keeping out of view all that might cause a change in his opinions, and if he only succeeds—basing his method, as he does, on two fundamental psychological laws—I do not see what can be said against his doing so. It would be an egotistical impertinence to object that his procedure is irrational, for that only amounts to saying that his method of settling belief is not ours. He does not propose to himself to be rational, and, indeed, will often talk with scorn of man's weak and illusive reason.* So let him think as he pleases.†

But this method of fixing belief, which may be called the method of tenacity,‡ will be unable to hold its ground in practice. The social impulse is against it. The man who adopts it will find that other men think differently from him, and it will be apt to occur to him, in some saner moment, that their opinions are quite as good as his own, and this will shake his confidence in his belief. This conception, that another man's thought or sentiment may be equivalent to one's own, is a distinctly new step, and a highly important one. It arises from an impulse too strong in man to be suppressed, without danger of destroying the human species. Unless we make ourselves hermits, we shall necessarily influence each other's opinions; so that the problem becomes how to fix belief, not in the individual merely, but in the community.[14]

Let the will of the state act, then, instead of that of the individual. Let an institution be created which shall have for its object to keep correct doctrines before the attention of the people, to reiterate them perpetually, and to teach them to the young; having at the same time power to prevent contrary doctrines from being taught, advocated, or expressed. Let all possible causes of a change of mind be removed from men's apprehensions. Let them be kept ignorant, lest they should learn of some reason to

*In the 1872 logic book, Peirce observes (W3:15): "It is a method which has been much practised and highly approved, especially by people whose experience has been that reasoning only leads from doubt to doubt."

†In an 1872–1873 MS Peirce writes instead (R 333:18): "In our free country he has a right to think as he pleases." In a deleted passage immediately following Peirce suggests calling the first method "the method of freedom."

‡In 1872, Peirce calls it "the method of obstinacy" (W3:25).

think otherwise than they do. Let their passions be enlisted, so that they may regard private and unusual opinions with hatred and horror. Then, let all men who reject the established belief be terrified into silence. Let the people turn out and tar-and-feather such men, or let inquisitions be made into the manner of thinking of suspected persons, and, when they are found guilty of forbidden beliefs, let them be subjected to some signal punishment. When complete agreement could not otherwise be reached, a general massacre of all who have not thought in a certain way has proved a very effective means of settling opinion in a country. If the power to do this be wanting, let a list of opinions be drawn up, to which no man of the least independence of thought can assent, and let the faithful be required to accept all these propositions, in order to segregate them as radically as possible from the influence of the rest of the world.

This method has, from the earliest times, been one of the chief means of upholding correct theological and political doctrines, and of preserving their universal or catholic character. In Rome, especially, it has been practised from the days of Numa Pompilius to those of Pius Nonus.[*][15] This is the most perfect example in history; but wherever there is a priesthood—and no religion has been without one[†]—this method has been more or less made use of. Wherever there is an aristocracy, or a guild, or any association of a class of men whose interests depend or are supposed to depend on certain propositions, there will be inevitably found some traces of this natural product[‡] of social feeling. Cruelties always accompany this system; and when it is consistently carried out, they become atrocities of the most horrible kind in the eyes of any rational man.[§] Nor should this occasion surprise, for the officer of a society does not feel justified in surrendering the interests of that society for the sake of mercy, as he might his own private interests. It is natural, therefore, that sympathy and fellowship should thus produce a most ruthless power.[**]

---

[*]In the French translation of 1878 Pope Pius Nonus, who had just died, is replaced with his successor, Pope Leo 13 (W3:348). Note, however, that Peirce does not make this change in 1894 when he rewrites the passage for HTR, possibly suggesting Peirce's support for Leo 13, a Thomist who denied that science and faith conflict, and who sought to reconcile the teachings of the Church with the demands of the modern world.

[†]The French translation of 1878 omits "and no religion has been without one".

[‡]In HTR (R 407:13) Peirce writes "some rudiments, at least of this natural fruit".

[§]In HTR (R 407:13) Peirce writes "in the eyes of any outsider."

[**]In HTR this sentence runs as follows (R 407:13): "Quite naturally, therefore, that which was born of sympathy and fellowship grows up into the most ruthless powers."

In judging this method of fixing belief, which may be called the method of authority,[*] we must, in the first place, allow its immeasurable mental and moral superiority to the method of tenacity. Its success is proportionately greater; and, in fact, it has over and over again worked the most majestic results.[†] The mere structures of stone which it has caused to be put together—in Siam, for example, in Egypt, and in Europe—have many of them a sublimity hardly more than rivaled by the greatest works of Nature. And, except the geological epochs, there are no periods of time so vast as those which are measured by some of these organized faiths.[‡] If we scrutinize the matter closely, we shall find that there has not been one of their creeds which has remained always the same; yet the change is so slow as to be imperceptible during one person's life, so that individual belief remains sensibly fixed. For the mass of mankind, then, there is perhaps no better method than this. If it is their highest impulse to be intellectual slaves, then slaves they ought to remain.[§]

But no institution can undertake to regulate opinions upon every subject. Only the most important ones can be attended to, and on the rest men's minds must be left to the action of natural causes. This imperfection will be no source of weakness so long as men are in such a state of culture that

---

[*] In 1872 Peirce calls it "the method of despotism" (W3:26) and the "method of persecution" (W3:17).

[†] In HTR Peirce replaces "we must, … the most majestic results." with (R 407:13–14):

> we put aside as irrelevant its immeasurable mental and moral superiority to the method of tenacity. These are interesting enough qualities; but this method is to be judged, like every other, by its success alone; and this success has been truly superhuman. Of all the achievements of man, this alone is majestic.

[‡] In HTR the remainder of this paragraph is replaced with the following (R 407:14):

> Unify them in the sense of Alexander Pope's *Universal Prayer,* and who is the individual whose conceit shall stand up and place his dictum against theirs? These faiths lay claim to divine authorship; and it is true that men have no more *invented* them, than the birds have invented their songs. It is a relapse toward the method of tenacity that segregates them and blinds the ecclesiastic to the value of anything but hatred. Every distinctive creed was as a historical fact invented to harm somebody. Still, the upshot has, on the whole, been success unparalleled. If slavery of opinion is natural and wholesome for man, then slaves they ought to remain.

[§] In an earlier draft (WMS 189), Peirce wrote "It is their highest impulse to be intellectual slaves, and slaves they ought to remain" (R 333:23), before changing it into the conditional statement given above.
In HTR the next two paragraphs are replaced with the following four (R 407:14–20):

> Every such system was first established by some individual legislator or prophet; and once established it grew of itself. But within this principle of growth lurk germs of decay.

one opinion does not influence another—that is, so long as they cannot put two and two together. But in the most priestridden states some individuals will be found who are raised above that condition. These men possess a wider sort of social feeling; they see that men in other countries and in other ages have held to very different doctrines from those which they themselves have been brought up to believe; and they cannot help seeing that it is the mere accident of their having been taught as they have, and of their having been surrounded with the manners and associations they have, that has caused them to believe as they do and not far differently. And their candor cannot resist the reflection that there is no reason to rate their own views at a higher value than those of other nations and other centuries; and this gives rise to doubts in their minds.

The power of individualism becomes extinct; the organization alone has life. Now, in the course of ages old questions pass out of mind: new questions become urgent. The sea advances or recedes; some horde which has always lived by conquest happens to make a conquest of consequence to the world at large. In one way or another, commerce is diverted from its ancient roads. Such change brings novel experiences and new ideas. Men begin to rebel at doings of the authorities to which in former times they would have submitted. Questions never before raised come up for decision; yet an individual legislator would no longer be listened to. Never had the instinct of rulers failed to see that the summoning of a council of the people was a measure fraught with peril to authority. Yet, however they strive to avoid it, they in effect invoke public opinion, which is a momentous appeal to a new method of settling opinion. Disturbances occur; knots of men discuss the state of affairs; and a suspicion is kindled which runs about like a train of gunpowder, that the Dicta men have been reverencing originated in caprice, in the pertinacity of some busybody, in the schemes of an ambitious man, or in other influences which are seen to edify a deliberative assembly. Men now begin to demand that, as the power which maintains the belief has become no longer capricious but public and methodical, so the propositions to be believed shall be determined in a public and methodical manner. Let the action of natural preferences be unimpeded, then, and under their influence, let men, conversing together and discussing their opinions, gradually develop such beliefs as are fittest to survive. This is the Method of Dialectic; in philosophy called the *a priori* method. It springs up from the humus of decayed religions. Greek philosophy first appeared when the myths began to shock people; and modern philosophy trod hard upon the heels of the Reformation.

Let us see in what manner a few of the greatest philosophers have undertaken to settle opinion, and what their success has been. Descartes, who would have a man begin by doubting everything, remarks that there is one thing he will find himself unable to doubt, and that is, that he does doubt; and when he reflects that he doubts, he can no longer doubt that he exists. Then, because he is all the while doubting whether there are any such things as shape and motion, Descartes thinks he must be persuaded that shape and motion do not belong to his nature, or anything else but consciousness. This is taking it for granted that nothing in his nature lies hidden beneath the surface. Next Descartes asks the doubter to remark that he has the idea of a Being, in the highest degree intelligent, powerful, and perfect. Now a Being would not have those qualities unless He existed necessarily and eternally. By existing necessarily he means existing by virtue of the existence of the idea. Consequently, all doubt as to the existence of this Being must cease. This Plainly supposes that belief is to be fixed by what men find in

They will further perceive that such doubts as these must exist in their minds with reference to every belief which seems to be determined by the caprice either of themselves or of those who originated the popular opinions. The willful adherence to a belief, and the arbitrary forcing of it upon others, must, therefore, both be given up, and a new method of settling opinions must be adopted, which shall not only produce an impulse to believe, but shall also decide what proposition it is which is to be believed. Let the action of natural preferences be unimpeded, then, and under their influence let men, conversing together and regarding matters in different lights, gradually develop beliefs in harmony with natural causes. This method resembles

---

their minds. He is reasoning like this: I find it written in the volume of my mind that there is something X, which is such a sort of thing that the moment it is written down it exists. Plainly, he is aiming at a kind of truth which saying so can make *[it]* to be so. He gives two further proofs of God's existence. Descartes makes God easier to know than anything else; for whatever we think He is, he is. He fails to remark that this is precisely the definition of a *figment*. In particular, God cannot be a deceiver; whence it follows, that whatever we quite clearly and distinctly think to be true about any subject, must *be* true. Accordingly, if people will thoroughly discuss a subject, and quite clearly and distinctly make up their minds what they think about it, the desired settlement of the question will be reached. I may remark that the world has pretty thoroughly deliberated upon that theory and has quite distinctly come to the conclusion that it is utter nonsense; whence that judgment is indisputably right.

Many critics have told me that I misrepresent the *a priori* philosophers, when I represent them as adopting whatever opinion there seems to be a natural inclination to adopt. But nobody can say the above does not accurately define the position of Descartes, and upon what does he repose except natural ways of thinking? Perhaps I shall be told, however, that since Kant that vice has been cured. Kant's great boast is that he critically examines into our natural inclinations toward certain opinions. An opinion that something is *universally* true clearly goes further than experience can warrant. An opinion that something is *necessarily* true (that is, not merely is true in the existing state of things; but would be true in every state of things) equally goes further than experience will warrant. Those remarks had been made by Leibniz and admitted by Hume; and Kant reiterates it. Though they are propositions of a nominalistic cast, they can hardly be denied. I may add that whatever is held to be *precisely* true goes further than experience can possibly warrant. Accepting those criteria of the origin of ideas, Kant proceeds to reason as follows: Geometrical propositions are held to be universally true. Hence, they are not given by experience. Consequently, it must be owing to an inward necessity of man's nature that he sees everything in space. Ergo, the sum of the angles of a triangle will be equal to two right angles for all the objects of our vision. Just that, and nothing more, is Kant's line of thought. But the dry-rot of reason in the seminaries has gone to the point where such stuff is held to be admirable argumentation. I might go through the *Critic of the Pure Reason*, section by section, and show that the thought throughout is precisely of this character. He everywhere shows that ordinary objects, such as trees and gold-pieces, involve elements not contained in the first presentations of sense. But we cannot persuade ourselves to give up the reality of trees and gold-pieces. There is a general inward insistence upon them, and that is the warrant for swallowing the entire bolus of general belief about them. This is merely accepting without question a belief as soon as it is shown to please a great many people very much. When he comes to the ideas of God, Freedom, and Immortality, he hesitates; because people who think only

that by which conceptions of art have been brought to maturity. The most perfect example of it is to be found in the history of metaphysical philosophy. Systems of this sort have not usually rested upon any observed facts, at least not in any great degree. They have been chiefly adopted because their fundamental propositions seemed "agreeable to reason." This is an apt expression; it does not mean that which agrees with experience, but that which we find ourselves inclined to believe. Plato, for example, finds it agreeable to reason that the distances of the celestial spheres from one another should be proportional to the different lengths of strings which produce harmonious chords. Many philosophers have been led to their main conclusions by considerations like this; but this is the lowest and least developed form which the method takes, for it is clear that another man might find Kepler's theory, that the celestial spheres are proportional to the

---

of bread and butter, pleasure and power, are indifferent to these ideas. He subjects these ideas to a different kind of examination, and finally admits them upon grounds which appear to the seminarists more or less suspicious, but which in the eyes of laboratorists are infinitely stronger than the grounds upon which he has accepted space, time, and causality. Those last grounds amount to nothing but this, that what there is a very decided and general inclination to believe must be true. Had Kant merely said, I shall adopt for the present the belief that the three angles of a triangle are equal to two right angles because nobody but brother Lambert and some Italian has ever called it in question, his attitude would be well enough. But on the contrary, he and those who today represent his school distinctly maintain the proposition is *proved*, and the Lambertists *refuted*, by what comes merely to the general disinclination to think with them.

As for Hegel, who led Germany for a generation, he recognizes clearly what he is about. He simply launches his boat into the current of thought and allows himself to be carried wherever that current leads. He himself calls his method *dialectic*, meaning that a frank discussion of the difficulties to which any opinion spontaneously gives rise will lead to modification after modification until a tenable position is attained. This is a distinct profession of faith in the method of inclinations.

Other philosophers appeal to "the test of inconceivability of the opposite," to "presuppositions" (by which they mean *Voraussetzungen*, properly translated, *postulates*), and other devices; but all these are but so many systems of rummaging the garret of the skull to find an enduring opinion about the universe.

On a draft sheet Peirce formulates the preceding passage thus (R 278b:219):

Rulers have never failed to see that the summoning of the States General was a measure fraught with danger to Authority. Men are thus led to think that the Dicta they have reverenced originated in caprice, in some busybody's pertinacity, or in somebody's pursuit of a personal end; for the deliberations of assemblies are never edifying. As the power which maintains the belief has become public and methodical, so men now demand that the propositions to be believed shall be determined in a public and methodical manner. Let the action of natural preferences be unimpeded, then, and under their influence let men, conversing together and discussing their opinions, gradually develop such as are fittest to survive. This is the method of Dialectic, called, in the sphere of philosophy the *a priori* method.

inscribed and circumscribed spheres of the different regular solids, more agreeable to *his* reason. But the shock of opinions will soon lead men to rest on preferences of a far more universal nature. Take, for example, the doctrine that man only acts selfishly—that is, from the consideration that acting in one way will afford him more pleasure than acting in another. This rests on no fact in the world, but it has had a wide acceptance as being the only reasonable theory.[*16]

---

*In HTR the next three paragraphs are replaced with the following (R 407:20–21):

> When we pass from the perusal of works upholding the method of authority to those of the philosophers, we not only find ourselves in a vastly higher intellectual atmosphere, but also in a clearer, freer, brighter, and more refreshing moral atmosphere. All this, however, is beside the one significant question of whether the method succeeds in fixing men's opinions. The projects of these authors are most persuasive. One dare swear they should succeed. But in point of fact, up to date they decidedly do not; and the outlook in this direction is most discouraging. The difficulty is that the opinions which today seem most unshakable are found tomorrow to be out of fashion. They are really far more changeable than they appear to a hasty reader to be; since the phrases made to dress out defunct opinions are worn at second hand by their successors.[1] Changes of opinion are brought about by events beyond human control. All mankind were so firmly of opinion that heavy bodies must fall faster than light ones, that any other view was scouted as absurd, eccentric, and probably insincere. Yet as soon as some of the absurd and eccentric men could succeed in inducing some of the adherents of common sense to look at their experiments,—no easy task,—it became apparent that nature would not follow human opinion, however unanimous. So there was nothing for it but human opinion must move to nature's position. That was a lesson in humility. A few men, the small band of laboratory men, began to see that they had to abandon the pride of an opinion assumed absolutely final in any respect, and to use all their endeavors to yield as unresistingly as possible to the overwhelming tide of experience, which must master them at last, and to listen to what nature seems to be telling us. The trial of this method of experience in natural science for these three centuries,—though bitterly detested by the majority of men,—encourages us to hope that we are approaching nearer and nearer to an opinion which is not destined to be broken down,—though we cannot expect ever quite to reach that ideal goal.
>
> [1] We still talk of "cause and effect" although, in the mechanical world, the opinion that phrase was meant to express has been shelved long ago. We now know that the acceleration of a particle at any instant depends upon its position relative to other particles at that same instant; while the old idea was that the past affects the future, while the future does not affect the past. So the "law of demand and supply" has utterly different meanings with different economists.

The *Collected Papers* editors insert at this point the following note they draw from an unidentified manuscript they date 1903 (CP 5.382n):

> An acceptance whose real support has been the opinion that pleasure is the only ultimate good. But this opinion, or even the opinion that pleasure *per se* is any *good* at all, is only tenable so long as he who holds it remains without any distinct idea of what he means by "good."

This method is far more intellectual and respectable from the point of view of reason than either of the others which we have noticed.* But† its failure has been the most manifest. It makes of inquiry something similar to the development of taste; but taste, unfortunately, is always more or less a matter of fashion, and accordingly metaphysicians have never come to any fixed agreement, but the pendulum has swung backward and forward between a more material and a more spiritual philosophy, from the earliest times to the latest. And so from this, which has been called the *a priori* method,‡[17] we are driven, in Lord Bacon's phrase, to a true induction.[18] We have examined into this *a priori* method as something which promised to deliver our opinions from their accidental and capricious element. But development, while it is a process which eliminates the effect of some casual circumstances, only magnifies that of others. This method, therefore, does not differ in a very essential way from that of authority. The government may not have lifted its finger to influence my convictions; I may have been left outwardly quite free to choose, we will say, between monogamy and polygamy, and, appealing to my conscience only, I may have concluded that the latter practice is in itself licentious. But when I come to see that the chief obstacle to the spread of Christianity among a people of as high culture as the Hindoos has been a conviction of the immorality of our way of treating women, I cannot help seeing that, though governments do not interfere, sentiments in their development will be very greatly determined by accidental causes. Now, there are some people, among whom I must suppose that my reader is to be found, who, when they see that any belief of theirs is determined by any circumstance extraneous to the facts, will from that moment not merely admit in words that that belief is doubtful, but will experience a real doubt of it, so that it ceases to be§ a belief.

To satisfy our doubts, therefore, it is necessary that a method should be found by which our beliefs may be caused by** nothing human, but by some external permanency—by something upon which our thinking

*In ER Peirce inserts at this point the following (R 334:49): "Indeed, as long as no better method can be applied, it ought to be followed, since it is then the expression of instinct, which must be the ultimate cause of belief in all cases."

†In ER (R334:49) Peirce replaces "But" with "Nevertheless,"

‡In 1872 Peirce also calls this "the method of public opinion" (W3:17).

§In ER (R 334:50) Peirce writes "ceases, in some degree at least, to be" for "ceases to be".

**In ER (R 334:50) Peirce writes "determined by" for "caused by".

has no effect.*[19] Some mystics imagine that they have such a method in a private inspiration from on high. But that is only a form of the method of tenacity, in which the conception of truth as something public is not yet developed. Our external permanency would not be external, in our sense, if it was restricted in its influence to one individual. It must be something which affects, or might affect, every man. And, though these affections are necessarily as various as are individual conditions, yet the method must be such that the ultimate conclusion of every man shall be the same.† Such is the method of science.‡ Its fundamental hypothesis, restated in more familiar language, is this: There are real things, whose characters are entirely independent of our opinions about them; those realities affect our senses according to regular laws, and, though our sensations are as different as our relations to the objects, yet, by taking advantage of the laws of perception,§ we can ascertain by reasoning how things really are, and any man, if he have sufficient experience and reason enough about it, will be led to the one true conclusion.** The new conception here involved is that of reality.[20] It may be asked how I know that there are any realities. If this hypothesis is the sole support of my method of inquiry, my method of inquiry must not be used to support my hypothesis. The reply is this: 1. If investigation

---

*The *Collected Papers* editors add here a note they draw from an unidentified manuscript they date 1903 (CP 5.384n): "But which, on the other hand, unceasingly tends to influence thought; or in other words, by something Real."

†The *Collected Papers* editors add here a note they draw from an unidentified manuscript they date 1903 (CP 5.384n): "Or would be the same if inquiry were sufficiently persisted in."

‡In 1872 Peirce names it "the scientific method" (W3:27).

§In WMS 189 Peirce writes (R 333:24) "laws which subsist" instead of "laws of perception".

**In 1905, in a "critical commentary" (R 289:2) on the first two *Popular Science Monthly* papers, entitled "Consequences of Pragmaticism," Peirce writes (R 289.17–18)

> I desire to express with empathic grief that in those original papers I fell into a grievous error. Looking into the future and thinking of what would take place, I overlooked the great truth that the future, as living, always is more or less general, and I thought that what *would be* could be resolved into what *will be*, this will be not being to my mind very different in its essential habit from what was. But in truth, it is what will be that is resolvable without difficulty in what *may be*, *might be*, or *would be* and belongs to quite a different character from what has hitherto been.

What Peirce believes he should have written is (R 289:17): "It thus appears that where there is any truth, inquiry would, under favorable circumstances, ultimately reach it."

cannot be regarded as proving that there are real things, it at least does not lead to a contrary conclusion; but the method and the conception on which it is based remain ever in harmony. No doubts of the method, therefore, necessarily arise from its practice, as is the case with all the others. 2. The feeling which gives rise to any method of fixing belief is a dissatisfaction at two repugnant propositions.[21] But here already is a vague concession that there is some *one* thing to which a proposition should conform.[*] Nobody, therefore, can really doubt that there are realities, or, if he did, doubt would not be a source of dissatisfaction. The hypothesis, therefore, is one which every mind admits. So that the social impulse does not cause me to doubt it.[†] 3. Everybody uses the scientific method about a great many things, and only ceases to use it when he does not know how to apply it. 4. Experience of the method has not led me to doubt it,[‡] but, on the contrary, scientific investigation has had the most wonderful triumphs in the way of settling opinion. These afford the explanation of my not doubting the method or the hypothesis which it supposes; and not having any doubt, nor believing that anybody else whom I could influence has, it would be the merest babble for me to say more about it. If there be anybody with a living doubt upon the subject, let him consider it.

To describe the method of scientific investigation is the object of this series of papers. At present[§] I have only room to notice some points of contrast between it and other methods of fixing belief.

This is the only one of the four methods which presents any distinction of a right and a wrong way.[22] If I adopt the method of tenacity and shut myself out from all influences, whatever I think necessary to doing this is necessary according to that method. So with the method of authority: the state may try to put down heresy by means which, from a scientific point of view, seem very ill-calculated to accomplish its purposes; but the only test *on that method* is what the state thinks, so that it cannot pursue the method wrongly. So with the *a priori* method.[**] The very essence of it is to think as one is inclined to think. All metaphysicians will be sure to do

[*] In the 1878 French version "to which a proposition should conform" is translated as "à quoi puisse être conforme une proposition" (W3:352).

[†] This sentence is absent in the French version (W3:352).

[‡] In ER Peirce writes (R 334:51): "led us to doubt it".

[§] In 1872 Peirce writes "this book. In this chapter" for "this series … present".

[**] In 1872 Peirce writes "If I endeavor to lay my susceptibilities of belief perfectly open to the influences which work upon them, I cannot on those principles go wrong" (W3:28) and immediately moves on to the scientific method. (W3:28).

that, however they may be inclined to judge each other to be perversely wrong. The Hegelian system recognizes every natural tendency of thought as logical, although it be certain to be abolished by counter-tendencies. Hegel thinks there is a regular system in the succession of these tendencies, in consequence of which, after drifting one way and the other for a long time, opinion will at last go right. And it is true that metaphysicians get the right ideas at last; Hegel's system of Nature represents tolerably the science of that day; and one may be sure that whatever scientific investigation has put out of doubt will presently receive *a priori* demonstration* on the part of the metaphysicians. But with the scientific method the case is different. I may start with known and observed facts to proceed to the unknown; and yet the rules which I follow in doing so may not be such as investigation would approve. The test of whether I am truly following the method is not an immediate appeal to my feelings and purposes, but, on the contrary, itself involves the application of the method. Hence it is that bad reasoning as well as good reasoning is possible; and this fact is the foundation of the practical side of logic.†

---

*In HTR Peirce writes (R 407:22): "receive the sanction of *a priori* demonstration".

†In 1872 Peirce here continues (W3:28) "When a man has once chosen the scientific method he positively violates it when he allows any weight *[MS breaks off]*" In HTR Peirce inserts the following paragraph (R 333:8–9):

> It is not to be supposed that it is practicable quite to abandon the use of the first three methods. Experience itself recommends a moderate and diffident reliance upon the natural light of reason, which, while it has never failed to go wrong if trusted too far, has frequently been of priceless value in furnishing suggestions capable of correction by experience. It was greatly relied upon by Galileo in making out the laws of motion, which otherwise though we fumbled for forever we might never have found; and in psychology it would be foolish to cover that light with any curtain. Experience, too, undoubtedly vindicates a limited adhesion to authority, especially in moral and religious questions. Experiments in the use of private reason as the safe guide in regard to social questions have been distinctly unsuccessful. The church may have fallen into numberless errors; but the question of whether a private man will do so wisely to consider her as infallible for practical purposes in the existing state of things is the main question for him. On the one hand, history is a sort of public experience in the light of which the teachings of the church have been pretty well sifted; on the other hand, it would be inconsistent for a man to resolve humbly to submit himself to the teachings of experience, and then to except from that rule the very experience which most truly deserves the name; that of his own inward life and heart. Even the method of tenacity has its uses. When a man has once publicly taken a position upon a question of interest to others, honor will not permit him lightly to abandon it until it has had its fair trial, and the judgment of experience has been pronounced. But all these temporary employments of the first three methods, in default of experience or its bidding, by a man ready to let them go the instant the method of experience shall prove directly available will be as far in their practical results from any adoption of those methods as in their theory and motive.

It is not to be supposed that the first three methods of settling opinion present no advantage whatever over the scientific method.* On the contrary, each has some peculiar convenience of its own. The *a priori* method is distinguished for its comfortable conclusions. It is the nature of the process to adopt whatever belief we are inclined to, and there are certain flatteries to the vanity of man which we all believe by nature, until we are awakened from our pleasing dream by some rough facts. The method of authority will always govern the mass of mankind; and those who wield the various forms of organized force in the state will never be convinced that dangerous reasoning ought not to be suppressed in some way. If liberty of speech is to be untrammeled from the grosser forms of constraint, then uniformity of opinion will be secured by a moral terrorism to which the respectability of society will give its thorough approval. Following the method of authority is the path of peace. Certain non-conformities are permitted; certain others (considered unsafe) are forbidden. These are different in different countries and in different ages; but, wherever you are, let it be known that you seriously hold a tabooed belief, and you may be perfectly sure of being treated with a cruelty less brutal but more refined than hunting you like a wolf. Thus, the greatest intellectual benefactors of mankind have never dared, and dare not now, to utter the whole of their thought; and thus a shade of *prima facie* doubt is cast upon every proposition which is considered essential to the security of society. Singularly enough, the persecution does not all come from without; but a man torments himself and is oftentimes most distressed at finding himself believing propositions which he has been brought up to regard with aversion. The peaceful and sympathetic man will, therefore, find it hard to resist the temptation to submit his opinions to authority. But most of all I admire the method of tenacity for its strength, simplicity, and directness. Men who pursue it are distinguished for their decision of character, which becomes very easy with such a mental rule. They do not waste time in trying to make up their minds what they want, but, fastening like lightning upon whatever alternative comes first, they hold to it to the end, whatever happens, without an instant's irresolution. This is one of the splendid qualities which generally accompany brilliant, unlasting success. It is impossible not to envy the man who can dismiss reason, although we know how it must turn out at last.

Such are the advantages which the other methods of settling opinion have over scientific investigation. A man should consider well of them; and

*In HTR the first sentence is changed into (R 407:23): "Not that the first three methods of settling opinions are every way at a disadvantage compared with that of experience."

then he should consider that, after all, he wishes his opinions to coincide with the fact, and that there is no reason why the results of these three methods should do so. To bring about this effect is the prerogative of the method of science. Upon such considerations he has to make his choice—a choice which is far more than the adoption of any intellectual opinion, which is one of the ruling decisions of his life, to which, when once made, he is bound to adhere. The force of habit will sometimes cause a man to hold on to old beliefs, after he is in a condition to see that they have no sound basis. But reflection upon the state of the case will overcome these habits, and he ought to allow reflection its full weight. People sometimes shrink from doing this, having an idea that beliefs are wholesome which they cannot help feeling rest on nothing. But let such persons suppose an analogous though different case from their own. Let them ask themselves what they would say to a reformed Mussulman who should hesitate to give up his old notions in regard to the relations of the sexes; or to a reformed Catholic who should still shrink from reading the Bible.* Would they not say that these persons ought to consider the matter fully, and clearly understand the new doctrine, and then ought to embrace it, in its entirety? But, above all, let it be considered that what is more wholesome than any particular belief is integrity of belief, and that to avoid looking into the support of any belief from a fear that it may turn out rotten is quite as immoral as it is disadvantageous. The person who confesses that there is such a thing as truth, which is distinguished from falsehood simply by this, that if acted on it will carry us to the point we aim at and not astray, and then, though convinced of this, dares not know the truth and seeks to avoid it, is in a sorry state of mind indeed.†

Yes, the other methods do have their merits: a clear logical conscience does cost something—just as any virtue, just as all that we cherish, costs us dear. But we should not desire it to be otherwise. The genius of a man's logical method should be loved and reverenced as his bride, whom he has chosen from all the world. He need not contemn the others; on the contrary, he may honor them deeply, and in doing so he only honors her the more. But she is the one that he has chosen, and he knows that he was right in making that choice. And having made it, he will work and fight for her, and will not complain that there are blows to take, hoping that there may

*The 1878 French translation omits the part of the sentence that follows the semi-colon (W3:355).

†The 1878 French translation ends here (W3:355). The *Collected Papers* editors add the following note (CP 5.387n): "Delete the remainder.—marginal note, 1893, 1903." Neither note was found, and the paragraph *is* included in HTR (R 407:25).

be as many and as hard to give, and will strive to be the worthy knight and champion of her from the blaze of whose splendors he draws his inspiration and his courage.[23]

Appendix

The following paragraphs comprise §50–53 of Peirce's 1894 How to Reason *(HTR), which includes "The Fixation of Belief" as Chapter V. The four sections, transcribed from R407:2–5, were written to replace the first three sections of "The Fixation of Belief."**

§50. Though nothing is more insipid than allegory, a work of art, to be interesting, must be an allegory, without the readers perceiving it. Don Quixotte and Sancho Panza exist within every man's breast. The *Trois Mousquetaires* are made the man of feeling, the man of action, and the man of thought, corresponding to the three categories of Chapter II.[24] The Bacchic train upon the Grecian vase is the train of thought.[25] The music of the dance is the particular habit, or association, which governs it. And,—what I meant to say,—the revel at Brussels in Childe Harold broken in upon by the awful sound of cannon, is the external stimulation which breaks up the play of free fancy, and tends to obliterate and discredit whatever association may happen to dominate it at the moment.[26]

Suppose that on any occasion I hesitate what to do,—whether, say, to pay a car-conductor a nickel or five coppers. The man stands before me, an urgent stimulus, and I am thrown into an agitation (an infinitesimal one, I grant, in the present example) until I have made my choice.

§51. Doubt, however, is not usually hesitancy about what is to be done then and there. It is anticipated hesitancy about what I shall do hereafter, or a

*An earlier version marks "The Fixation of Belief" as Chapter IV, and begins thus (R 407:27):

> I have to wait in a railway-station; and by way of passtime I read the advertisements on the walls, I compare the advantages of different trains and different routes which I never expect to take, merely fancying myself to be in a state of hesitancy, because I am bored with having nothing to trouble me. Feigned hesitancy, whether feigned for mere amusement or with a lofty purpose, plays a great part in the production of scientific inquiry. However the doubt may originate, it stimulates the mind to an activity which may be slight or energetic, calm or turbulent. Images pass rapidly through consciousness, one incessantly melting into another, until at last, when all is over,—it may be in a fraction of a second or after long years,—we find ourselves decided as to how we should act under such circumstances as those which occasioned this hesitation. We have formed a habit; and this habit is a belief.

feigned hesitancy about a fictitious state of things. It is the power of making believe we hesitate, together with the pregnant fact that the decision upon the merely make-believe dilemma goes toward forming a bona fide habit that will be operative in a real emergency.[27] It is these two things in combination that constitutes us intellectual beings.

Every answer to a question that has any meaning is a decision as to how we would act under imagined circumstances, or how the world would be expected to react upon our senses. Thus, suppose I am told that if two straight lines in one plane are cut by a third making the sum of the internal angles on one side less than two right angles, then those lines if sufficiently produced will meet on the side on which the said sum is less than two right angles. This means to me that if I had two lines drawn on a plane and wished to find where they should meet, I could draw a third line cutting them and ascertaining on which side the sum of the two internal angles was less than two right angles, and should lengthen the lines on that side. In a like manner, all doubt is a state of hesitancy about an imagined state of things.

§52. Doubt is an uneasy condition from which we struggle to free ourselves. In this, it is like any other stimulus. It is true that just as men may, for the sake of pleasures of the table, like to be hungry and take means to make themselves so, although hunger always involves a desire to fill the stomach, so for the sake of the pleasure of inquiry men may like to seek out doubts. Yet, for all that, doubt essentially involves a struggle to escape it.

A belief, on the other hand, is a habit according to which we shall behave in a certain way on occasions of a certain description. The Assassins, or followers of the Old Man of the Mountain, used to rush into death at his least command, because they believed that obedience to him would ensure everlasting felicity. Had they doubted this, they would not have acted as they did. So it is with every belief, according to its degree.

§54. But belief is something more than a habit; for we generally know when we wish to ask a question and when we wish to pronounce a judgment. To believe, therefore, feels differently from doubting. Not only that, but we know pretty nearly what we believe. Let us recall the nature of a sign and ask ourselves how we can know that a feeling of any sort is a sign that we have a habit implanted within us.[28]

We can understand one habit by likening it to another habit. But to understand what any habit is there must be some habit of which we are directly conscious in its generality. That is to say, we must have a certain generality in our direct consciousness. Bishop Berkeley and a great many clear thinkers laugh at the idea of our being able to imagine a triangle that is neither equilateral, isosceles, nor scalene.[29] They seem to think the object of imagination must be precisely determinate in every respect. But it seems certain that something general we must imagine. I do not intend, in this book, to go into questions of psychology. It is not necessary for us to know in detail how our thinking is done, but only how it *can* be done. Still, I may as well say, at once, that I think our direct consciousness covers a duration of time, although only an infinitely brief duration.[30] At any rate, I can see no way of escaping the proposition that to attach any general significance to a sign and to know that we do attach a general significance to it, we must have a direct imagination of something not in all respects determinate.

NOTES

1. Peirce opens with a direct attack on René Descartes (1596–1650), who began his 1637 *Discourse on Method* as follows:

> The most widely shared thing in the world is good sense, for everyone thinks he is so well provided with it that even those who are the most difficult to satisfy in everything else do not usually desire to have more good sense than they have. In this matter it is not likely that everyone is wrong. But this is rather a testimony to the fact that the power of judging well and distinguishing what is true from what is false, which is really what we call good sense or reason, is naturally equal in all men, and thus the diversity of our opinions does not arise because some people are more reasonable than others, but only because we conduct our thoughts by different routes and do not consider the same things. For it is not enough to have a good mind. The main thing is to apply it well.

2. In the next paper Peirce argues that the concept of transubstantiation is meaningless.

3. Close to the end of his letter to Paul Carus (chapter 9 below) Peirce refers to this as "a very false and foolish remark about Kepler." Much inspired by Robert Small's *An Account of the Astronomical Discoveries of Kepler* (London, 1804), Peirce gives a far more appreciative and more elaborate discussion of Kepler's discovery in "Keppler" (W8, sel. 49), and in his 1892–93 Lowell lectures (W9, sel. 36). The *Collected Papers* editors insert at this point a note (CP 5.362n) reproducing Peirce's discussion of Kepler in his August 1910 letter to Carus, and erroneously date it 1893.

4. The maxim translates as: read, read, read, work, pray, and read again.

5. The seed for this lies undoubtedly in Peirce's experience with experimental chemistry. At the age of twelve his uncle helped him put together a chemistry

laboratory at home designed around Liebig's method. On this method the student receives a set of bottles each marked with a letter. The student is then asked to analyze the contents of each bottle, using as the sole guide an introductory textbook in qualitative analysis. The Liebig method is a very practical way of learning chemical analysis, one where the difference in the contents of the bottles is determined by the practical consequences of the various operations performed upon them. Hence, Peirce's early experience with Liebig's method likely inspired his pragmatism. It also inspired Peirce general view of logic. As he writes in *How to Reason* (R 411.18–19):

> All laboratory inquiry consists in manipulating things, making experiments, observing results, and thus finding what the active rule (or intellect) of nature prescribes; all closet-inquiry, or reasoning, consists in manipulating diagrams, experimenting upon them, observing the results, and thus finding what secret depths of Thought (which with colossal infatuation I call *my* mind) has for its order.

6. The *Origin of Species* appeared in 1859, when Peirce was 21, and came to play a central role in his discussions with Chauncey Wright (1830–1875), an avid defender of John Stuart Mill's nominalism. See also chapter 7, with notes.

7. Rudolf Clausius (1822–1888), German physicist and mathematician; James Clerk Maxwell (1831–1879), Scottish physicist and mathematician. Both played a central role in the development of the kinetic theory of gasses, which is the first theory to rely on statistical laws, in this case to describe macroscopic phenomena (such as pressure, volume, or temperature) in terms of microscopic phenomena (the behavior of molecules).

8. Note that Peirce further down uses this notion of "the long run," as well as a focus on the community rather than the individual, when describing his method of science.

9. Also called "leading principle," which is defined by Peirce in Baldwin's *Dictionary of Philosophy and Psychology* (1901), 2:1–2.

10. In a short unfinished manuscript entitled "Logic," written 2 November 1910, Peirce gives an insightful account of doubt, comparing it with ignorance (R 828:2–4):

> Doubt is not the same as ignorance, nor as the consciousness of being ignorant; for if one does not care to know one cannot be said to be in doubt. Even if one is ignorant and does wish to know, that state will not, of itself, constitute doubt. If, for example, I wish to know how many days there are in that eclipse-cycle called a *saros*, knowing that it is nearly 8 hours more than 18 calendar years and 10 or 11 days, according as there are 5 or 4 leap-years in the interval, I shall simply multiply 365 by 18 and add 15⅓ to the product, which I can easily do in my head (for it is 7300⅓ less 715); and consequently I shall never feel any doubt; and this reflection causes me to recognize that what we call "doubt" is an *emotion*. It is particularly likely to set in when one's uncertainty is between two alternatives; for it seems to consist in an uneasy passage from one state of emotional expectation to another, though (in my own case at least)

> very little of the emotional element is required in the alternative imaginations to render the doubt very appreciably disagreeable. I find that, to me, the disagreeableness of doubt seems to be due to my imagining the question to become a practical one to me. I find, for instance, that my doubt whether a lancelet ought to be ranked with the vertebrates or not, is much less annoying than my doubt whether that group of elements in Mendeléef's Table is real that at present is represented by Manganese alone; and I account for this by the circumstance that I have made a serious study of chemistry, but not of zoology. *[...]*

11. Reflecting upon the discussions within the Metaphysical Club, Peirce writes that Nicholas St. John Green (CP 5.12), "often urged the importance of applying Bain's definition of belief, as that upon which a man is prepared to act. From this definition, pragmatism is scarce more than a corollary." The reference is to Alexander Bain (1818–1903), especially his 1859 *The Emotions and the Will.* In 1870 Peirce reviews Bain's *Logic* for the *Nation* (W2, item 43). For a further reflection on Green's contributions to the Metaphysical Club, see chapter 7.

12. The reference is clearly to René Descartes's 1641 *Meditations on First Philosophy*, especially the first meditation; see also CP 6.498. When reworking the paper for *How to Reason* (see the appendix at the end of this chapter) Peirce has come to embrace artificial doubt as a possible motive for inquiry.

13. Based on W3:18 (1872) it may be conjectured that Peirce is thinking of the *Advertisers' Gazette*, published by George P. Rowell and edited by Charles N. Kent, which was published from November 1866 until April 1876.

14. In a draft of his 1893 reply to Paul Carus, Peirce writes (R 958:145): "The essay on the Fixation of Belief was written to draw attention to that agapastic or social theory of epistemology, which I first announced in a paper called 'Some Consequences of Four Incapacities'." Peirce discusses agapasm in his 1892 "Evolutionary Love" (W8:194) and seeks to apply it to the history of science in his 1892–93 Lowell Lectures.

15. It was under Pius Nonus, or Giovanni Mastai Ferretti, who was pope from 1846 till 1878, that the doctrine of papal infallibility was accepted—an event that greatly impressed Peirce. In its stead Peirce advocated a doctrine he called fallibilism (see e.g., CP 2.75, 1902).

16. In a draft reply to Paul Carus's "Mr. Charles S. Peirce's Onslaught on the Doctrine of Necessity" (*Monist* 2 [July 1892]: 560–82), Peirce wrote (R 958:141): "In Section III. Dr. Carus calls attention to views expressed by me in 1877 in an article on the Fixation of Belief. Of these, Dr. Carus expresses an approval which is very gratifying to me, except as to what I there said of the *a priori* method. I acknowledge that that passage requires modification." The entire text is forthcoming in W9 as "Reply to the Necessitarians *[*Rejoinder to Carus's First Article*]*."

17. See e.g. Samuel Tyler, *A Discourse of the Baconian Philosophy* (New York: Barker and Scribner, 1850), 148–56.

18. This is also Tyler's view, ibid., 156; see also *New Organon*, Book I, Aphorism 14.

19. This means a rejection of Roger Bacon's notion that we can obtain synthetic knowledge through "interior illumination" (see above).

20. A more extensive discussion of reality is found in the next chapter where the pragmatic maxim is applied to the concept.

21. In a draft for "The Neglected Argument for the Reality of God" (1908), Peirce makes the following observation (R 842:10–11):

> Now belief is essentially satisfactory; and the state of Cartesian doubt is mere pretence or self-deception. For real doubt is most distressing, and nobody can pass from the naturally satisfied state of belief to a state of distressing doubt, unless some still stronger belief forces him to doubt. We cannot pass from satisfaction to dissatisfaction from sheer dissatisfaction with satisfaction.

22. In *How to Reason* and elsewhere Peirce draws a direct connection with self-control (R 407:27): "Reasoning, or inference, is something which can be good or bad, because it is within the domain of self-control."

23. This is possibly a veiled reference to Peirce's troubled marital situation. Zina accompanied him on his 1875 trip to Europe—the same trip during which he met the publisher William H. Appleton (see the introduction)—but deserted him while they were in Europe. Zina separated from Peirce in October the following year.

24. This is most likely a reference to R 403, titled "The Categories," although this is the first and not the second chapter of *How to Reason,* which is largely drawn from Peirce's 1868 "On a New List of Categories" (W2, sel. 4). R 1583:2 identifies "List of Categories" as the second chapter in the 1893 precursor to *How to Reason*, titled *Search for a Method.*

25. Peirce uses the same metaphor in the introduction to *How to Reason* (R 402:2):

> In the absence of external impressions, thoughts dance through the mind, each one gently leading in another, like a train of Bacchants on a Grecian vase. After a while, the clear train of thought is broken, the ideas remain scattered for a time, and then reconcentrate in another train.

In a later revision of the introduction (R 400:7) Peirce attributes the metaphor to Hegel.

26. Lord Byron, *Childe Harold's Pilgrimage* (London: J. Murray, 1816), Canto 3, Stanza 21–22.

27. Once, at a dinner party Peirce's younger brother Herbert swiftly extinguished a lady's dress when it caught fire. The incident greatly impressed the still young Peirce, and when he asked his brother about it the latter replied that after Mrs. Longfellow had died from a similar accident "he had often run over in imagination all the details of what ought to be done in such an emergency" (CP5.487n1).

28. This is most likely a reference to chapter 2 of *How to Reason*, entitled "What is a sign?" (EP2:4–10).

29. See John Locke, *An Essay Concerning Human Understanding*, IV.vii.9, and George Berkeley, *The Principles of Human Knowledge,* Introduction, §13.

30. Peirce here rejects the idea of a knife-edge present, siding with William James who argues in its stead for a specious present. See James's *Principles of Psychology* (New York: H. Holt & Co., 1890), 609–10.

Chapter 2

HOW TO MAKE OUR IDEAS CLEAR

William James famously identified this second article of the Illustrations series as the birthplace of Pragmatism. Building on the theory of inquiry developed in the first article, Peirce develops a theory of meaning, distinguishing three grades of clearness—the third of which is later called the Principle of Pragmatism, or the Pragmatic Maxim. The term pragmatism itself does not occur in the article, as Peirce only began using the term after James had used it publicly to describe Peirce's view, beginning in 1898. After introducing the pragmatic maxim, Peirce gave several examples, mostly from physics, to show how the maxim is to be applied. The first is the application to the concept of hardness. In this discussion Peirce gave a decidedly nominalistic interpretation to the maxim. This came to haunt him later, as the early pragmatists (mostly following James and Schiller) embraced a nominalist interpretation of pragmatism. Consequently, after the turn of century Peirce saw himself forced to spend much time and effort trying to distance himself from the more popular pragmatists, up to renaming his own brand of pragmatism "pragmaticism."

The second article appeared in Popular Science Monthly *12 (January 1878): 286–302. The text below is as it was printed in the* Popular Science Monthly. *It is only minimally edited and italic brackets surround text added by the editor. A French translation appeared as "Comment rendre nos idées claires" in* Revue Philosophique de la France et de L'Étranger *7*

(January 1879): 39–57. The footnotes contain discrepancies with the French version, alterations Peirce made while preparing the text for inclusion as essay 9 of the 1893 A Search for a Method*—which shortly thereafter was reused as chapter 16 of the 1894 logic book* How to Reason *under the title "Clearness of Apprehension" (R 422, hereafter referred to as HTR)—and several notes possibly related to the* Essays on Reasoning *of circa 1910, where this article was to be included as the second part of the first essay (R 334, hereafter referred to as ER). However, whereas Peirce's annotated version of "The Fixation of Belief," has been recovered, no such document could be identified for "How to Make Our Ideas Clear." Instead only a few notes were retrieved, and without any indication as to where Peirce meant to insert these surviving notes in the text. Evidence suggests that the* Collected Papers *editors had access to some annotated version (either a typescript or an offprint) that they date 1903. Given their earlier misdating (see chapter 1 and the introduction) it is conjectured that this may have been the missing ER offprint mentioned above, and it is treated as such in the footnotes. Where known the date of composition is given. A discussion of the history of the article, its French translation, and the various attempts at rewriting it, is found in the introduction.*

I.

Whoever has looked into a modern treatise on logic of the common sort,* will doubtless remember the two distinctions between *clear* and *obscure* conceptions, and between *distinct* and *confused* conceptions. They have lain in the books now for nigh two centuries, unimproved and unmodified, and are generally reckoned by logicians as among the gems of their doctrine.[1]

A clear idea is defined as one which is so apprehended that it will be recognized wherever it is met with, and so that no other will be mistaken for it. If it fails of this clearness, it is said to be obscure.[2]

This is rather a neat bit of philosophical terminology; yet, since it is clearness that they were defining, I wish the logicians had made their definition a little more plain. Never to fail to recognize an idea, and under no circumstances to mistake another for it, let it come in how recondite a form it may, would indeed imply such prodigious force and clearness of intellect as is seldom met with in this world. On the other hand, merely to have such an acquaintance with the idea as to have become familiar with

*In HTR Peirce writes instead (R 422:2): has looked into one of the treatises upon logic dating from *L'Art de penser* of the Port-Royalists down to very recent times,

it, and to have lost all hesitancy in recognizing it in ordinary cases, hardly seems to deserve the name of clearness of apprehension, since after all it only amounts to a subjective feeling of mastery which may be entirely mistaken. I take it, however, that when the logicians speak of "clearness," they mean nothing more than such a familiarity with an idea, since they regard the quality as but a small merit, which needs to be supplemented by another, which they call *distinctness.*

A distinct idea is defined as one which contains nothing which is not clear.[3] This is technical language; by the *contents* of an idea logicians understand whatever is contained in its definition. So that an idea is *distinctly* apprehended, according to them, when we can give a precise definition of it, in abstract terms. Here the professional logicians leave the subject; and I would not have troubled the reader with what they have to say, if it were not such a striking example of how they have been slumbering through ages of intellectual activity, listlessly disregarding the enginery of modern thought, and never dreaming of applying its lessons to the improvement of logic. It is easy to show that the doctrine that familiar use and abstract distinctness make the perfection of apprehension has its only true place in philosophies which have long been extinct; and it is now time to formulate the method of attaining to a more perfect clearness of thought, such as we see and admire in the thinkers of our own time.

When Descartes set about the reconstruction of philosophy,* his first step was to (theoretically) permit skepticism and to discard the practice of the schoolmen of looking to authority as the ultimate source of truth. That done, he sought a more natural fountain of true principles, and professed

---

*Peirce wrote the following note, marked H, for ER, which was most likely to be inserted here (R 334:38–39):

> the one κτῆμα ἐς ἀεί (everlasting boon,) that he conferred upon our race was in setting up an indestructible barrier to the transmission of the doctrine of most of the scholastic doctors that all certainty of truth ultimately rests upon external authority. In place of that he wishes to substitute the internal authority of consciousness of self, so crossing at one side the threshold between "the Method of Authority" and "the Method of Apriority," as they have been described in the first part of this essay.
>
> His *Méditations* are still read and perused by every student of philosophy and by every student of literature, and very profitably by each. Nevertheless, it would be hard to imagine reasoning more ~~worthless~~ objectionable than his; and his whole quest is for the mythical Eldorado of absolute certainty instead of for such an approach toward the truth as arduous endeavor, through one man's life after another's, can reasonably hope to attain. He talks about doubting everything except his own existence; but that he never really did so is made plain enough in many ways, among which a few may here be instanced. He never recognizes that doubt has degrees, as he must have done if he had been describing with strict veracity his own actual state of mind. He says it is *certain* that nothing *[MS breaks off]*

to find it in the human mind; thus passing, in the directest way, from the method of authority to that of apriority, as described in my first paper. Self-consciousness was to furnish us with our fundamental truths, and to decide what was agreeable to reason. But since, evidently, not all ideas are true, he was led to note, as the first condition of infallibility, that they must be clear. The distinction between an idea *seeming* clear and really being so, never occurred to him. Trusting to introspection, as he did, even for a knowledge of external things, why should he question its testimony in respect to the contents of our own minds? But then, I suppose, seeing men, who seemed to be quite clear and positive, holding opposite opinions upon fundamental principles, he was further led to say that clearness of ideas is not sufficient, but that they need also to be distinct, i.e., to have nothing unclear about them. What he probably meant by this (for he did not explain himself with precision) was, that they must sustain the test of dialectical examination; that they must not only seem clear at the outset, but that discussion must never be able to bring to light points of obscurity connected with them.

Such was the distinction of Descartes, and one sees that it was precisely on the level of his philosophy. It was somewhat developed by Leibnitz.[4] This great and singular genius was as remarkable for what he failed to see as for what he saw. That a piece of mechanism could not do work perpetually without being fed with power in some form, was a thing perfectly apparent to him; yet he did not understand that the machinery of the mind can only transform knowledge, but never originate it, unless it be fed with facts of observation.[5] He thus missed the most essential point of the Cartesian philosophy, which is, that to accept propositions which seem perfectly evident to us is a thing which, whether it be logical or illogical, we cannot help doing. Instead of regarding the matter in this way, he sought to reduce the first principles of science to formulas which cannot be denied without self-contradiction,* and was apparently unaware of the great difference between his position and that of Descartes.† So he reverted to the old formalities of logic, and, above all,

*In HTR (R 422:3) Peirce writes instead: he sought to reduce the first principles of science to two classes, those which cannot be denied without self-contradiction and those which result from the principle of sufficient reason (of which more anon),

†The *Collected Papers* editors insert at this point the following note that they date 1903 (CP 5.392n):

> He was, however, above all, one of the minds that grow; while at first he was an extreme nominalist, like Hobbes, and dabbled in the nonsensical and impotent *Ars magna* of Raymond Lully, he subsequently embraced the law of continuity and other doctrines opposed to nominalism. I speak here of his earlier views.

abstract definitions played a great part in his philosophy. It was quite natural, therefore, that on observing that the method of Descartes labored under the difficulty that we may seem to ourselves to have clear apprehensions of ideas which in truth are very hazy, no better remedy occurred to him than to require an abstract definition of every important term. Accordingly, in adopting the distinction of *clear* and *distinct* notions, he described the latter quality as the clear apprehension of everything contained in the definition; and the books have ever since copied his words.[6] There is no danger that his chimerical scheme will ever again be overvalued. Nothing new can ever be learned by analyzing definitions. Nevertheless, our existing beliefs can be set in order by this process, and order is an essential element of intellectual economy, as of every other. It may be acknowledged, therefore, that the books are right in making familiarity with a notion the first step toward clearness of apprehension, and the defining of it the second. But in omitting all mention of any higher perspicuity of thought, they simply mirror a philosophy which was exploded a hundred years ago. That much-admired "ornament of logic"—the doctrine of clearness and distinctness—may be pretty enough, but it is high time to relegate to our cabinet of curiosities the antique *bijou,* and to wear about us something better adapted to modern uses.*

*The *Collected Papers* editors insert at this point the following note that they date 1903 (CP 5.393n): "Delete this Paragraph." Most likely this comes from a now lost annotated offprint to which belong some of the notes surviving in R 334. Possibly the deleted paragraph was to be replaced with the following note, marked F (R 334:25–27):

> Nevertheless, it behooves an old student of such subjects to testify, for the benefit of those who have not delved into discarded books, that many a time it has happened that a reasoning has appeared justly convincing, not to giddy heads only but to the best exercized and consequently the most cautious, and has for generations encountered no objection from good reasoners, and yet at last has been proved to contain a flaw that destroys its value. There have, moreover, been many instances of a manifestly just inference not being drawn; but some instances of this second class may have been due to attention not having been called to the arguments, while in others the fault was due to a fallacious argument against the just conclusion, i.e., to a fault of the first class. Yet unless we are prepared to accuse all the ancient civilized peoples either of laziness or of stupidity, the numerous cases in which easy experiments would have refuted their opinions ought to incline us toward the supposition that the Instinct of Just Reasoning had not in those days attained as full a development in the human genus as it has in ours; and this supposition receives some colour of support from the fact that the Greeks depicted in the Iliad, who perhaps did not antecede the tragedians by a greater interval of time than King Alfred antecedes us, were absolutely devoid of that Moral Instinct upon which the plots of Sophocles are pivoted. It cannot be denied that each of the three human Instincts of Beauty, of Morality, and of Just Reasoning has undergone strange fluctuations in the course of history; and yet, if we are really descended from anything like apes, there has certainly been an upward growth of each on the whole.

The very first lesson that we have a right to demand that logic shall teach us is, how to make our ideas clear; and a most important one it is, depreciated only by minds who stand in need of it. To know what we think, to be masters of our own meaning, will make a solid foundation for great and weighty thought. It is most easily learned by those whose ideas are meagre and restricted; and far happier they than such as wallow helplessly in a rich mud of conceptions. A nation, it is true, may, in the course of generations, overcome the disadvantage of an excessive wealth of language and its natural concomitant, a vast, unfathomable deep of ideas. We may see it in history, slowly perfecting its literary forms, sloughing at length its metaphysics, and, by virtue of the untirable patience which is often a compensation, attaining great excellence in every branch of mental acquirement. The page of history is not yet unrolled which is to tell us whether such a people will or will not in the long-run prevail over one whose ideas (like the words of their language) are few, but which possesses a wonderful mastery over those which it has. For an individual, however, there can be no question that a few clear ideas are worth more than many confused ones. A young man would hardly be persuaded to sacrifice the greater part of his thoughts to save the rest; and the muddled head is the least apt to see the necessity of such a sacrifice. Him we can usually only commiserate, as a person with a congenital defect. Time will help him, but intellectual maturity with regard to clearness comes rather late,* an unfortunate arrangement of Nature, inasmuch as clearness is of less use to a man settled in life, whose errors have in great measure had their effect, than it would be to one whose path lies† before him. It is terrible to see how a single unclear idea, a single formula without meaning, lurking in a young man's head, will sometimes act like an obstruction of inert matter in an artery, hindering the nutrition of the brain, and condemning its victim to pine away in the fullness of his intellectual vigor and in the midst of intellectual plenty. Many a man has cherished for years as his hobby some vague shadow of an idea, too meaningless to be positively false; he has, nevertheless, passionately loved it, has made it his companion by day and by night, and has given to it his strength and his life, leaving all other occupations for its sake, and in short has lived with it and for it, until it has become, as it were, flesh of his flesh and bone of his bone; and then he has waked up some bright morning to find it gone, clean vanished away like the beautiful Melusina of the

*In ER (on authority of CP5.393n) Peirce changes "comes rather late," into "is apt to come rather late. This seems".

†In ER (on authority of CP5.393n) Peirce changes "lies" into "lay".

fable,[*] and the essence of his life gone with it.[7] I have myself known such a man; and who can tell how many histories of circle-squarers, metaphysicians, astrologers, and what not, may not be told in the old German story?

II.

The principles set forth in the first of these papers[†] lead, at once, to a method of reaching a clearness of thought of a far[‡] higher grade than the "distinctness" of the logicians. We have there found[§] that the action of thought is excited by the irritation of doubt, and ceases when belief is attained; so that the production of belief is the sole function of thought.[8] All these words, however, are too strong for my purpose. It is as if I had described the phenomena as they appear under a mental microscope. Doubt and Belief, as the words are commonly employed, relate to religious or other grave discussions. But here I use them to designate the starting of any question, no matter how small or how great, and the resolution of it. If, for instance, in a horse-car, I pull out my purse and find a five-cent nickel and five coppers, I decide, while my hand is going to the purse, in which way I will pay my fare. To call such a question Doubt, and my decision Belief, is certainly to use words very disproportionate to the occasion. To speak of such a doubt as causing an irritation which needs to be appeased, suggests a temper which is uncomfortable to the verge of insanity. Yet, looking at the matter minutely, it must be admitted that, if there is the least hesitation as to whether I shall pay the five coppers or the nickel (as there will be sure to be, unless I act from some previously contracted habit in the matter), though irritation is too strong a word, yet I am excited to such small mental activity as may be necessary to deciding how I shall act. Most frequently doubts arise from some indecision, however momentary, in our action. Sometimes it is not so. I have, for example, to wait in a railway-station,

---

[*]The 1879 French translation concludes the paragraph as follows (W3:359):

> comme Mélusine, la belle fée, et toute sa vie s'était envolée avec elle. J'ai connu moi-même un de ces hommes. Qui pourrait compter tous les les quadrateurs de cercle, métaphysiciens, astrologues, que sais-je encore, dont les annales de la vieille Allemagne pourraient nous redire l'histoire?

[†]In ER (on authority of CP5.394n) Peirce changes "of these papers" into "part of this essay".

[‡]In ER (on authority of CP5.394n) Peirce deletes "a far".

[§]In ER (on authority of CP5.394n) Peirce changes "We have there found" into "It was there noticed".

and to pass the time I read the advertisements on the walls, I compare the advantages of different trains and different routes which I never expect to take, merely fancying myself to be in a state of hesitancy, because I am bored with having nothing to trouble me. Feigned hesitancy, whether feigned for mere amusement or with a lofty purpose, plays a great part in the production of scientific inquiry.[9] However the doubt may originate, it stimulates the mind to an activity which may be slight or energetic, calm or turbulent. Images pass rapidly through consciousness, one incessantly melting into another, until at last, when all is over—it may be in a fraction of a second, in an hour, or after long years—we find ourselves decided as to how we should act under such circumstances as those which occasioned our hesitation. In other words, we have attained belief.

In this process we observe two sorts of elements of consciousness, the distinction between which may best be made clear by means of an illustration. In a piece of music there are the separate notes, and there is the air. A single tone may be prolonged for an hour or a day, and it exists as perfectly in each second of that time as in the whole taken together; so that, as long as it is sounding, it might be present to a sense from which everything in the past was as completely absent as the future itself. But it is different with the air, the performance of which occupies a certain time, during the portions of which only portions of it are played. It consists in an orderliness in the succession of sounds which strike the ear at different times; and to perceive it there must be some continuity of consciousness which makes the events of a lapse of time present to us. We certainly only perceive the air by hearing the separate notes; yet we cannot be said to directly hear it, for we hear only what is present at the instant, and an orderliness of succession cannot exist in an instant. These two sorts of objects, what we are *immediately* conscious of and what we are *mediately* conscious of, are found in all consciousness. Some elements (the sensations) are completely present at every instant so long as they last, while others (like thought) are actions having beginning, middle, and end, and consist in a congruence in the succession of sensations which flow through the mind. They cannot be immediately present to us, but must cover some portion of the past or future. Thought is a thread of melody running through the succession of our sensations.[10]

We may add that just as a piece of music may be written in parts, each part having its own air, so various systems of relationship of succession subsist together between the same sensations. These different systems are distinguished by having different motives, ideas, or functions. Thought is only one such system, for its sole motive, idea, and function, is to produce belief, and whatever does not concern that purpose belongs to some other system of relations. The action of thinking may incidentally have other results; it may serve to amuse us, for example, and among *dilettanti* it is not rare to find those who have so perverted thought to the purposes of pleasure that it seems to vex them to think that the questions upon which they delight to exercise it may ever get finally settled; and a positive discovery which takes a favorite subject out of the arena of literary debate is met with ill-concealed dislike. This disposition is the very debauchery of thought. But the soul and meaning of thought, abstracted from the other elements which accompany it, though it may be voluntarily thwarted, can never be made to direct itself toward anything but the production of belief. Thought in action has for its only possible motive the attainment of thought at rest; and whatever does not refer to belief is no part of the thought itself.

And what, then, is belief? It is the demi-cadence which closes a musical phrase in the symphony of our intellectual life. We have seen that it has just three properties: First, it is something that we are aware of; second, it appeases the irritation of doubt; and, third, it involves the establishment in our nature of a rule of action, or, say for short, a *habit.* As it appeases the irritation of doubt, which is the motive for thinking, thought relaxes, and comes to rest for a moment when belief is reached. But, since belief is a rule for action, the application of which involves further doubt and further thought, at the same time that it is a stopping-place, it is also a new starting-place for thought. That is why I have permitted myself to call it thought at rest, although thought is essentially an action. The *final* upshot of thinking is the exercise of volition, and of this thought no longer forms a part; but belief is only a stadium of mental action, an effect upon our nature due to thought, which will influence future thinking.

The essence of belief is the establishment of a habit, and different beliefs are distinguished by the different modes of action to which they give rise. If beliefs do not differ in this respect, if they appease the same doubt by producing the same rule of action, then no mere differences in the manner of consciousness of them can make them different beliefs, any more than playing a tune in different keys is playing different tunes. Imaginary distinctions are often drawn between beliefs which differ only in their mode of expression;—the wrangling which ensues is real enough, however. To believe that any objects are arranged* as in Fig. 1, and to believe that they are arranged in Fig. 2, are one and the same belief; yet it is conceivable that a man should assert one proposition and deny the other. Such false distinctions do as much harm as the confusion of beliefs really different, and are among the pitfalls of which we ought constantly to beware, especially when

*In ER (on authority of CP5.398n) Peirce inserts "among themselves".

we are upon metaphysical ground. One singular deception of this sort, which often occurs, is to mistake the sensation produced by our own unclearness of thought for a character of the object we are thinking. Instead of perceiving that the obscurity is purely subjective, we fancy that we contemplate a quality of the object which is essentially mysterious; and if our conception be afterward presented to us in a clear form we do not recognize it as the same, owing to the absence of the feeling of unintelligibility. So long as this deception lasts, it obviously puts an impassable barrier in the way of perspicuous thinking; so that it equally interests the opponents of rational thought to perpetuate it, and its adherents to guard against it.

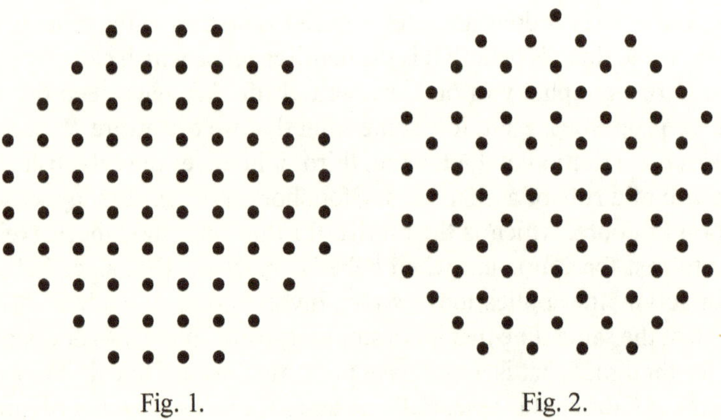

Fig. 1. Fig. 2.

Another such deception is to mistake a mere difference in the grammatical construction of two words for a distinction between the ideas they express. In this pedantic age, when the general mob of writers attend so much more to words than to things, this error is common enough. When I just said that thought is an *action,* and that it consists in a *relation,* although a person performs an action but not a relation, which can only be the result of an action, yet there was no inconsistency in what I said, but only a grammatical vagueness.

From all these sophisms we shall be perfectly safe so long as we reflect that the whole function of thought is to produce habits of action; and that whatever there is connected with a thought, but irrelevant to its purpose, is an accretion to it, but no part of it. If there be a unity among our sensations which has no reference to how we shall act on a given occasion, as when we listen to a piece of music, why we do not call that thinking. To develop its meaning, we have, therefore, simply to determine what habits it produces, for what a thing means is simply what habits it involves. Now, the identity of a habit depends on how it might lead us to act, not merely under such circumstances as are likely to arise, but under such as might possibly occur, no matter how improbable they may be.[*] What the habit is depends on *when* and *how* it causes us to act. As for the *when,* every stimulus to action is derived from perception; as for the *how,* every purpose of action is to produce some sensible result. Thus, we come down to what is tangible and practical,[†] as the root of every real distinction of thought, no matter how subtile it may be; and there is no distinction of meaning so fine as to consist in anything but a possible difference of practice.

To see what this principle leads to, consider in the light of it such a doctrine as that of transubstantiation. The Protestant churches[‡] generally hold that the elements of the sacrament are flesh and blood only in a tropical sense; they nourish our souls as meat and the juice of it would our bodies. But the Catholics maintain that they are literally just that;[§] although they possess all the sensible qualities of wafer-cakes and diluted wine. But we can have no conception of wine except what may enter into a belief, either—

1. That this, that, or the other, is wine; or,
2. That wine possesses certain properties.

Such beliefs are nothing but self-notifications that we should, upon occasion, act in regard to such things as we believe to be wine according to the qualities which we believe wine to possess. The occasion of such action would be some sensible perception, the motive of it to produce some sensible result. Thus our action has exclusive reference to what affects the senses, our habit has the same bearing as our action, our belief the same as our habit, our conception the same as our belief; and we can consequently mean nothing by wine but what has certain effects, direct or indirect, upon our senses; and to talk of something as having all the sensible characters of wine, yet being in reality blood, is senseless jargon. Now, it is not my object to pursue the theological question; and having used it as a logical example I drop it, without caring to anticipate the theologian's reply. I only desire to point out how impossible it is that we should have an idea in our minds which relates to anything but conceived sensible effects of things. Our idea of anything *is* our idea of its sensible effects; and if we fancy that we have any other we deceive ourselves, and mistake a mere sensation accompanying the thought for a part of the thought itself. It is absurd to say that thought has any meaning unrelated to its only function. It is foolish

[*] In HTR Peirce adds (R 422:7): —no matter if contrary to all previous experience.

[†] In HTR Peirce writes (R 422:7): tangible and conceivably practical

[‡] In HTR Peirce writes instead (R 422:7): Most Protestant churches

[§] In ER (on authority of CP5.401n) Peirce writes "meat and blood;" for "that;".

for Catholics and Protestants to fancy themselves in disagreement about the elements of the sacrament, if they agree in regard to all their sensible effects, here or hereafter.*

It appears, then, that the rule for attaining the third grade of clearness of apprehension is as follows: Consider what effects, which† might conceivably have practical bearings, we conceive the object of our conception to have. Then, our conception of these effects is the whole of our conception of the object.‡[11]

*In HTR Peirce writes (R 422:7): here and hereafter.

†In ER (on authority of CP5.402n) Peirce writes "that" for "which".

‡In HTR Peirce inserts the following three paragraphs (R 422:8–9):

> Before we undertake to apply this rule, let us reflect a little upon what it implies. It has been said to be a sceptical and materialistic principle. But it is only an application of the sole principle of logic which was recommended by Jesus; "Ye may know them by their fruits"; and it is very intimately allied with the ideas of the gospel. We must certainly guard ourselves against understanding this rule in too individualistic a sense. To say that man accomplishes nothing but that to which his endeavors are directed would be a cruel condemnation of the great bulk of mankind, who never have leisure to labor for anything but the necessities of life for themselves and their families. But, without directly striving for it, far less comprehending it, they perform all that civilization requires, and bring forth another generation to advance history another step. Their fruit is, therefore, collective; it is the achievement of the whole people. What is it, then, that the whole people is about, what is this civilization that is the outcome of history, but is never completed? We cannot expect to attain a complete conception of it; but we can see that it is a gradual process, that it involves a realization of ideas in man's consciousness and in his works, and that it takes place by virtue of man's capacity for learning, and by experience continually pouring upon him ideas he has not yet acquired. We may say that it is the process whereby man, with all his miserable littleness, becomes gradually more and more imbued with the Spirit of God, in which Nature and History are rife. We are also told to believe in a world to come; but the idea is itself too vague to contribute much to the perspicuity of ordinary ideas. It is a common observation that those who dwell continually upon their expectations are apt to become oblivious to the requirements of their actual station. The great principle of logic is self-surrender, which does not mean that self is to lay low for the sake of an ultimate triumph. It may turn out so; but that must not be the governing purpose.
>
> When we come to study the great principle of continuity and see how all is fluid and every point directly partakes the being of every other, it will appear that individualism and falsity are one and the same. Meantime, we know that man is not whole as long as he is single, that he is essentially a possible member of society. Especially, one man's experience is nothing, if it stands alone. If he sees what others cannot, we call it hallucination. It is not "my" experience, but "our" experience that has to be thought of; and this "us" has indefinite possibilities.
>
> Neither must we understand the practical in any low and sordid sense. Individual action is a means and not our end. Individual pleasure is not our end; we are all putting our shoulders to the wheel for an end that none of us can catch more than a glimpse at,—that which the generations are working out. But we can see that the development of embodied ideas is what it will consist in.

III.

Let us illustrate this rule* by some examples; and, to begin with the simplest one possible, let us ask what we mean by calling a thing *hard.* Evidently that it will not be scratched by many other substances. The whole conception of this quality, as of every other, lies in its conceived effects.† There is absolutely no difference between a hard thing and a soft thing so long as they are not brought to the test.[12] Suppose, then, that a diamond could be crystallized in the midst of a cushion of soft cotton, and should remain there until it was finally burned up. Would it be false to say that that diamond was soft? This seems a foolish question, and would be so, in fact, except in the realm of logic. There such questions are often of the greatest utility as serving to bring logical principles into sharper relief than real discussions ever could. In studying logic we must not put them aside with hasty answers, but must consider them with attentive care, in order to make out the principles involved. We may, in the present case, modify our question, and ask what prevents us from saying that all hard bodies remain perfectly soft until they are touched, when their hardness increases with the pressure until they are scratched. Reflection will show that the reply is this: there would be no *falsity* in such modes of speech. They would involve a modification of our present usage of speech with regard to the words hard and soft, but not of their meanings. For they represent no fact to be different from what it is; only they involve arrangements of facts which would be exceedingly maladroit. This leads us to remark that the question of what would occur under circumstances which do not actually arise is not a question of fact, but only of the most perspicuous arrangement of them. For example, the

---

*In HTR Peirce writes (R 422:10): our rule for attaining clearness of thought

†In 1905, in a "critical commentary" (R 289:2) on the first two *Popular Science Monthly* papers, entitled "Consequences of Pragmaticism," Peirce writes (R 289:12–13):

> At this point an error emerges in my original paper which is repeated in later passages; and in view of the enormities into which some of the neopragmatists have been drawn whether by incautiously assenting to what I said, or, as is more likely, by similar crookedness in their independent thought, I am bound to call it a damnable error. Namely, I imagine a diamond to be formed in a bed of cotton wool and there to be burned up before any hard object ever touched it; whereupon I say that it is a mere question of the usage of speech whether that diamond be called hard or soft. But to say this is to represent the being of one habit or of the reverse habit as constituted by actual occurrences. In other words, it is to represent a general as constrained by the existence of individuals of a given kind instead of by the invariable existence of individuals of the opposite kind. The true account of such a case is precisely the account that commonsense would give of it.

question of free-will and fate in its simplest form, stripped of verbiage, is something like this: I have done something of which I am ashamed; could I, by an effort of the will, have resisted the temptation, and done otherwise? The philosophical reply is, that this is not a question of fact, but only of the arrangement of facts. Arranging them so as to exhibit what is particularly pertinent to my question—namely, that I ought to blame myself for having done wrong—it is perfectly true to say that, if I had willed to do otherwise than I did, I should have done otherwise. On the other hand, arranging the facts so as to exhibit another important consideration, it is equally true that, when a temptation has once been allowed to work, it will, if it has a certain force, produce its effect, let me struggle how I may. There is no objection to a contradiction in what would result from a false supposition. The *reductio ad absurdum* consists in showing that contradictory results would follow from a hypothesis which is consequently judged to be false. Many questions are involved in the free-will discussion, and I am far from desiring to say that both sides are equally right. On the contrary, I am of opinion that one side denies important facts, and that the other does not. But what I do say is, that the above single question was the origin of the whole doubt; that, had it not been for this question, the controversy would never have arisen; and that this question is perfectly solved in the manner which I have indicated.

Let us next seek a clear idea of Weight.[*13] This is another very easy case. To say that a body is heavy means simply that, in the absence of opposing force, it will fall. This (neglecting certain specifications of how it will fall, etc., which exist in the mind of the physicist who uses the word) is evidently the whole conception of weight. It is a fair question whether some particular facts may not *account* for gravity; but what we mean by the force itself is completely involved in its effects.

This leads us to undertake an account of the idea of Force in general. This is the great conception which, developed in the early part of the seventeenth century from the rude idea of a cause, and constantly improved

---

[*] In the 1905 "Consequences of Pragmaticism" Peirce writes (R 289:19):

> The next examples in the original paper were designed, not to contribute anything to the philosophical understanding of the pragmatistic principle, but merely to practice the reader in making use of it. They explicate the idea of physical force. Had space been unlimited I might here have added an explanation of the intellectual force of probability, as I did on p.129f of the same volume of the *Pop. Sci. Monthly*. But it was here omitted for the reason that in the *North American Review* of eleven years before, July 1867 (Vol. CV. p. 317) I have said all that seemed necessary in a notice of Venn's *Logic of Chance [*W2:98–102*]*.

upon since,* has shown us how to explain all the changes of motion which bodies experience, and how to think about all physical phenomena; which has given birth to modern science, and changed the face of the globe; and which, aside from its more special uses, has played a principal part in directing the course of modern thought, and in furthering modern social development. It is, therefore, worth some pains to comprehend it.[14] According to our rule, we must begin by asking what is the immediate use of thinking about force; and the answer is, that we thus account for changes of motion. If bodies were left to themselves, without the intervention of forces, every motion would continue unchanged both in velocity and in direction. Furthermore, change of motion never takes place abruptly; if its direction is changed, it is always through a curve without angles; if its velocity alters, it is by degrees. The gradual changes which are constantly taking place are conceived by geometers to be compounded together according to the rules of the parallelogram of forces. If the reader does not already know what this is, he will find it, I hope, to his advantage to endeavor to follow the following explanation; but if mathematics are insupportable to him, pray let him skip three paragraphs rather than that we should part company here.†

A *path* is a line whose beginning and end are distinguished. Two paths are considered to be equivalent, which, beginning at the same point, lead to the same point. Thus the two paths, *ABCDE* and *AFGHE,* are equivalent. Paths which do *not* begin at the same point are considered to be equivalent, provided that, on moving either of them without turning it, but keeping it always parallel to its original position, when its beginning coincides with that of the other path, the ends also coincide. Paths are considered as geometrically added together, when one begins where the other ends; thus the path *AE* is conceived to be a sum of *AB, BC, CD,* and *DE.* In the parallelogram of Fig. 4 the diagonal *AC* is the sum of *AB* and *BC;* or, since *AD* is geometrically equivalent to *BC, AC* is the geometrical sum of *AB* and *AD.*

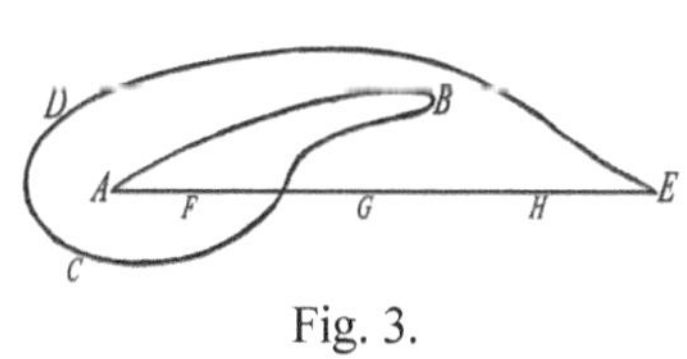

Fig. 3.

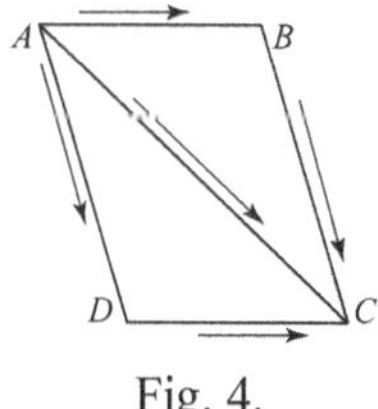

Fig. 4.

*In a draft sheet for HTR (RS 104:145, 1893), Peirce replaces "and constantly improved upon since," with: —itself a development of an instinctive notion which no race of animals can be without which makes any effort to seize its prey,—and constantly improved under scientific discussion,

†What follows after the semicolon is omitted in the 1879 French translation (W3:367).

All this is purely conventional. It simply amounts to this: that we choose to call paths having the relations I have described equal or added. But, though it is a convention, it is a convention with a good reason. The rule for geometrical addition may be applied not only to paths, but to any other things which can be represented by paths. Now, as a path is determined by the varying direction and distance of the point which moves over it from the starting-point, it follows that anything which from its beginning to its end is determined by a varying direction* and a varying magnitude is capable of being represented by a line. Accordingly, *velocities* may be represented by lines, for they have only directions and rates. The same thing is true of *accelerations,* or changes of velocities. This is evident enough in the case of velocities; and it becomes evident for accelerations if we consider that precisely what velocities are to positions—namely, states of change of them—that accelerations are to velocities.

The so-called "parallelogram of forces" is simply a rule for compounding accelerations. The rule is, to represent the accelerations by paths, and then to geometrically add the paths.† The geometers, however, not only use the "parallelogram of forces" to compound different accelerations, but also to resolve one acceleration into a sum of several. Let *AB* (Fig. 5) be the path which represents a certain acceleration—say, such a change in the motion of a body that at the end of one second the body will, under the influence of that change, be in a position different from what it would have had if its motion had continued unchanged such that a path equivalent to *AB* would lead from the latter position to the former. This acceleration may be considered as the sum of the accelerations represented by *AC* and *CB.* It may also be considered as the sum of the very different accelerations represented by *AD* and *DB,* where *AD* is almost the opposite of *AC.* And it is clear that there is an immense variety of ways in which *AB* might be resolved into the sum of two accelerations.

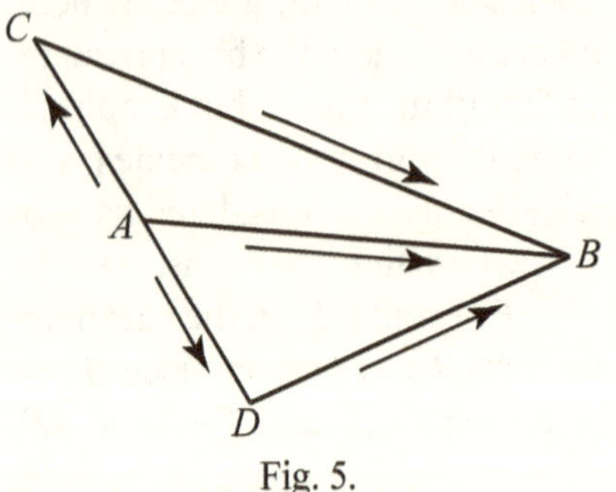

Fig. 5.

After this tedious explanation, which I hope, in view of the extraordinary interest of the conception of force, may not have exhausted the reader's

* In HTR Peirce inserts at this point (R 422:11): and distance which moves over it from the starting-point, it follows that anything which from its beginning to its end is determined by a varying distance

† What follows after the last comma is omitted in the 1879 French translation (W3:367).

patience, we are prepared at last to state the grand fact which this conception embodies. This fact is that if the actual changes of motion which the different particles of bodies experience are each resolved in its appropriate way, each component acceleration is precisely such as is prescribed by a certain law of Nature, according to which bodies in the relative positions which the bodies in question actually have at the moment,* always receive certain accelerations, which, being compounded by geometrical addition, give the acceleration which the body actually experiences.[15]

This is the only fact which the idea of force represents, and whoever will take the trouble clearly to apprehend what this fact is, perfectly comprehends what force is. Whether we ought to say that a force *is* an acceleration, or that it *causes* an acceleration, is a mere question of propriety of language, which has no more to do with our real meaning than the difference between the French idiom *"Il fait froid"* and its English equivalent *"It is cold."* Yet it is surprising to see how this simple affair has muddled men's minds.[16] In how many profound treatises is not force spoken of as a "mysterious entity," which seems to be only a way of confessing that the author despairs of ever getting a clear notion of what the word means! In a recent admired work on "Analytic Mechanics" it is stated that we understand precisely the effect of force, but what force itself is we do not understand! This is simply a self-contradiction.[17] The idea which the word force excites in our minds has no other function than to affect our actions, and these actions can have no reference to force otherwise than through its effects. Consequently, if we know what the effects of force are, we are acquainted with every fact which is implied in saying that a force exists, and there is nothing more to know. The truth is, there is some vague notion afloat that a question may mean something which the mind cannot conceive; and when some hair-splitting philosophers have been confronted with the absurdity of such a view, they have invented an empty distinction between positive and negative conceptions, in the attempt to give their non-idea a form not obviously nonsensical. The nullity of it is sufficiently plain from the considerations given a few pages back; and, apart from those considerations, the quibbling character of the distinction must have struck every mind accustomed to real thinking.

IV.

Let us now approach the subject of logic, and consider a conception which particularly concerns it, that of *reality*.[18] Taking clearness in the sense of

*[Original Peirce footnote, omitted in the French translation:] Possibly the velocities also have to be taken into account.

familiarity, no idea could be clearer than this. Every child uses it with perfect confidence, never dreaming that he does not understand it. As for clearness in its second grade, however, it would probably puzzle most men, even among those of a reflective turn of mind, to give an abstract definition of the real. Yet such a definition may perhaps be reached by considering the points of difference between reality and its opposite, fiction. A figment is a product of somebody's imagination; it has such characters as his thought impresses upon it. That whose characters are independent of how you or I think is an external reality. There are, however, phenomena within our own minds, dependent upon our thought, which are at the same time real in the sense that we really think them. But though their characters depend on how we think, they do not depend on what we think those characters to be. Thus, a dream has a real existence as a mental phenomenon, if somebody has really dreamt it; that he dreamt so and so, does not depend on what anybody thinks was dreamt, but is completely independent of all opinion on the subject. On the other hand, considering, not the fact of dreaming, but the thing dreamt, it retains its peculiarities by virtue of no other fact than that it was dreamt to possess them. Thus we may define the real as that whose characters are independent of what anybody may think them to be.[19]

But, however satisfactory such a definition may be found, it would be a great mistake to suppose that it makes the idea of reality perfectly clear. Here, then, let us apply our rules. According to them, reality, like every other quality, consists in the peculiar sensible effects which things partaking of it produce. The only effect which real things have is to cause belief, for all the sensations which they excite emerge into consciousness in the form of beliefs. The question therefore is, how is true belief (or belief in the real) distinguished from false belief (or belief in fiction). Now, as we have seen in the former paper, the ideas of truth and falsehood, in their full development, appertain exclusively to the scientific method of settling opinion.* A person who arbitrarily chooses the propositions which he will adopt can use the word truth only to emphasize the expression of his determination to hold on to his choice. Of course, the method of tenacity never prevailed exclusively; reason is too natural to men for that. But in the literature of the dark ages we find some fine examples of it. When Scotus Erigena is commenting upon a poetical passage in which hellebore is spoken of as having caused the death of Socrates, he does not hesitate to inform the inquiring reader that Helleborus and Socrates were two eminent Greek philosophers, and that the latter having been overcome in

*In HTR Peirce writes (R 422:13): experiential method of settling opinion.

argument by the former took the matter to heart and died of it![20] What sort of an idea of truth could a man have who could adopt and teach, without the qualification of a perhaps, an opinion taken so entirely at random? The real spirit of Socrates, who I hope would have been delighted to have been "overcome in argument," because he would have learned something by it, is in curious contrast with the naïve idea of the glossist, for whom* discussion would seem to have been simply a struggle. When philosophy began to awake from its long slumber, and before theology completely dominated it, the practice seems to have been for each professor to seize upon any philosophical position he found unoccupied and which seemed a strong one, to intrench himself in it, and to sally forth from time to time to give battle to the others. Thus, even the scanty records we possess of those disputes enable us to make out a dozen or more opinions held by different teachers at one time concerning the question of nominalism and realism. Read the opening part of the *Historia Calamitatum* of Abelard, who was certainly as philosophical as any of his contemporaries, and see the spirit of combat which it breathes.[21] For him, the truth is simply his particular stronghold. When the method of authority prevailed, the truth meant little more than the Catholic faith. All the efforts of the scholastic doctors are directed toward harmonizing their faith in Aristotle and their faith in the Church, and one may search their ponderous folios through without finding an argument which goes any further. It is noticeable that where different faiths flourish side by side, renegades are looked upon with contempt even by the party whose belief they adopt; so completely has the idea of loyalty replaced that of truth-seeking. Since the time of Descartes, the defect in the conception of truth has been less apparent. Still, it will sometimes strike a scientific man that the philosophers have been less intent on finding out what the facts are, than on inquiring what belief is most in harmony with their system. It is hard to convince a follower of the *a priori* method by adducing facts; but show him that an opinion he is defending is inconsistent with what he has laid down elsewhere, and he will be very apt to retract it. These minds do not seem to believe that disputation is ever to cease; they seem to think that the opinion which is natural for one man is not so for another, and that belief will, consequently, never be settled. In contenting themselves with fixing their own opinions by a method which would lead another man to a different result, they betray their feeble hold of the conception of what truth is.

---

*In HTR Peirce inserts in between commas (R 422:13): as for the "born missionary" of today

On the other hand, all the followers of science are fully persuaded[*] that the processes of investigation, if only pushed far enough, will give one certain solution to every question to which they can be applied.[†] One man may investigate the velocity of light by studying the transits of Venus and the aberration of the stars; another by the oppositions of Mars and the eclipses of Jupiter's satellites; a third by the method of Fizeau; a fourth by that of Foucault; a fifth by the motions of the curves of Lissajoux; a sixth, a seventh, an eighth, and a ninth, may follow the different methods of comparing the measures of statical and dynamical electricity. They may at first obtain different results, but, as each perfects his method and his processes, the results will move steadily[‡] together toward a destined centre. So with all scientific research. Different minds may set out with the most antagonistic views, but the progress of investigation carries them by a force outside of themselves to one and the same conclusion. This activity of thought by which we are carried, not where we wish, but to a foreordained goal, is like the operation of destiny. No modification of the point of view taken, no selection of other facts for study, no natural bent of mind even, can enable a man to escape the predestinate opinion. This great law[§] is embodied in the conception of truth and reality. The opinion which is fated[**] to be ultimately agreed to by all who investigate, is what we mean by the truth, and the object represented in this opinion is the real. That is the way I would explain reality.[22]

But it may be said that this view is directly opposed to the abstract definition which we have given of reality, inasmuch as it makes the characters of the real to depend on what is ultimately thought about them. But the answer to this is that, on the one hand, reality is independent, not necessarily of thought in general, but only of what you or I or any finite number of men may think about it; and that, on the other hand, though the object of the final opinion depends on what that opinion is, yet what that opinion is does not depend on what you or I or any man thinks. Our perversity and that of others may indefinitely postpone the settlement of opinion; it might even conceivably cause an arbitrary proposition to be universally accepted as long as the human race should last. Yet even that would not change the

---

[*] In HTR Peirce writes instead (R 422:14): All the followers of science are animated by a cheerful hope

[†] In HTR Peirce writes instead (R 422:14): to each question to which they apply it.

[‡] In HTR Peirce replaces (R 422:14): "will move steadily" with "are found to move"

[§] In HTR Peirce writes (R 422:14): This great hope

[**] *[Original Peirce footnote:]* Fate means merely that which is sure to come true, and can nohow be avoided. It is a superstition to suppose that a certain sort of events are ever fated, and it is another to suppose that the word fate can never be freed from its superstitious taint. We are all fated to die.

nature of the belief, which alone could be the result of investigation carried sufficiently far; and if, after the extinction of our race, another should arise with faculties and disposition for investigation, that true opinion must be the one which they would ultimately come to. "Truth crushed to earth shall rise again,"[23] and the opinion which would finally result from investigation does not depend on how anybody may actually think. But the reality of that which is real does depend on the real fact that investigation is destined to lead, at last, if continued long enough, to a belief in it.

But I may be asked what I have to say to all the minute facts of history, forgotten never to be recovered, to the lost books of the ancients, to the buried secrets.[24]

> "Full many a gem of purest ray serene
> The dark, unfathomed caves of ocean bear;
> Full many a flower is born to blush unseen,
> And waste its sweetness on the desert air."[25]

Do these things not really exist because they are hopelessly beyond the reach of our knowledge? And then, after the universe is dead (according to the prediction of some scientists), and all life has ceased forever, will not the shock of atoms continue though there will be no mind to know it? To this I reply that, though in no possible state of knowledge can any number be great enough to express the relation between the amount of what rests unknown* to the amount of the known, yet it is unphilosophical to suppose that, with regard to any given question (which has any clear meaning), investigation would not bring forth a solution of it, if it were carried far enough. Who would have said, a few years ago, that we could ever know of what substances stars are made whose light may have been longer in reaching us than the human race has existed?[26] Who can be sure of what we shall not know in a few hundred years? Who can guess what would be the result of continuing the pursuit of science for ten thousand years, with the activity of the last hundred?† And if it were to go on for a million, or a billion, or any number of years you please, how is it possible to say that there is any question which might not ultimately be solved?

---

*In HTR Peirce writes (R422:15): what remains unknown

†In HTR the remainder is replaced with the following (R422:15–19):

> I refrain from inquiring about billions and vigintillions.
> But then we may as well grant that we do not know that there are not questions that will remain forever open. We do not know that in regard to any question about a continuous quantity, the value of that quantity will ever be precisely determined; on the contrary, it would seem scarcely possible it ever should be. Still, if out of any given class of questions infinitely numerous, as time goes on, the proportion that remains

But it may be objected, "Why make so much of these remote considerations, especially when it is in your principle that only practical distinctions have a meaning?" Well, I must confess that it makes very little difference whether we say that a stone on the bottom of the ocean, in complete darkness, is brilliant or not—that is to say, that it *probably* makes no difference, remembering always that that stone *may* be fished up to-morrow. But that there are gems at the bottom of the sea, flowers in the untraveled desert, etc., are propositions which, like that about a diamond being hard when it is not pressed, concern much more the arrangement of our language than they do the meaning of our ideas.

---

unanswered approximates indefinitely to zero, or if the uncertainty of the value is destined to diminish indefinitely, then we may claim that our conception of reality is substantially verified. Perhaps it might be shown that there was some ground for believing that such was the case. But I prefer to assume that there is no such reason. We *hope* that the particular question into which we are inquiring is susceptible of an approximate answer in a reasonable time; and if things do not possess strict and precise reality, we shall see further on that there is reason to think they do so sufficiently to warrant our retaining this way of thinking about them.

Observe that the conception we have developed is that the reality of the real consists in a real unanimity being destined to be reached, that is beyond human control. We are therefore assigning to ideas, to their history, and to the agreement of many minds a reality more direct and primary than the reality of any *thing*. Are we to say that the fact that there is such a thing as an opinion consists in the fact that there is an opinion that there is an opinion? We plainly must hold that the reality of an idea is a reality *ipso facto*. If I see a red light, that light is red to itself and by itself, and no appeal to any other person to say how it looks to me can help the reality of its appearance one whit. There is the light; I see it; it is over against me! We shall find, too, when we come to examine the matter, that the reality of time is in part directly perceived. That all time hangs consistently together, though I will be sleeping or inattentive, I gather in various ways. But for a little I can perceive a gradual change in the light before me. The color of the light appears gradually to change, and as the light is nothing but appearance, there seems little room to doubt that the appearance really does appear gradually to change. I appear to perceive a direct appearance of change, and so of time. That time is then directly real, like the light, or is real by and in the light. If then various reasons lead me to think this time connects appearances I do not directly perceive to be so connected, those are reasons for thinking that time had a reality in the appearances quite apart from any and all opinions that might be held about it. By considerations such as this, we shall be step by step brought to admit that ideas, their history is time, and their spread is space, have self-reality. So that all that we have been saying comes to this that the reality of things that are *not* ideas consists in the reality and permanence of the ideas in which those things are represented. But if it can be shown that the Great Pyramid, for example, whose reality we make to consist in the persistence of it as an object of opinion, is itself at bottom a living idea,—and not merely the *object* of an idea,—then, and then alone, it may possess as idea a self-reality. It is the same with my own existence. I have an instinctive idea of myself, but no direct perception of myself, that is, of the living being behind (or before) my perceptions. I know myself as I know another man, only much more familiarly. My reality is quite distinct from the reality of the ideas I perceive; and unless it can be shown that I am an idea or a real connection

It seems to me, however, that we have, by the application of our rule, reached so clear an apprehension of what we mean by reality, and of the fact which the idea rests on, that we should not, perhaps, be making a pretension so presumptuous as it would be singular, if we were to offer a metaphysical theory of existence for universal acceptance among those who employ the scientific method of fixing belief. However, as metaphysics is a subject much more curious than useful, the knowledge of which, like that of a sunken reef, serves chiefly to enable us to keep clear of it, I will not trouble the reader with any more Ontology at this moment.[27] I have already been led much further into that path than I should have desired; and I have given the reader such a dose of mathematics, psychology, and all that is most abstruse, that I fear he may already have left me, and that what I am now writing is for the compositor and proof-reader exclusively. I trusted to the importance of the subject. There is no royal road to logic, and really valuable ideas can only be had at the price of close attention.[28] But I know that in the matter of ideas the public prefer the cheap and nasty; and in my next paper I am going to return to the easily intelligible, and not wander from it again. The reader who has been at the pains of wading through this month's paper, shall be rewarded in the next one by seeing how beautifully what has been developed in this tedious way can be applied to the ascertainment of the rules of scientific reasoning.

We have, hitherto, not crossed the threshold of scientific logic. It is certainly important to know how to make our ideas clear, but they may be ever so clear without being true. How to make them so, we have next to study. How to give birth to those vital and procreative ideas which multiply into a

---

of ideas the only thing that can be meant by my reality is the consistency of the final opinion that I am what I am. But many psychologists would say that I am precisely a coördination of ideas; just as many metaphysicians would hold that the Great Pyramid is a living idea, and not merely the object of an idea.

Let us further remark that it is the lower order of clearness to which the higher is obliged to conform. If our abstract definition of the real, that it is that whose characters are independent of what any individual or any given individuals may think them to be, if this were in decided conflict with the familiar use of the idea, it would be the abstract definition which would have to be altered. And again, when we seek to make this more clear by translating the whole into what is directly known to us, that is ideas, if this translation did not fit the ~~exact~~\abstract definition, it is the translation that would have to give way. If it does fit, it is because of the intractable character of the destined final opinion which none of us can in the least modify by any willful effort. Whatever withstands us is real. Though we can only apprehend ideas, and to have a clear meaning must mean something about ideas, yet it is a certain intractableness of certain ideas which can make them representative of the real. Hence, the reality of external things is a reality, because those things do not yield but withstand us, that is to say, the idea of them does this, because they persistently and consistently bear down every opinion to the contrary.

thousand forms and diffuse themselves everywhere, advancing civilization and making the dignity of man, is an art not yet reduced to rules, but of the secret of which the history of science affords some hints.

NOTES

1. *A Treatise on Logic* (Cambridge, 1864) by Francis Bowen, who was Peirce's philosophy professor at Harvard, contains an extensive discussion of the subject (77–86). See also Isaac Watts's *Logic: Or, the Right Use of Reason in the Inquiry After Truth*, I.iii.4. First published in 1724, Watts' *Logic* became a very popular textbook in the English-speaking world. The terminology goes back to Descartes. See e.g. his *Discourse on Method,* Pt. 4, and his *Principles of Philosophy* Pt.1 §§45–46 (see also W2:505). See also John Locke's *Essay Concerning Human Understanding*, Bk. II, Ch. xxix, "Of Clear and Obscure, Distinct and Confused Ideas" with Leibniz's response in *New Essays on Human Understanding.* See also Leibniz's eighth letter to Burnet (W2:505). In the *Century Dictionary* Peirce explains (CD:1037):

> [The words clear and distinct] were first used technically as applied to vision by writers on optics. *Clear* vision occurs where there is sufficient light; *distinct* vision, where the parts of the objects seen can be recognized. Descartes extended the terms to the mental apprehension of truth, which he considered analogous to vision. Leibnitz gave more technically logical definitions, especially of the term *distinct* ... and added the term *adequate*.

When defining *distinct*, he writes in a similar vein (CD:1693):

> The distinction made by writers on vision between imperfection of vision due to want of light (obscurity) and that owing to distance (confusion) was transferred to psychology by Descartes. With him a distinct idea is one which resists dialectical criticism. Later writers, adhering more closely to the optical metaphor, make a clear idea to be one distinguishable from others, and a *distinct* idea to be one whose parts can be distinguished from one another; hence, one which can be abstractly defined.

2. This is the first grade of clearness. In a manuscript called "Of the Nature of Measurement" (ca.1895) Peirce describes it as "containing no element which perfect familiarity does not enable us to use with entire confidence," adding that this first grade "is not sufficient for precise statement, and logical security" (R 254:5). Peirce subsequently distinguishes sensuous clearness from logical clearness, observing that having sensuous clearness may not in the least contribute to logical clearness (R 254:6).

3. Distinctness marks the second grade of clearness. See note above, for Peirce's definition of the term in the *Century Dictionary.*

4. See e.g. Leibniz's *Discourse on Metaphysics* §24, and *New Essays on Human Understanding* II.xxix. Leibniz adds the notion *adequate*, which Peirce defines as (CD:71): "(a) A cognition involving no notion which is not perfectly clear and distinct. (b) A cognition at once precise and complete."

5. This is a criticism of Leibniz's view that we come to know adequate ideas through intuition.

6. Leibniz, *Discourse on Metaphysics* §24; also *New Essays on Human Understanding* II.xxix.15.

7. As with the last paragraph of "The Fixation of Belief," this too may be a veiled reference to the marital problems between Peirce and his first wife Melusina Fay. There are various Melusina legends. Most involve a beautiful fairy that marries some nobleman on the promise that he never enters her private quarters. Invariably the nobleman breaks his promise, upon which the fairy suddenly changes into a dragon and disappears. In some variants she changes into a serpent, or a serpent from the waist down.

8. In a 1905 *Nation* review of Wilhelm Wundt's *Principles of Physiological Psychology,* Peirce writes (CP 8.201): "the sole function of thought is to regulate motor reactions."

9. See also "The Fixation of Belief," Sect. IV, with notes, where Peirce argues against feigned doubt, insisting that doubt be a real source of discomfort.

10. Here we find an early expression of the idea of a stream of consciousness; see also W2:34.

11. Commonly referred to as the pragmatic maxim, this rule was brought into the limelight by William James in his 1898 Berkeley Union Address "Philosophical Conceptions and Practical Results" as "the principle of pragmatism" See John J. McDermott, ed., *The Writings of William James: A Comprehensive Edition* (Chicago: University of Chicago Press, 1977), 348. Peirce begins to refer to it as a maxim of pragmatism after the turn of the century. The interpretation James gives to the maxim, however, is a nominalistic interpretation for which the paper gives some credence (especially in the example of hardness below) but which Peirce rejects. James's interpretation of pragmatism, however, catches on, which forces Peirce to call his own reading of pragmatism "pragmaticism."

12. The following 1905 manuscript shows the extent to which Peirce comes to disagree with how he treated the example in "How to Make Our Ideas Clear" (RS 290:3–6, 1905; for a variant of this passage, see EP2:356f):

> Now let us take up the case of the diamond formed and afterward consumed upon cushion of cotton-wool, without ever having had its hardness brought to the test. Now it must not be forgotten that there are laws of nature; and although a Tychist may believe that they are affected by a more than microscopic vagueness, yet from the first day a diamond was ever formed their precision must have been very nearly perfect. Now since this is called a diamond we know that it was a mass of pure carbon in the form of a crystal brittle and of facile octahedral cleavage, if not peculiarly twinned, took the form (omitting minutiae of crystallography) of anoctohedron, apparently regular, but with grooved edges, *some faces being curved*; without being touched it could be shown to be insoluble, transparent, *very highly refractive*, showing a peculiar bluish phosphorescence under "dark light," X rays, and especially radium rays, of *as high specific gravity as realgar, orpiment, or chloritoid*, and *giving off less heat on combustion than any other form of carbon.* Posterity will very likely find that those of these properties that are italicized are due to the same condition of the diamond's atoms to which its

adamantine hardness is due, perhaps the high polymerization of its molecule. At any rate the hardness can be nothing but one of the manifestations of *some* condition of the diamond's atoms. At any rate, it is monstrous to say that a mere accidental circumstance, such as the crystal of corundum which had been sent for to test the diamond's hardness having arrived too late, can affect at all the real quality of the gem.

In short it is impossible not to believe that the hardness of the diamond is a quality which it possesses at all times whether it has been tested to our knowledge, or without our knowledge, or has not been tested at all. It is really hard.

On the other hand, we cannot but recognize that to whatever character of the relationship between the atoms the quality of hardness or any similar quality may be due, that character, when discovered would be found to consist in the truth of some conditional proposition expressing that if certain conditions be fulfilled certain phenomena must result. It is impossible to imagine of what other nature it possibly could be.

We are thus forced to admit that the truth of a conditional proposition stating that some state of things will accompany some hypothetical state of things that *may* be or *may not* be, can constitute a real fact about an individual thing. In fact all the general phenomena of physics are of that nature.

13. The remainder of the section is devoted to the concept of force. Especially in the first half of the nineteenth century the Newtonian concept of force, which many considered an obscure metaphysical notion, was hotly debated. Its rejection led to the development of a pure geometry of motion called kinematics of which what follows is a very rudimentary sketch. See e.g., Alberto A Martínez, *Kinematics: The Lost Origins of Einstein's Relativity* (Baltimore: Johns Hopkins University Press, 2009).

14. For a detailed discussion of Peirce's views on causation, see Menno Hulswit, *From Cause to Causation: A Peircean Perspective* (Dordrecht: Kluwer, 2002).

15. This is in essence the view taken by the new kinematics and is very close to what e.g. Gustav Kirchhoff writes in the preface to his *Vorlesungen über mathematische Physik* (Leipzig: B.G. Teubner, 1876).

16. An early criticism of this passage is found in Thomas Hubbart Musick, *A Brief on the Doctrine of the Conservation of Force* (Mexico, MO: Mexico Union Print, 1878), 5.

17. Peirce's source is not identified. A passage that comes close to it is found in A. Privat Deschanel, *Elementary Treatise on Natural Philosophy*, 4 vols., translated and expanded by J.D. Everett (New York: Appleton, 1876), I:9.

18. For earlier substantive discussions of the notion of reality, which plays a key role in the fourth method of fixing belief (chapter 1), see "Some Consequences of Four Incapacities" (W2, item 22), and Peirce's 1871 review of Fraser's edition of the works of Berkeley (W2, item 48).

19. Peirce routinely attributes this abstract definition of reality to John Duns Scotus, and often argues that the ethics of terminology demands that we only use technical terms with the meaning initially given to it. In November 1909, in a draft of the never published "Significs and Logic" (R 642:11), Peirce states that the closest Scotus comes to giving a definition of reality—Peirce calls it a quasi definition—is in his *Opus Oxoniense, Distinctio* VIII, *Quaestio* iv., where Scotus

writes: "Ens reale quod distinguitur contra ens rationis, est illud quod ex se habet esse circumscripto omni operae intellectus, ut intellectus est." Reality, as defined here, Peirce remarks, is not independent of *all* thought, but only of "intellectus, ut intellectus est." Peirce takes the last phrase to be a misreading for "intellectus, ut intellectum est," on the ground that otherwise Scotus "would be denying the reality of states of mind and the like" (ibid), which he thinks is absurd. If one fantasizes about a mermaid, the mermaid may not be real, but the fact that one is fantasizing about it will be. Peirce does not give a translation of this quasi definition, but the following is a reasonable translation: A real being as contrasted with a rational being is that which has its being of itself, assuming that we set apart every activity of the understanding insofar as it belongs to the understanding. This allows for an "activity of the understanding insofar as it belongs to the understanding," to be interpreted as a fiction—i.e., as a product of the understanding which is nothing but a product of the understanding. As we now know, the *Opus Oxoniense* is neither written by Scotus, nor written in Oxford; it is a posthumous edition of Scotus's principal work compiled by his students. In the critical edition of Scotus's work, which appeared in 1956, the *Opus Oxoniense* is included as the *Ordinatio.* In this edition the quotation given by Peirce surfaces in a slightly modified form as *Ordinatio* I 8. 177.

20. Johannes Scotus Erigena (c.815–c.877), Irish theologian and Neoplatonist philosopher.

21. The *Historia Calamitatum,* or *A History of My Calamities*, is an autobiographical work written in the form of a letter by Peter Abelard (1079–1142).

22. In the concluding paragraph of "The Probability of Induction" (see chapter 4) Peirce derives the rule of induction from "the principle that reality is only the object of the final opinion to which sufficient investigation would lead."

23. From "The Battle-Field" (1839) by William Cullen Bryant, stanza 9:

> Truth, crushed to earth, shall rise again;
> The eternal years of God are hers;
> But Error, wounded, writhes with pain,
> And dies among his worshippers.

24. In his August 1910 letter to Paul Carus (chapter 9), Peirce identifies this passage, with its reference to Gray's elegy, as one that illustrates "the principal positive error" of the first two articles, namely their nominalism.

25. Thomas Gray, "Elegy Written in a Country Churchyard" (1751), Stanza 14. Included in *The Poetical Works of Thomas Gray,* which appeared in numerous editions.

26. This is a popular example of Peirce. The reference is to Auguste Comte's unfortunate prediction in the *Cours de la Philosophie Positive* (1835) that we would never be able to know the chemical composition of stars. It took only fourteen years before Gustav Kirchhoff found that one can derive the chemical composition of a gas from its electromagnetic spectrum, and another fifteen till William Huggins began studying the stars with a spectrograph attached to a telescope.

27. Recall Youmans' complaint that Peirce's first paper was too metaphysical (see the introduction).

28. This echoes Euclid's response to Ptolemy who asked him whether there was a shorter road to geometry than through the *Elements* (Proclus, *Commentary on the First Book of Euclid's* Elements, Bk. II, Ch. 4), and reflects back to the opening of "The Fixation of Belief." While revising these first two papers for later publication Peirce spends much time and effort justifying this point.

Chapter 3

THE DOCTRINE OF CHANCES

The third Illustrations article proved too large and had to be divided into two (W3:528). "The Doctrine of Chances" forms the first part; "The Probability of Induction" forms the second part. Starting from the observation that "a science first begins to be exact when it is quantitatively treated," Peirce began his discussion of the logic of science with a discussion of quantitative logic, postponing a discussion of deduction (which was typically considered the essence of logic) till the sixth article. Later, when working on a new edition of the Illustrations for the* Open Court, *Peirce spent considerable time and effort further developing his view, especially*

---

*The following outline, spanning both papers, survives as WMS 316 (R 762:1):

> I. That a science begins to be really accurate from the moment when it receives a quantitative treatment. II. The theory of probabilities is logic quantitatively considered. III. The vagueness, uncertainty, & error which have infested this theory. IV. The conception of probability made clear. V. The rules of the calculus. VI. All the difficulties disappear. VII. It is not a formularization of natural good sense, but is superior thereto. VIII. The use of scientific reasoning. Indeterminacy of the probability. This is essential; when the probability is determined we have nothing but a species of deduction. IX. How are synthetical judgments à priori possible? The question which exploded the Leibnizian philosophy. The question how are synthetical judgments possible? makes another great step in philosophy. X. Practical importance of the question. Abbé Gratry makes the thing miraculous. XI. Solution of the question. XII. New conception of Knowledge. XIII. Rule of reasoning.

by contrasting it with the views expressed by Laplace in his Théorie analytique des probabilités *and John Stuart Mill in his* System of Logic, *both of which he rejected (see chapters 7 and 9 below).*

"The Doctrine of Chances" first appeared in Popular Science Monthly *12 (March 1878): 604–15. The text that follows is as it appeared in* Popular Science Monthly. *It is only minimally edited and italic brackets surround any text added by the editor. "The Doctrine of Chances" is listed as chapter 10 of* Search for a Method, *and was included under the title "Clearness of Apprehension" as chapter 18 of the 1894 logic book* How to Reason, *hereafter referred to as HTR. However, in the surviving typescript (R 424) Peirce made only a few minor changes. The* Collected Papers *editors include four notes they date 1893 that are not found in the annotated typescript for* How to Reason *(R 424). This suggests that they had access to a different version of the text, possibly intended for* A Search for a Method. *They did not include alterations made by Peirce in the R 424 typescript. Manuscripts R 703 and R 705, which were written between 10 and 15 August 1910, contain notes for the third* Popular Science Monthly *article. Most of this is reproduced, in part as footnotes and in part as an appendix to the current chapter. A more extensive discussion of the history of the article is found in the introduction.*

I.

It is a common observation that a science first begins to be exact when it is quantitatively treated.* What are called the exact sciences are no others

---

*R 706 (1909) elaborates as follows (R 706:28–29; dated January 29, 1909):

> In the original papers in the *Popular Science Monthly* of March and April, 1878, of which the present essay is a revision, I contented myself with applying the "common observation" that "a science first begins to be exact when it gets quantitatively treated." Now this is, as I called it, "an observation." But observation only supplies us with individual facts: it does not, by itself alone, supply a reason for the necessity of the facts observed. It may be a fact that any science takes a higher science as soon as it becomes quantitative, or, what comes to the same thing, so far as observation can show, that any science as it matures is led to pay more and more heed to quantities; for naturally we first remark the rough discriminations of quality and later the finer differences of quantity. Each of these two forms of interpreting the observable facts represents, I do not doubt, a side of the truth that the other leaves unexpressed. When we come to slight degrees of difference we resort to quantity, simply as a convenient means of expressing and of finding slight dissimilarities; but moreover quantitative expressions always tend to stimulate our attention to small distinctions. But it certainly is not a universal truth that progress in a science will render it more attentive to quantity; for it has not been so with nineteenth-century mathematics. *[...]*

than the mathematical ones. Chemists reasoned vaguely until Lavoisier showed them how to apply the balance to the verification of their theories, when chemistry leaped suddenly into the position of the most perfect of the classificatory sciences.[1] It has thus become so precise and certain that we usually think of it along with optics, thermotics, and electrics. But these are studies of general laws, while chemistry considers merely the relations and classification of certain objects; and belongs, in reality, in the same category as systematic botany and zoölogy. Compare it with these last, however, and the advantage that it derives from its quantitative treatment is very evident.[*2]

The rudest numerical scales, such as that by which the mineralogists distinguish the different degrees of hardness, are found useful. The mere counting of pistils and stamens sufficed to bring botany out of total chaos into some kind of form. It is not, however, so much from *counting* as from *measuring,* not so much from the conception of number as from that of continuous quantity, that the advantage of mathematical treatment comes. Number, after all, only serves to pin us down to a precision in our thoughts which, however beneficial, can seldom lead to lofty conceptions, and frequently descends to pettiness. Of those two faculties of which Bacon speaks, that which marks differences and that which notes resemblances,[3] the employment of number can only aid the lesser one; and the excessive use of it must tend to narrow the powers of the mind. But the conception of continuous quantity has a great office to fulfill, independently of any attempt at precision. Far from tending to the exaggeration of differences, it is the direct instrument of the finest generalizations.[4] When a naturalist wishes

---

*R 704 (1910) contains the following note, with instructions on where to insert it as a footnote (R 704:2):

> Such studies have been made in our generation of the general laws of chemistry that the reader may well be surprised to find it spoken of as a merely classificatory science. Yet that the chemists could not be so affected at the time of the original publication of this Essay is shown by the fact that only a few months previously van 't Hoff[2] had put forward the general law of massaction as a new discovery of his own. The fact that it had really been several times already announced as a newly discovered law only fortifies the evidence that chemists were not yet accustomed to think of general laws as belonging to their science, although of course there had been a few such since the early years of modern *[science.]*

In a draft of the same note, Peirce writes (R 703:36):

> I am satisfied by considerable search after pertinent facts that no distinction between different allied sciences can represent any truth of fact other than a difference between what habitually passes in the minds, and moves the investigations of the two general bodies of cultivators of those sciences at the time to which the distinction refers.

to study a species, he collects a considerable number of specimens more or less similar. In contemplating them, he observes certain ones which are more or less alike in some particular respect. They all have, for instance, a certain S-shaped marking. He observes that they are not *precisely* alike, in this respect; the S has not precisely the same shape, but the differences are such as to lead him to believe that forms could be found intermediate between any two of those he possesses. He, now, finds other forms apparently quite dissimilar—say a marking in the form of a C—and the question is, whether he can find intermediate ones which will connect these latter with the others. This he often succeeds in doing in cases where it would at first be thought impossible; whereas, he sometimes finds those which differ, at first glance, much less, to be separated in Nature by the non-occurrence of intermediaries. In this way, he builds up from the study of Nature a new general conception of the character in question. He obtains, for example, an idea of a leaf which includes every part of the flower, and an idea of a vertebra which includes the skull. I surely need not say much to show what a logical engine there is here. It is the essence of the method of the naturalist. How he applies it first to one character, and then to another, and finally obtains a notion of a species of animals, the differences between whose members, however great, are confined within limits, is a matter which does not here concern us. The whole method of classification must be considered later;[*5] but, at present, I only desire to point out[†] that it is by taking advantage of the idea of continuity, or the passage from one form to another by insensible degrees,[‡] that the naturalist builds his conceptions. Now, the naturalists are the great builders of conceptions; there is no other branch of science where so much of this work is done as in theirs; and we must, in great measure, take them for our teachers in this important part of logic. And it will be found everywhere that the idea of continuity[§] is a powerful aid to the formation of true and fruitful conceptions. By means of it, the greatest differences are broken down and resolved into differ-

---

*In HTR Peirce writes (R 424:2): "has been considered in another chapter." This other chapter is possibly a reference to chapter IX of HTR (R 413).

†In HTR Peirce replaces "point out" with "remind the student" (R 424:2).

‡CP 2.646 contains the following note which the editors date 1893: "Or rather of an idea that continuity suggests—that of limitless intermediation; i.e., of a series between every two members of which there is another member of it"—to be substituted for the phrase "or … degrees." The *Collected Papers* editors add a total of four notes that they date 1893. These notes are not found in R 424, which is the version of "The Doctrine of Chances" that Peirce prepared for *How to Reason.*

§CP 2.646 contains the following note that the editors date 1893: For "continuity" substitute "went of close study of these concepts."—1893

ences of degree, and the incessant application of it is the greatest value in broadening our conceptions. I propose to make a great use of this idea[*] in the present series of papers; and the particular series of important fallacies, which, arising from a neglect of it,[†] have desolated philosophy, must further on be closely studied. At present, I simply call the reader's attention to the utility of this conception.

In studies of numbers, the idea of continuity is so indispensable, that it is perpetually introduced even where there is no continuity in fact, as where we say that there are in the United States 10.7 inhabitants per square mile, or that in New York 14.72 persons live in the average house.[‡] Another example is that law of the distribution of errors which Quetelet, Galton, and others, have applied with so much success to the study of biological and social matters.[6] This application of continuity to cases where it does not really exist illustrates, also, another point which will hereafter demand a separate study, namely, the great utility which fictions sometimes have in science.

II.

The theory of probabilities is simply the science of logic quantitatively treated. There are two conceivable certainties with reference to any hypothesis, the certainty of its truth and the certainty of its falsity. The numbers *one* and *zero* are appropriated, in this calculus, to marking these extremes of knowledge; while fractions having values intermediate between them indicate, as we may vaguely say, the degrees in which the evidence leans toward one or the other. The general problem of probabilities is, from a given state of facts, to determine the numerical probability of a possible fact. This is the same as to inquire how much the given facts are worth, considered as evidence to prove the possible fact. Thus the problem of probabilities is simply the general problem of logic.

Probability is a continuous quantity, so that great advantages may be expected from this mode of studying logic. Some writers have gone so far

[*]CP 2.464 contains the following note that the editors date 1893: "And others that are involved in that of continuity."—1893

[†]CP 5.646 contains the following note that the editors date 1893: For "neglect of it" substitute "want of close study of these concepts"—1893

[‡]*[Original Peirce footnote:]* This mode of thought is so familiarly associated with all exact numerical consideration, that the phrase appropriate to it is imitated by shallow writers in order to produce the appearance of exactitude where none exists. Certain newspapers which affect a learned tone talk of "the average man," when they simply mean *most men,* and have no idea of striking an average.

as to maintain that, by means of the calculus of chances, every solid inference may be represented by legitimate arithmetical operations upon the numbers given in the premises. If this be, indeed, true, the great problem of logic, how it is that the observation of one fact can give us knowledge of another independent fact, is reduced to a mere question of arithmetic. It seems proper to examine this pretension before undertaking any more recondite solution of the paradox.

But, unfortunately, writers on probabilities are not agreed in regard to this result. This branch of mathematics is the only one, I believe, in which good writers frequently get results entirely erroneous. In elementary geometry the reasoning is frequently fallacious, but erroneous conclusions are avoided; but it may be doubted if there is a single extensive treatise on probabilities in existence which does not contain solutions absolutely indefensible. This is partly owing to the want of any regular method of procedure; for the subject involves too many subtilties to make it easy to put its problems into equations without such an aid. But, beyond this, the fundamental principles of its calculus are more or less in dispute. In regard to that class of questions to which it is chiefly applied for practical purposes, there is comparatively little doubt; but in regard to others to which it has been sought to extend it, opinion is somewhat unsettled.

This last class of difficulties can only be entirely overcome by making the idea of probability perfectly clear in our minds in the way set forth in our last paper.[7]

III.

To get a clear idea of what we mean by probability,[8] we have to consider what real and sensible difference there is between one degree of probability and another.

The character of probability belongs primarily, without doubt, to certain inferences. Locke explains it as follows: After remarking that the mathematician positively knows that the sum of the three angles of a triangle is equal to two right angles because he apprehends the geometrical proof, he thus continues: "But another man who never took the pains to observe the demonstration, hearing a mathematician, a man of credit, affirm the three angles of a triangle to be equal to two right ones, *assents* to it; i.e., receives it for true. In which case the foundation of his assent is the probability of the thing, the proof being such as, for the most part, carries truth with it; the man on whose testimony he receives it not being wont to affirm anything contrary to, or besides his knowledge, especially in matters of this kind."[9]

The celebrated *Essay concerning Humane Understanding* contains many passages which, like this one, make the first steps in profound analyses which are not further developed. It was shown in the first of these papers that the validity of an inference does not depend on any tendency of the mind to accept it, however strong such tendency may be; but consists in the real fact that, when premises like those of the argument in question are true, conclusions related to them like that of this argument are also true. It was remarked that in a logical mind an argument is always conceived as a member of a *genus* of arguments all constructed in the same way, and such that, when their premises are real facts, their conclusions are so also. If the argument is demonstrative, then this is always so; if it is only probable, then it is for the most part so. As Locke says, the probable argument is "*such as* for the most part carries truth with it."[10]

According to this, that real and sensible difference between one degree of probability and another, in which the meaning of the distinction lies, is that in the frequent employment of two different modes of inference, one will carry truth with it oftener than the other. It is evident that this is the only difference there is in the existing fact. Having certain premises, a man draws a certain conclusion, and as far as this inference alone is concerned the only possible practical question is whether that conclusion is true or not, and between existence and non-existence there is no middle term. "Being only is and nothing is altogether not," said Parmenides; and this is in strict accordance with the analysis of the conception of reality given in the last paper.[11] For we found that the distinction of reality and fiction depends on the supposition that sufficient investigation would cause one opinion to be universally received and all others to be rejected. That presupposition involved in the very conceptions of reality and figment involves a complete sundering of the two. It is the heaven-and-hell idea in the domain of thought. But, in the long run, there is a real fact which corresponds to the idea of probability, and it is that a given mode of inference sometimes proves successful and sometimes not, and that in a ratio ultimately fixed. As we go on drawing inference after inference of the given kind, during the first ten or hundred cases the ratio of successes may be expected to show considerable fluctuations; but when we come into the thousands and millions, these fluctuations become less and less; and if we continue long enough, the ratio will approximate toward a fixed limit. We may therefore define the probability of a mode of argument as the proportion of cases in which it carries truth with it.

The inference from the premise, A, to the conclusion, B, depends, as we have seen, on the guiding principle, that if a fact of the class A is true, a fact of the class B is true. The probability consists of the fraction whose numerator is the number of times in which both A and B are true, and whose denominator is the total number times in which A is true, whether B is so or not. Instead of speaking of this as the probability of the inference, there is not the slightest objection to calling it the probability that, if A happens, B happens. But to speak of the probability of the event B, without naming the condition, really has no meaning at all. It is true that when it is perfectly obvious what condition is meant, the ellipsis may be permitted. But we should avoid contracting the habit of using language in this way (universal as the habit is), because it gives rise to a vague way of thinking, as if the action of causation might either determine an event to happen or determine it not to happen, or leave it more or less free to happen or not, so as to give rise to an *inherent* chance in regard to its occurrence. It is quite clear to me that some of the worst and most persistent errors in the use of the doctrine of chances have arisen from this vicious mode of expression.*[12]

IV.

But there remains an important point to be cleared up. According to what has been said, the idea of probability essentially belongs to a kind of inference which is repeated indefinitely. An individual inference must be either true or false, and can show no effect of probability; and, therefore, in reference to a single case considered in itself, probability can have no meaning.[13] Yet if a man had to choose between drawing a card from a pack containing twenty-five red cards and a black one, or from a pack containing twenty-five black cards and a red one, and if the drawing of a red card were destined to transport him to eternal felicity, and that of a black one to consign him to everlasting woe, it would be folly to deny that he ought to prefer the pack containing the larger proportion of red cards, although, from the nature of the risk, it could not be repeated. It is not easy to reconcile this with our analysis of the conception of chance. But suppose he should choose the red pack, and should draw the wrong card, what consolation would he have? He might say that he had acted in accordance with reason, but that would only show that his reason was absolutely worthless. And if he should choose the right card, how could he regard it as anything but a happy accident? He could not say that if he had drawn from the other pack, he might have drawn the wrong one, because an hypothetical proposition such as, "if A, then B,"

*[Original Peirce footnote:] The conception of probability here set forth is substantially that first developed by Mr. Venn, in his *Logic of Chance*.[12] Of course, a vague apprehension of the idea had always existed, but the problem was to make it perfectly clear, and to him belongs the credit of first doing this.

means nothing with reference to a single case. Truth consists in the existence of a real fact corresponding to the true proposition. Corresponding to the proposition, "if A, then B," there may be the fact that *whenever* such an event as A happens such an event as B happens. But in the case supposed, which has no parallel as far as this man is concerned, there would be no real fact whose existence could give any truth to the statement that, if he had drawn from the other pack, he might have drawn a black card. Indeed, since the validity of an inference consists in the truth of the hypothetical proposition that *if* the premises be true the conclusion will also be true, and since the only real fact which can correspond to such a proposition is that whenever the antecedent is true the consequent is so also, it follows that there can be no sense in reasoning in an isolated case, at all.

These considerations appear, at first sight, to dispose of the difficulty mentioned. Yet the case of the other side is not yet exhausted. Although probability will probably manifest its effect in, say, a thousand risks, by a certain proportion between the numbers of successes and failures, yet this, as we have seen, is only to say that it certainly will, at length, do so. Now the number of risks, the number of probable inferences, which a man draws in his whole life, is a finite one, and he cannot be absolutely *certain* that the mean result will accord with the probabilities at all. Taking all his risks collectively, then, it cannot be certain that they will not fail, and his case does not differ, except in degree, from the one last supposed. It is an indubitable result of the theory of probabilities that every gambler, if he continues long enough, must ultimately be ruined. Suppose he tries the martingale, which some believe infallible, and which is, as I am informed, disallowed in the gambling-houses.[14] In this method of playing, he first bets say $1; if he loses it he bets $2; if he loses that he bets $4; if he loses that he bets $8; if he then gains he has lost 1 + 2 + 4 = 7, and he has gained $1 more; and no matter how many bets he loses, the first one he gains will make him $1 richer than he was in the beginning. In that way, he will probably gain at first; but, at last, the time will come when the run of luck is so against him that he will not have money enough to double, and must therefore let his bet go. This will *probably* happen before he has won as much as he had in the first place, so that this run against him will leave him poorer than he began; some time or other it will be sure to happen. It is true that there is always a possibility of his winning any sum the bank can pay, and we thus come upon a celebrated paradox that, though he is certain to be ruined, the value of his expectation calculated according to the usual rules (which omit this consideration) is large. But, whether a gambler plays in this way or any other, the same thing is true, namely, that he if he plays long enough he will be sure some time to have such a run against him as to exhaust his entire fortune. The same thing is true of an insurance company. Let the directors take the utmost pains to be independent of great conflagrations and pestilences, their actuaries can tell them that, according to the doctrine of chances, the time must come, at last, when their losses will bring them to a stop. They may tide over such a crisis by extraordinary means, but then they will start again in a weakened state, and the same thing will happen again all the sooner. An actuary might be inclined to deny this, because he knows that the expectation of his company is large, or perhaps (neglecting the interest upon money) is infinite. But calculations of expectations leave out of account the circumstance now under consideration, which reverses the whole thing. However, I must not be understood as saying that insurance is on this account unsound, more than other kinds of business. All human affairs rest upon probabilities, and the same thing is true everywhere. If man were immortal he could be perfectly sure of seeing the day when everything in which he had trusted should betray his trust, and, in short, of coming eventually to hopeless misery. He would break down, at last, as every great fortune, as every dynasty, as every civilization does. In place of this we have death.

But what, without death, would happen to every man, with death must happen to some man. At the same time, death makes the number of our risks, of our inferences, finite, and so makes their mean result uncertain. The very idea of probability and of reasoning rests on the assumption that this number is indefinitely great. We are thus landed in the same difficulty as before, and I can see but one solution of it. It seems to me that we are driven to this, that logicality inexorably requires that our interests shall *not* be limited. They must not stop at our own fate, but must embrace the whole community. This community, again, must not be limited, but must extend to all races of beings with whom we can come into immediate or mediate intellectual relation. It must reach, however vaguely, beyond this geological epoch, beyond all bounds. He who would not sacrifice his own soul to save the whole world, is, as it seems to me, illogical in all his inferences, collectively. Logic is rooted in the social principle.[15]

To be logical men should not be selfish; and, in point of fact, they are not so selfish as they are thought.[16] The willful prosecution of one's desires is a different thing from selfishness. The miser is not selfish; his money does him no good, and he cares for what shall become of it after his death. We are constantly speaking of *our* possessions on the Pacific, and of *our* destiny as a republic, where no personal interests are involved, in a way which shows that we have wider ones. We discuss with anxiety the possible exhaustion of coal in some hundreds of years, or the cooling-off of the sun in some millions, and show in the most popular of all religious tenets that we can conceive the possibility of a man's descending into hell for the salvation of his fellows.

Now, it is not necessary for logicality that a man should himself be capable of the heroism of self-sacrifice. It is sufficient that he should recognize the possibility of it, should perceive that only that man's inferences who has it are really logical, and should consequently regard his own as being only so far valid as they would be accepted by the hero. So far as he thus refers his inferences to that standard, he becomes identified with such a mind.

This makes logicality attainable enough. Sometimes we can personally attain to heroism. The soldier who runs to scale a wall knows that he will probably be shot, but that is not all he cares for. He also knows that if all the regiment, with whom in feeling he identifies himself, rush forward at once, the fort will be taken. In other cases we can only imitate the virtue. The man whom we have supposed as having to draw from the two packs, who if he is not a logician will draw from the red pack from mere habit, will see, if he is logician enough, that he cannot be logical so long as he is concerned only with his own fate, but that that man who should care equally for what was to happen in all possible cases of the sort could act logically, and would draw from the pack with the most red cards, and thus, though incapable himself of such sublimity, our logician would imitate the effect of that man's courage in order to share his logicality.

But all this requires a conceived identification of one's interests with those of an unlimited community. Now, there exist no reasons, and a later discussion will show that there can be no reasons, for thinking that the human race, or any intellectual race, will exist forever. On the other hand, there can be no reason against it;* and, fortunately, as the whole requirement is that we should have certain sentiments, there is nothing in the facts to forbid our having a *hope*, or calm and cheerful wish, that the community may last beyond any assignable date.

It may seem strange that I should put forward three sentiments, namely, interest in an indefinite community, recognition of the possibility of this interest being made supreme, and hope in the unlimited continuance of intellectual activity, as indispensable requirements of logic. Yet, when we consider that logic depends on a mere struggle to escape doubt, which, as it terminates in action, must begin in emotion, and that, furthermore, the only cause of our planting ourselves on reason is that other methods of escaping doubt fail on account of the social impulse, why should we wonder

[Original Peirce footnote:] I do not here admit an absolutely unknowable. Evidence could show us what would probably be the case after any given lapse of time; and though a subsequent time might be assigned which that evidence might not cover, yet further evidence would cover it.

to find social sentiment presupposed in reasoning? As for the other two sentiments which I find necessary, they are so only as supports and accessories of that. It interests me to notice that these three sentiments seem to be pretty much the same as that famous trio of Charity, Faith, and Hope, which, in the estimation of St. Paul, are the finest and greatest of spiritual gifts.[17] Neither Old nor New Testament is a text-book of the logic of science, but the latter is certainly the highest existing authority in regard to the dispositions of heart which a man ought to have.

V.

Such average statistical numbers as the number of inhabitants per square mile, the average number of deaths per week, the number of convictions per indictment, or, generally speaking, the number of *x*'s per *y*, where the *x*'s are a class of things some or all of which are connected with another class of things, their *y*'s, I term *relative numbers.* Of the two classes of things to which a relative number refers, that one of which it is a number may be called its *relate*, and that one *per* which the numeration is made may be called its *correlate.*

Probability is a kind of relative number; namely, it is the ratio of the number of arguments of a certain genus which carry truth with them to the total number of arguments of that genus, and the rules for the calculation of probabilities are very easily derived from this consideration. They may all be given here, since they are extremely simple, and it is sometimes convenient to know something of the elementary rules of calculation of chances.

RULE I. *Direct Calculation.*—To calculate, directly, any relative number, say for instance the number of passengers in the average trip of a street-car, we must proceed as follows:

Count the number of passengers for each trip; add all these numbers, and divide by the number of trips. There are cases in which this rule may be simplified. Suppose we wish to know the number of inhabitants to a dwelling in New York. The same person cannot inhabit two dwellings. If he divide his time between two dwellings he ought to be counted a half-inhabitant of each. In this case we have only to divide the total number of the inhabitants of New York by the number of their dwellings, without the necessity of counting separately those which inhabit each one. A similar proceeding will apply wherever each individual relate belongs to one individual correlate exclusively. If we want the number of *x*'s per *y*, and no *x* belongs to more than one *y*, we have only to divide the whole number of *x*'s of *y*'s by the number of *y*'s. Such a method would, of course, fail if applied to finding the average number of street-car passengers per trip. We could not divide the total number of travelers by the number of trips, since many of them would have made many passages.

To find the probability that from a given class of premises, A, a given class of conclusions, B, follow, it is simply necessary to ascertain what proportion of the times in which premises of that class are true, the appropriate conclusions are also true. In other words, it is the number of cases of the occurrence of both the events A and B, divided by the total number of cases of the occurrence of the event A.

RULE II. *Addition of Relative Numbers.*—Given two relative numbers having the same correlate, say the number of x's per y, and the number of z's per y; it is required to find the number of x's and z's together per y. If there is nothing which is at once an x and a z to the same y, the sum of the two given numbers would give the required number. Suppose, for example, that we had given the average number of friends that men have, and the average number of enemies, the sum of these two is the average number of persons interested in a man. On the other hand, it plainly would not do to add the average number of persons having constitutional diseases to the average number over military age, and to the average number exempted by each special cause from military service, in order to get the average number exempt in any way, since many are exempt in two or more ways at once.

This rule applies directly to probabilities. Given the probability that two different and mutually exclusive events will happen under the same supposed set of circumstances. Given, for instance, the probability that if A then B, and also the probability that if A then C, then the sum of these two probabilities is the probability that if A then either B or C, so long as there is no event which belongs at once to the two classes B and C.

RULE III. *Multiplication of Relative Numbers.*—Suppose that we have given the relative number of x's per y; also the relative number of z's per x of y; or, to take a concrete example, suppose that we have given, first, the average number of children in families living in New York; and, second, the average number of teeth in the head of a New York child—then the product of these two numbers would give the average number of children's teeth in a New York family. But this mode of reckoning will only apply in general under two restrictions. In the first place, it would not be true if the same child could belong to different families, for in that case those children who belonged to several different families might have an exceptionally large or small number of teeth, which would affect the average number of children's teeth in a family more than it would affect the average number of teeth in a child's head. In the second place, the rule would not be true if different children could share the same teeth, the average number of children's teeth being in that case evidently something different from the average number of teeth belonging to a child.

In order to apply this rule to probabilities, we must proceed as follows: Suppose that we have given the probability that the conclusion B follows from the premise A, B and A representing as usual certain classes of propositions. Suppose that we also knew the probability of an inference in which B should be the premise, and a proposition of a third kind, C, the conclusion. Here, then, we have the materials for the application of this rule. We have, first, the relative number of B's per A. We next should have the relative number of C's per B following from A. But the classes of propositions being so selected that the probability of C following from any B in general is just the same as the probability of C's following from one of those B's which is deducible from an A, the two probabilities may be multiplied together, in order to give the probability of C following from A. The same restrictions exist as before. It might happen that the probability that B follows from A was affected by certain propositions of the class B following from several different propositions of the class A. But, practically speaking, all these restrictions are of very little consequence, and it is usually recognized as a principle universally true that the probability that, if A is true, B is, multiplied by the probability that, if B is true, C is, gives the probability that, if A is true, C is.

There is a rule supplementary to this, of which great use is made. It is not universally valid, and the greatest caution has to be exercised in making use of it—a double care, first, never to use it when it will involve serious error; and, second, never to fail to take advantage of it in cases in which it can be employed. This rule depends upon the fact that in very many cases the probability that C is true if B is, is substantially the same as the probability that C is true if A is. Suppose, for example, we have the average number of males among the children born in New York; suppose that we also have the average number of children born in the winter months among those born in New York. Now, we may assume without doubt, at least as a closely approximate proposition (and no very nice calculation would be in place in regard to probabilities), that the proportion of males among all the children born in New York is the same as the proportion of males born in summer in New York, and, therefore, if the names of all the children born during a year were put into an urn, we might multiply the probability that any name drawn would be the name of a male child by the probability that it would be the name of a child born in summer, in order to obtain the probability that it would be the name of a male child born in summer. The questions of probability, in the treatises upon the subject, have usually been such as to relate to balls drawn from urns, and games of cards, and so on, in which the question of the *independence* of events, as it is called—that is to say, the question of whether the probability of C, under the hypothesis B, is the same as its probability under the hypothesis A, has been very simple; but, in the application of probabilities to the ordinary questions of life, it is often an exceedingly nice question whether two events may be considered as independent with sufficient accuracy or not. In all calculations about cards it is assumed that the cards are thoroughly shuffled, which makes one deal quite independent of another. In point of fact the cards seldom are, in practice, shuffled sufficiently to make this true; thus, in a game of whist, in which the cards have fallen in suits of four of the same suit, and are so gathered up, they will lie more or less in sets of four of the same suit, and this will be true even after they are shuffled. At least some traces of this arrangement will remain, in consequence of which the number of "short suits," as they are called—that is to say, the number of hands in which the cards are very unequally divided in regard to suits—is smaller than the calculation would make it to be; so that, when there is a misdeal, where the cards, being thrown about the table, get very thoroughly shuffled, it is a common saying that in the hands next dealt out there are generally short suits. A few years ago a friend of mine, who plays whist a great deal, was so good as to count the number of spades dealt to him in 165 hands, in which the cards had been, if anything, shuffled better than usual. According to calculation, there should have been 85 of these hands in which my friend held either three or four spades, but in point of fact there were 94, showing the influence of imperfect shuffling.

According to the view here taken, these are the only fundamental rules for the calculation of chances. An additional one, derived from a different conception of probability, is given in some treatises, which if it be sound might be made the basis of a theory of reasoning. Being, as I believe it is, absolutely absurd, the consideration of it serves to bring us to the true theory; and it is for the sake of this discussion, which must be postponed to the next number, that I have brought the doctrine of chances to the reader's attention at this early stage of our studies of the logic of science.

NOTE*

1910 Aug.† On reperusing this article after the lapse of a full generation, it strikes me as making two points that were worth making. The better made

*The following note, which is preserved as R 703:2–31, was written by Peirce on August 11–15, 1910; it is reproduced in the *Collected Papers* immediately following "The Doctrine of Chances" (CP 2.661–68).

†Peirce dated each sheet separately and for each indicated the time he began writing. The first sheet is dated August 11, 1910, 4:00 PM; the last sheet is dated August 14, 1910, 5:45 PM.

of the two had been still better made ten years before in my three articles in the *Speculative Review,* Vol II. (W. T. Harris, Editor).[18] This point is that no man can be logical whose supreme desire is the well-being of himself or of any other existing person or collection of persons.

The other good point is that probability never properly refers immediately to a single event, but exclusively to the happening of a given kind of event on any occasion of a given kind. So far all is well. But when I come to define probability, I repeatedly say that it is the quotient of the *number* of occurrences of the event divided by the *number* of occurrences of the occasion. Now this is manifestly wrong; for probability relates to the future; and how can I say how many times a given die will be thrown in the future? To be sure I might, immediately after my throw, put the die in strong nitric acid, and dissolve it; but this suggestion only puts the preposterous character of the definition in a still stronger light. For it is plain, that if probability be the ratio of the occurrences of the specific event to the occurrences of the generic occasion, it is the ratio that there *would be* in the long run, and has nothing to do with any supposed cessation of the occasions. This long run can be nothing but an endlessly long run; and even if it be correct to speak of an infinite "number," yet (infinity divided by infinity) has certainly, *in itself,* no definite value.

But we have not yet come to the end of the flaws in the definition, since no notice whatever has been taken of two conditions which require the strictest precautions in all experiments to determine the probability of a specific event on a generic occasion.* Namely in the first place we must limit our endeavours strictly to counting occurrences to the right genus of occasion and carefully resist all other motives for counting them, and strive to take them just as they would ordinarily occur. In the next place, it must be known that the occurrence of the specific event on one occasion will have no tendency to produce or to prevent the occurrence of the same event upon any other of the occurrences of the generic occasion. In the third place, after the probability has been ascertained, we must remember that this probability cannot be relied upon at any future time unless we have adequate grounds for believing that it has not too much changed in this interval.

I will now give over jeering at my former inaccuracies, committed when I had been a student of logic for only about a quarter of a century, and was naturally not so well-versed in it as now, and will proceed to define probability. I must promise that we, all of us, use this word with a degree of laxity which corrupts and rots our reasoning to a degree that very few

---

**Peirce writes:* occasions.

of us are at all awake to. When I say our "reasoning," I mean not formal reasonings only but our thoughts in general, so far as they are concerned with any of those approaches toward knowledge which we confound with probability. The result is that we not only fall into the falsest ways of thinking, but, what is often still worse, we give up sundry problems as beyond our powers,—problems of gravest concern, too,—when, in fact, we should find they were not in a bit so, if we only rightly discriminated between the different kinds of imperfection of certitude, and if we had only once acquainted ourselves with their different natures. I shall in these notes endeavor to mark the three ways of falling short of certainty by the three terms *probability, verisimilitude* or *likelihood,* and *plausibility.*[19] Just at present I propose to deal only with Probability; but I will so far characterize *Verisimilitude* and *Plausibility* as mark them off as being entirely different from Probability. Beginning with Plausibility, I will first endeavour to give an example of an idea which shall be strikingly marked by its very low degree of this quality. Eusapia Palladino had been proved to be a very clever prestigiateuse and cheat, and was visited by a Mr. Carrington, who I suppose to be so clever in finding out how tricks are done, that it is highly improbable that any given trick should long baffle him.[20] In point of fact he has often caught the Palladino creature in acts of fraud. Some of her performances, however, he cannot explain; and thereupon he urges the theory that these are supernatural, or, as he prefers to phrase it, "supernormal." Well, I know how it is that when a man has been long intensely exercized and over-fatigued by an enigma, his common-sense will sometimes desert him; but as for me, it seems to me that the Palladino has simply been too clever for him, as no doubt she would be for me. The theory that there is anything "supernormal," or *super* anything but *super-chérie* in the case, seems to me as needless as any theory I ever came across. That is to say, granted that it is not yet *proved* that women who deceive for gain receive aid from the spiritual world, I think it more plausible that there are tricks that can deceive Mr. Carrington than that the Palladino woman has received such aid. By Plausible, I mean that a theory that has not yet been subjected to any test, although more or less surprising phenomena have occurred which it would explain if it were true, is in itself of such a character as to recommend it for further examination or, if it be *highly* plausible, justify us in seriously inclining toward belief in it, as long as the phenomena be inexplicable otherwise.

I will now give an idea of what I mean by *likely* or *verisimilar.* It is to be understood that I am only endeavouring so far to explain the meanings I attach to 'plausible' and to 'likely' as this may be an assistance to the reader in understanding the meaning I attach to *probable.* I call that theory *likely* which is not yet proved but is supported by such evidence, that if the rest of the conceivably possible evidence *should* turn out upon examination to be of a *similar* character, the theory would be conclusively proved. Strictly speaking, matters of fact never can be demonstrably proved, since it will always remain conceivable that there should be some mistake about it. For instance, I regard it as *sufficiently* proved that my name is Charles Peirce and that I was born in Cambridge, Massachusetts in a stone colored wooden house in Mason Street.[21] But even of the part of this of which I am most assured—that of my name, —there is a certain small probability that I am in an abnormal condition and have got it wrong. I am conscious myself of occasional lapses of memory about other things; and though I well remember,—or think I do,—living in that house at a tender age, I do not in the least remember being born there, impressive as such a first experience might be expected to be. Indeed, I cannot specify any date on which any certain person informed me I had been born there; and it certainly would have been easy to deceive me in the matter had there been any serious reason for doing so; and how can I be so sure as I surely am that no such reason did exist. It would be a theory without plausibility; that is all.

The history of science, particularly physical science, in contradistinction to natural science,—or, as I usually, though inadequately phrase the distinction, the history of nomological in contradistinction to classificatory science,—this history ever since I first seriously set myself, at the age of thirteen, in 1852, in the study of logic,[22] shows only too grievously how great a boon could be any way determining and expressing by numbers the degree of likelihood that a theory had attained,—any general recognition, even among leading men of science, of the true degree of significance of a given fact, and of the proper method of determining it. I hope my writings may, at any rate, awaken a few to the enormous waste of effort it would save. But any numerical determination of likelihood is more than I can expect.

The only kind of reasoning which can render our conclusions certain,—and even this kind can do so only under the proviso that no blunder had been committed in the process,—attains this certainty by limiting the conclusion (as Kant virtually said, and others before him) to facts already expressed and accepted in the premisses. This is called necessary, or syllogistic reasoning. Syllogism, not confined to the kind that Aristotle and Theophrastus studied, is merely an artificial form in which it may be expressed, and it is not its best form, from any point of view. But the kind of reasoning which creates likelihoods by virtue of observations may render a likelihood *practically* certain,—as certain as that a stone let loose from the clutch will, under circumstances not obviously exceptional, fall to the ground,[23]—and this conclusion may be that under a certain general condition, easily verified, a certain actuality will be *probable,* that is to say, will come to pass once in so often in the long run. One such familiar conclusion, for example, is that a die thrown from a dice box will with a *probability* of one-third, that is, once in three times in the long run, turn up a number (either *tray* or *size*) that is divisible by three. But this can be affirmed with practical certainty only if by a "long run" be meant an endless series of trials, and (as just said) infinity divided by infinity gives of itself an entirely indefinite quotient. It is therefore necessary to define the phrase. I might give the definition with reference to the probability, p, where p is any vulgar fraction, and in reference to a generic condition, m, and a specific kind of event, n. But I think the reader will follow me more readily, if in place of the letter, m, (which in itself is but a certain letter, to which is attached a peculiar meaning, that of the fulfillment of some generic condition,) I put instead the supposition that a die is thrown from a dice-box; and this special supposition will be as readily understood by the reader to be replaceable by any other general condition along with a simultaneous replacement of the *event*, that a number divisible by three is turned up, and at the same time with the replacement of one third by whatever other vulgar fraction may be called for when some different example of a probability is before us. I am, then, to define the meaning of the statement that the *probability* that if a die be thrown from a dice box it will turn up a number divisible by three is one third. The statement means that the die has a certain "would-be;" and to say that a die has a "would-be" is to say that it has a property, quite analogous to any *habit* that a man might have. Only the "would-be" of the die is presumably as much simpler and more definite than the man's habit, as the die's homogeneous composition and cubical shape is simpler than the nature of the man's nervous system and soul; and just as it would be necessary, in order to define a man's habit, it would be needful to describe how it would lead him to behave and upon what sort of occasion; albeit this statement would by no means imply that this habit *consists* in that action, so to define the die's "would-be," it is necessary to say how it would lead the die to behave on an occasion that would bring out the full consequences of the would-be; and this statement will not of itself imply that the "would-be" of the die *consists* in such behaviour.

Now in order that the full effect of the die's "would-be" may find expression, it is necessary that the die should undergo in an endless series of throws from the dice-box, the result of no throw having the slightest influence upon the result of any other throw, or, as we express it, the throws must be *independent* each of every other.

It will be no objection to our considering the consequences of the supposition that the die is thrown in an endless succession of times, and that with a finite pause after each throw, that such an endless series of events is impossible, for the reason that the impossibility is merely a physical, and not a logical, impossibility, as was well illustrated in that famous sporting event in which Achilles succeeded in overtaking the champion tortoise, in spite of his giving the latter the start of a whole *stadion*. For it having been ascertained, by delicate measurement, between a mathematical point between the shoulder blades of Achilles (marked on the limit between a red, a green, and a violet sector of a stained disk) and a similar point on the carapace of the tortoise, that when Achilles arrived where the tortoise started, the latter was just 60 feet 8 inches and $\frac{1}{10}$ inch further on, which is just one tenth of a stadion, and that when Achilles reached that point the tortoise was still 6 feet and $\frac{81}{100}$ inch in advance of him, and finally that, both advancing at a perfectly uniform rate, the tortoise had run just 67 feet 5 inches when he was overtaken by Achilles, it follows that the tortoise progressed at just one tenth the speed of Achilles, the latter running a distance in *stadia* of 1.111111111, so that he had to traverse the sum of an infinite multitude of finite distances, each in a finite time, and yet covered the stadion and one ninth in a finite time. No contradiction, therefore, is involved in the idea of an endless series of finite times or spaces having but a finite sum, provided there is no *fixed* finite quantity which every member of an endless part of that series must each and every one *exceed*.

The reader must pardon me for occupying any of his time with such puerile stuff as that $0.1111 = \frac{1}{9}$; for astounding as it seems, it has more than once happened to me that a gentleman has come to me, every one of them not merely educated men, but highly accomplished, men who might well enough be famous over the civilized world, if fame were anything to the purpose, but men whose studies had been such that one would have expected to find each of them an adept in the accurate statement of arguments, and yet each has come and has undertaken to prove to me that the old catch of Achilles and the tortoise is a sound argument. If I tell you what after listening to them by the hour, I have always ended by saying, it may serve your turn on a similar occasion. I have said, "I suppose you do not mean to say that you really believe that a fast runner cannot, as a matter of fact, overtake a slow one. I therefore conclude that the argument which you have been unable to state, either syllogistically or in any other intelligible form, is intended to show that Zeno's reasoning about Achilles and the tortoise is sound according to some system of logic which admits that sound necessary reasoning may lead from true premisses to a false conclusion. But in my system of logic what I mean by bad necessary reasoning is precisely an argument which might lead from true premisses to a false conclusion, —just that and nothing else. If you prefer to call such reasoning a sound necessary argument, I have no objection in the world to your doing so; and you will kindly allow me to employ my different nomenclature. For I am such a plain, uncultured soul that when I reason I aim at nothing else than just to find out the truth.

To get back, then, to the die and its habit,—its "would-be"—I really know no other way of defining a habit than by describing the kind of behaviour in which the habit becomes actualized. So I am obliged to define the statement that there is a probability of one third that the die when thrown will turn up either a three or a six by stating how the numbers will run when the die is thrown.

But my purpose in doing so is to explain what *probability*, as I use the word, consists in. Now it would be no explanation at all to say that it consists in something being *probable*, so I must avoid using that word or any synonym of it. If I were to use such an expression, you would very properly turn upon me and say, "I either know what it is to be *probable*, in your sense of the term, or I do not. If I don't how can I be expected to understand you until you have explained yourself; and if I do what is the use of the explanation?" But the fact that the probability of the die turning up a three or a six is not *sure* to produce any determinate upon the run of the numbers thrown in any *finite* series of throws. It is only when the series is endless that we can be *sure* that it will have a particular character.

Even when there is an endless series of throws, there is no syllogistic certainty,—no "mathematical" certainty (if you are more familiar with this latter phrase,)—that the die will not turn up a six obstinately at every single throw. It might be that if in the course of the endless series, some friends should borrow the die to make a pair for a game of backgammon, there might be nothing unusual in the behaviour of the lent die, and yet when it was returned and our experimental series was resumed where it had been interrupted, the die might return to turning up nothing but six every time. I say it *might*, in the sense that it would not violate the principle of contradiction if it did. It surely *would not*, however, unless a miracle were performed; and moreover if such miracle *were* worked, I should say (since it is my use of the term probability that we have supposed to be in question,) that during this experimental series of throws, the die took on an abnormal,—a miraculous,—habit. For I should think that the performance of a certain line of behaviour throughout an endless succession of occasions, without exception, very decidedly *constituted* a habit. There may be some doubt about this; for owing to our not being accustomed to reason in this way about successions of events which are endless *in the sequence* and yet are completed *in time*, it is hard for me quite to satisfy myself what I ought to say in such a case. But I have reflected seriously upon it, and though I am not perfectly sure of my ground, (and I am a cautious reasoner,) yet I am more that what you would understand by "pretty confident," that supposing one to be in a condition to assert what *would surely be* the behaviour *in any single determinate respect,* of any subject throughout an endless series of occasions of a stated kind, he *ipso facto* knows a "would-be," or habit, of that subject. It is very true, mind you, that *no* collection whatever of single acts, though it were ever so many grades greater than a simple endless series, can constitute a would-be, nor can the knowledge of single acts, whatever their multitude, tell us for *sure* of a would-be. But there are two remarks to be made; 1st, that in the case under consideration a person is supposed to be in a condition to assert what surely *would be* the behaviour of the subject throughout the endless series of occasions,—a knowledge which cannot have been derived from reasoning from its behaviour on the single occasions; and 2nd, that that which in our case renders it true, as stated, that the person supposed "*ipso facto* knows a would-be of that subject," as not the occurrence of the single acts, but the fact that the person supposed "was in condition to assert what *would surely be* the behaviour of the subject throughout an endless series of occasions."*

I will now describe the behaviour of the die during the endless series of throws, in respect to turning up numbers divisible by three. It would be perfectly possible to construct a machine that would automatically throw the die and pick it up, and continue doing so as long as it was supplied with energy. It would further be still easier to design the plan of an arrangement whereby a hand should after each throw move over an arc graduated so as indicate the value of the quotient of the number of throws of three or six that had been known since the beginning of the experiment, divided by the total number of throws since the beginning. It is true that the mechanical difficulties would become quite insuperable before the die had been thrown many times; but fortunately a general description of the way the hand would move will answer our purposes much better than would the actual machine, were it ever so perfect.

After the first throw, the hand will go either to $0 = {}^0/_1$ or to $1 = {}^1/_1$; and there it may stay for several throws. But when it once moves, it will move after every throw, without exception, since the denominator of the fraction at whose value it points will always increase by 1, and consequently the value of the fraction will be diminished if the numerator remains unchanged, as

*[Original Peirce footnote:] Meantime it may be remarked that, though an endless series of acts is not a habit, nor a would-be, it does present the first of an endless series of steps toward the full nature of a would-be. Compare what I wrote nineteen years ago, in §13 of an article on the logic of relatives: *Monist*, Vol. VII. pp. 205–217.[24]

it will be increased in case the numerator is increased by 1, these two being the only possible cases. The behaviour of the hand may be described as an excessively irregular oscillation, back and forth, from one side of $^1/_3$ to the other. I will actually throw a die, record the throws and put dots on squared paper, each square one way representing a unit of the numerator and each square the other way representing a unit of the denominator. Then straight lines to the origin or corner of the squared surface will represent the positions of the hand, so that you will have a sort of picture of its motion*[.]**

NOTES

1. Peirce opens the sixth article, "Deduction, Induction, and Hypothesis," by stating "the chief business of the logician is to classify arguments."

2. Jacobus Henricus van 't Hoff (1852–1911), Dutch chemist.

3. Francis Bacon, *Novum Organum,* Bk. 1, Aph. 55 and Bk. 2, Aph. 27.

4. The (mathematical) theme of continuity plays a central role throughout Peirce's philosophical career.

5. Peirce discusses the issue of the classification of arguments at the beginning of the sixth Illustrations paper (see chapter 6 below).

6. The Law of Error originated with Karl Gauss. Lambert Adolphe Quetelet (1797–1874) applied the law to his theory of the average man in *Sur l'homme et le développement de se facultés: ou Essai de physique sociale* (Paris: Bachelier, 1835). Quetelet's views influenced Sir Francis Galton in *Hereditary Genius: An Inquiry Into Its Laws and Consequences* (London: Macmillan, 1869); see esp. appendix. Later, in 1886, Galton would famously say that the Law of Error: "reigns with serenity and in complete self-effacement amidst the wildest confusion. The huger the mob and the greater the apparent anarchy, the more perfect is its sway. It is the supreme law of Unreason" (*Natural Inheritance* [London: Macmillan, 1889], 66). For Peirce's recent technical work in this area, see his 1870 "On the Theory of Errors of Observations" (W3, sel., 42; see also sel. 41).

7. In a 26 January 1909 letter to William Edward Story, Peirce writes:

> Here is a definition which pleases me: To say that a certain event has a probability, *p*, means that in an endless series of independent occasions, on each of which the event either occurs or does not, if a tally be kept and a new value of the proportion of occasions on which the event has occurred since the beginning of the series be calculated at every new occasion, then, if there be one value and one value only of this proportion which never ceases to recur, while every other value will some time occur for the last time, that value is the probability of the event. Of course it is assumed that the general conditions do not alter during the whole series. The probability is the value that *would* never cease to recur if the general conditions were to remain the same.

*It is quite clear from the content, and perhaps also from the absent punctuation, that the manuscript is unfinished.

8. The entries for the first three meanings of the term probability in the *Century Dictionary* are Peirce's (CD:4741).

9. John Locke, *An Essay Concerning Human Understanding*, Bk. 4, Ch. 15 "Of Probability," sect. 1.

10. Ibid.

11. Peirce found Parmenides's statement in Bk 1, Pt 1, Ch 1, §C1, *Anmerkung* 1 of Hegel's *Wissenshaft der Logik*. See also W2:317, where Peirce criticizes Scotus Erigena's binary approach in *De divisione naturae,* by using Parmenides's claim.

12. John Venn, *Logic of Chance* (London: Macmillan,1866). For Peirce's review of that book, see W2, sel. 8.

13. Hence, for Peirce, no probabilistic argument can be given for the origin of the universe—universes, as he puts it in "The Probability of Induction," are simply not as plentiful as blackberries. Nonetheless, Peirce develops a view where the universe develops from (absolute) chance. See esp. "Evolutionary Love" (W8, sel. 30).

14. The Martingale or doubling-up system is a betting strategy that became popular in eighteenth-century France; it also inspired the Gamblers Fallacy: the mistaken assumption that departures from what occurs on average or in the long run will be corrected in the short run.

15. See Peirce's discussion of the method of science in "The Fixation of Belief," with notes.

16. In "Evolutionary Love" (W8, sel. 30) Peirce criticizes the doctrine of self-interest espoused by economists, dismissively calling it "the gospel of greed," while developing an alternative.

17. See 1 *Corinthians* 13. Paul uses the Greek *agape* for charity—the unselfish and unconditional love for others. For Peirce's view on agape, and his doctrine of agapism, see esp. "Evolutionary Love" (W8, sel. 30).

18. These are the three papers Peirce published in 1868 in the *Journal of Speculative Philosophy*, edited by William Torrey Harris (W2, sel. 22–23).

19. Peirce makes the same trifold distinction in his letter to Paul Carus written later that month, which is included below as chapter 9.

20. Eusapia Palladino (1854–1918) was an Italian spiritualist medium. In 1908, the American Society for Psychical Research sent Hereward Carrington (1880–1958) to Naples to investigate her psychic abilities. Carrington described his encounter in *Eusapia Palladino and her Phenomena* (New York: B.W. Dodge & Co., 1909), and invited her to the US where he organized a tour for her.

21. The Peirce family resided from 1834 till 1845 at 11 Mason Street (Peirce was born in 1839). For a more extensive autobiographical account, see e.g. W8, sel. 12 and the accompanying editorial annotations.

22. Peirce is here referring to his discovery of his brother's copy of Richard Whately's *Elements of Logic.*

23. In the fourth Lowell Lecture of 1903, Peirce uses this as an example to show that thirdness is really operative in nature (CP 5.93–95).

24. Peirce is referring to "The Logic of Relatives," *Monist* 7 (January 1897): 161–217; note that this was written fourteen years earlier, not nineteen years earlier.

Chapter 4

THE PROBABILITY OF INDUCTION

In "The Probability of Induction" Peirce continued the theory he set out to develop in "The Doctrine of Chances." Peirce followed John Venn in distinguishing the conceptualistic *view of probability, which refers probabilities to events, from the* materialistic *or frequency view of probability, which makes probability the ratio of the frequency of favorable cases to all cases. Peirce charged that most writers on the subject confound the two, and argued against the first. Peirce, for instance, famously rejected a probabilistic theory of the universe on the ground that universes are not as plentiful as blackberries. In the second part of the article Peirce connects his views on probability with synthetic reasoning, especially induction, and discussed the Kantian notion of synthetic a priori judgments.*

The text that follows is as it originally appeared in Popular Science Monthly *12 (April 1878): 705–18. It is only minimally edited and italic brackets surround text added by the editor. The article is listed as the eleventh essay in the table of contents of the 1893* Search for a Method *and was presumably to be the nineteenth chapter of Peirce's 1894 logic book* How to Reason *(hereafter HTR), for which no manuscript material survives but which is listed in the table of contents as "Induction etc." Alternatively, it is possible that Peirce sought to recombine both papers as chapter 18, as in* Popular Science Monthly *they got separated against his intention, reserving chapter 19 for a different text. Whether Peirce actually prepared a version of "The Probability of Induction" for HTR is doubtful given the half-hearted attempt with which he revised "The Doctrine of Chances." A discussion of the history of the article, with an account of the various attempts at rewriting it, is found in the introduction.*

I.

We have found that every argument derives its force from the general truth of the class of inferences to which it belongs; and that probability is the proportion of arguments carrying truth with them among those of any *genus.*[1] This is most conveniently expressed in the nomenclature of the mediæval logicians. They called the fact expressed by a premise an *antecedent,* and that which follows from it its *consequent;* while the leading principle, that every (or almost every) such antecedent is followed by such a consequent, they termed the *consequence.* Using this language, we may say that probability belongs exclusively to *consequences,* and the probability of any consequence is the number of times in which antecedent and consequent both occur divided by the number of all the times in which the antecedent occurs. From this definition are deduced the following rules for the addition and multiplication of probabilities:

Rule for the Addition of Probabilities.—Given the separate probabilities of two consequences having the same antecedent and incompatible consequents. Then the sum of these two numbers is the probability of the consequence, that from the same antecedent one or other of those consequents follows.

Rule for the Multiplication of Probabilities.—Given the separate probabilities of the two consequences, "If A then B," and "If both A and B, then C." Then the product of these two numbers is the probability of the consequence, "If A, then both B and C."

Special Rule for the Multiplication of Independent Probabilities.—Given the separate probabilities of two consequences having the same antecedents, "If A, then B," and "If A, then C." Suppose that these consequences are such that the probability of the second is equal to the probability of the consequence, "If both A and B, then C." Then the product of the two given numbers is equal to the probability of the consequence, "If A, then both B and C."

To show the working of these rules we may examine the probabilities in regard to throwing dice. What is the probability of throwing a six with one die? The antecedent here is the event of throwing a die; the consequent, its turning up a six. As the die has six sides, all of which are turned up with equal frequency, the probability of turning up any one is $^1/_6$. Suppose two dice are thrown, what is the probability of throwing sixes? The probability of either coming up six is obviously the same when both are thrown as when one is thrown—namely, $^1/_6$. The probability that either will come up six when the other does is also the same as that of its coming up six whether the other does or not. The probabilities are, therefore, independent; and, by our rule, the probability that both events will happen together is the product of their several probabilities, or $^1/_6 \times {}^1/_6$. What is the probability of throwing deuce-ace? The probability that the first die will turn up ace and the second deuce is the same as the probability that both will turn up sixes—namely, $^1/_{36}$; the probability that the *second* will turn up ace and the *first* deuce is likewise $^1/_{36}$; these two events—first, ace; second, deuce; and, second, ace; first, deuce—are incompatible. Hence the rule for addition holds, and the probability that either will come up ace and the other deuce is $^1/_{36} + {}^1/_{36}$, or $^1/_{18}$.

In this way all problems about dice, etc., may be solved. When the number of dice thrown is supposed very large, mathematics (which may be defined as the art of making groups to facilitate numeration)[2] comes to our aid with certain devices to reduce the difficulties.

II.

The conception of probability as a matter of *fact,* i.e., as the proportion of times in which an occurrence of one kind is accompanied by an occurrence of another kind, is termed by Mr. Venn the materialistic view of the subject.[3] But probability has often been regarded as being simply the degree of belief which ought to attach to a proposition; and this mode of explaining the idea is termed by Venn the conceptualistic view.[4] Most writers have mixed the two conceptions together. They, first, define the probability of an event as the reason we have to believe that it has taken place, which is conceptualistic; but shortly after they state that it is the ratio of the number of cases favorable to the event to the total number of cases favorable or contrary, and all equally possible. Except that this introduces the thoroughly unclear idea of cases equally possible in place of cases equally frequent, this is a tolerable statement of the materialistic view. The pure conceptualistic theory has been best expounded by Mr. De Morgan in his *Formal Logic: or, the Calculus of Inference, Necessary and Probable*.[5]

The great difference between the two analyses is, that the conceptualists refer probability to an event, while the materialists make it the ratio of frequency of events of a *species* to those of a *genus* over that *species,* thus *giving it two terms instead of one.* The opposition may be made to appear as follows:

Suppose that we have two rules of inference, such that, of all the questions to the solution of which both can be applied, the first yields correct answers to $\frac{81}{100}$, and incorrect answers to the remaining $\frac{19}{100}$; while the second yields correct answers to $\frac{93}{100}$, and incorrect answers to the remaining $\frac{7}{100}$. Suppose, further, that the two rules are entirely independent as to their truth, so that the second answers correctly $\frac{93}{100}$ of the questions which the first answers correctly, and also $\frac{93}{100}$ of the questions which the first answers incorrectly, and answers incorrectly the remaining $\frac{7}{100}$ of the questions which the first answers correctly, and also the remaining $\frac{7}{100}$ of the questions which the first answers incorrectly. Then, of all the questions to the solution of which both rules can be applied—

both answer correctly.. $\frac{93}{100}$ of $\frac{81}{100}$, or $\frac{93 \times 81}{100 \times 100}$;

the second answers correctly and the first incorrectly.... $\frac{93}{100}$ of $\frac{19}{100}$, or $\frac{93 \times 19}{100 \times 100}$;

the second answers incorrectly and the first correctly.... $\frac{7}{100}$ of $\frac{81}{100}$, or $\frac{7 \times 81}{100 \times 100}$;

and both answer correctly... $\frac{7}{100}$ of $\frac{19}{100}$, or $\frac{7 \times 19}{100 \times 100}$;

Suppose, now, that, in reference to any question, both give the same answer. Then (the questions being always such as are to be answered by *yes* or *no*), those in reference to which their answers agree are the same as those which both answer correctly together with those which both answer falsely, or $\frac{93 \times 81}{100 \times 100} + \frac{7 \times 19}{100 \times 100}$ of all. The proportion of those which both answer correctly out of those their answers to which agree is, therefore—

$$\frac{\frac{93 \times 81}{100 \times 100}}{\frac{93 \times 81}{100 \times 100} + \frac{7 \times 19}{100 \times 100}} \quad \text{or} \quad \frac{93 \times 81}{(93 \times 81) + (7 \times 19)}$$

This is, therefore, the probability that, if both modes of inference yield the same result, that result is correct. We may here conveniently make use of another mode of expression. *Probability* is the ratio of the favorable cases to all the cases. Instead of expressing our result in terms of this ratio, we may make use of another—the ratio of favorable to unfavorable cases. This last ratio may be called the *chance* of an event. Then the chance of a true answer by the first mode of inference is $\frac{81}{19}$ and by the second is $\frac{93}{7}$; and the chance of a correct answer from both, when they agree, is—

$$\frac{81 \times 93}{19 \times 7} \text{ or } \frac{81}{19} \times \frac{93}{7},$$

or the product of the chances of each singly yielding a true answer.

It will be seen that a chance is a quantity which may have any magnitude, however great. An event in whose favor there is an even chance, or $^1/_1$, has a probability of $^1/_2$. An argument having an even chance can do nothing toward reënforcing others, since according to the rule its combination with another would only multiply the chance of the latter by 1.

Probability and chance undoubtedly belong primarily to consequences, and are relative to premises; but we may, nevertheless, speak of the chance of an event absolutely, meaning by that the chance of the combination of all arguments in reference to it which exist for us in the given state of our knowledge. Taken in this sense it is incontestable that the chance of an event has an intimate connection with the degree of our belief in it. Belief is certainly something more than a mere feeling; yet there is a feeling of believing, and this feeling does and ought to vary with the chance of the thing believed, as deduced from all the arguments. Any quantity which varies with the chance might, therefore, it would seem, serve as a thermometer for the proper intensity of belief. Among all such quantities there is one which is peculiarly appropriate. When there is a very great chance, the feeling of belief ought to be very intense. Absolute certainty, or an infinite chance, can never be attained by mortals, and this may be represented appropriately by an infinite belief. As the chance diminishes the feeling of believing should diminish, until an even chance is reached, where it should completely vanish and not incline either toward or away from the proposition. When the chance becomes less, then a contrary belief should spring up and should increase in intensity as the chance diminishes, and as the chance almost vanishes (which it can never quite do) the contrary belief should tend toward an infinite intensity. Now, there is one quantity which, more simply than any other, fulfills these conditions; it is the *logarithm* of the chance. But there is another consideration which must, if admitted, fix us to this choice for our thermometer. It is that our belief ought to be proportional to the weight of evidence, in this sense, that two arguments which are entirely independent, neither weakening nor strengthening each other, ought, when they concur, to produce a belief equal to the sum of the intensities of belief which either would produce separately. Now, we have seen that the chances of independent concurrent arguments are to be multiplied together to get the chance of their combination, and therefore the quantities which best express the intensities of belief should be such that they are to be *added* when the *chances* are multiplied in order to produce the quantity which corresponds to the combined chance. Now, the logarithm is the only quantity which fulfills this condition.[6] There is a general law of sensibility, called Fechner's psychophysical law. It is that the intensity of any sensation is proportional to the logarithm of the external force which produces it.[7] It is entirely in harmony with this law that the feeling of belief should be as the logarithm of the chance, this latter being the expression of the state of facts which produces the belief.

The rule for the combination of independent concurrent arguments takes a very simple form when expressed in terms of the intensity of belief, measured in the proposed way. It is this: Take the sum of all the feelings of belief which would be produced separately by all the arguments *pro,* subtract from that the similar sum for arguments *con,* and the remainder is the feeling of belief which we ought to have on the whole. This is a proceeding which men often resort to, under the name of *balancing reasons.*[8]

These considerations constitute an argument in favor of the conceptualistic view. The kernel of it is that the conjoint probability of all the arguments in our possession, with reference to any fact, must be intimately connected with the just degree of our belief in that fact; and this point is supplemented by various others showing the consistency of the theory with itself and with the rest of our knowledge.

But probability, to have any value at all, must express a fact. It is, therefore, a thing to be inferred upon evidence. Let us, then, consider for a moment the formation of a belief of probability. Suppose we have a large bag of beans from which one has been secretly taken at random and hidden under a thimble. We are now to form a probable judgment of the color of that bean, by drawing others singly from the bag and looking at them, each one to be thrown back, and the whole well mixed up after each drawing. Suppose the first drawing is white and the next black. We conclude that there is not an immense preponderance of either color, and that there is something like an even chance that the bean under the thimble is black. But this judgment may be altered by the next few drawings. When we have drawn ten times, if 4, 5, or 6, are white, we have more confidence that the chance is even. When we have drawn a thousand times, if about half have been white, we have great confidence in this result. We now feel pretty sure that, if we were to make a large number of bets upon the color of single beans drawn from the bag, we could approximately insure ourselves in the long run by betting each time upon the white, a confidence which would be entirely wanting if, instead of sampling the bag by 1,000 drawings, we had done so by only two. Now, as the whole utility of probability is to insure us in the long run, and as that assurance depends, not merely on the value of the chance, but also on the accuracy of the evaluation, it follows that we ought not to have the same feeling of belief in reference to all events of which the chance is even. In short, to express the proper state of our belief, not *one* number but *two* are requisite, the first depending on the inferred probability, the second on the amount of knowledge on which that probability is based.* It is true that when our knowledge is very precise, when we have made many drawings from the bag, or, as in most of the examples in the books, when the total contents of the bag are absolutely known, the number which expresses the uncertainty of the assumed probability and its liability to be changed by further experience may become insignificant, or utterly vanish. But, when our knowledge is very slight, this number may be even more important than the probability itself; and when we have no knowledge at all this completely overwhelms the other, so that there is no sense in saying that the chance of the totally unknown event is even (for what expresses absolutely no fact has absolutely no meaning), and what ought to be said is that the chance is entirely indefinite. We thus perceive that the conceptualistic view, though answering well enough in some cases, is quite inadequate.

Suppose that the first bean which we drew from our bag were black. That would constitute an argument, no matter how slender, that the bean under the thimble was also black. If the second bean were also to turn out black, that would be a second independent argument reënforcing the first. If the whole of the first twenty beans drawn should prove black, our confidence that the hidden bean was black would justly attain considerable strength. But suppose the twenty-first bean were to be white and that we were to go on drawing until we found that we had drawn 1,010 black beans and 990 white ones. We should conclude that our first twenty beans being black was simply an extraordinary accident, and that in fact the proportion of white beans to black was sensibly equal, and that it was an even chance that the hidden bean was black. Yet according to the rule of *balancing reasons,* since all the drawings of black beans are so many independent arguments in favor of the one under the thimble being black, and all the white drawings so many against it, an excess of twenty black beans ought to produce the same degree of belief that the hidden bean was black, whatever the total number drawn.

In the conceptualistic view of probability, complete ignorance, where the judgment ought not to swerve either toward or away from the hypothesis, is represented by the probability $^1/_2$.†

---

*[Original Peirce footnote:] Strictly we should need an infinite series of numbers each depending on the probable error of the last.

†[Original Peirce footnote:] "Perfect indecision, belief inclining neither way, an even chance."—DE MORGAN, p. 182.[9]

But let us suppose that we are totally ignorant what colored hair the inhabitants of Saturn have. Let us, then, take a color-chart in which all possible colors are shown shading into one another by imperceptible degrees. In such a chart the relative areas occupied by different classes of colors are perfectly arbitrary. Let us inclose such an area with a closed line, and ask what is the chance on conceptualistic principles that the color of the hair of the inhabitants of Saturn falls within that area? The answer cannot be indeterminate because we must be in some state of belief; and, indeed, conceptualistic writers do not admit indeterminate probabilities. As there is no certainty in the matter, the answer lies between *zero* and *unity*. As no numerical value is afforded by the data, the number must be determined by the nature of the scale of probability itself, and not by calculation from the data. The answer can, therefore, only be one-half, since the judgment should neither favor nor oppose the hypothesis. What is true of this area is true of any other one; and it will equally be true of a third area which embraces the other two. But the probability for each of the smaller areas being one-half, that for the larger should be at least unity, which is absurd.

III.

All of our reasonings are of two kinds: 1. *Explicative, analytic,* or *deductive;* 2. *Amplificative, synthetic,* or (loosely speaking) *inductive.*[10] In explicative reasoning, certain facts are first laid down in the premises. These facts are, in every case, an inexhaustible multitude, but they may often be summed up in one simple proposition by means of some regularity which runs through them all. Thus, take the proposition that Socrates was a man; this implies (to go no further) that during every fraction of a second of his whole life (or, if you please, during the greater part of them) he was a man. He did not at one instant appear as a tree and at another as a dog; he did not flow into water, or appear in two places at once; you could not put your finger through him as if he were an optical image, etc. Now, the facts being thus laid down, some order among some of them, not particularly made use of for the purpose of stating them, may perhaps be discovered; and this will enable us to throw part or all of them into a new statement, the possibility of which might have escaped attention. Such a statement will be the conclusion of an analytic inference. Of this sort are all mathematical demonstrations.[11] But synthetic reasoning is of another kind. In this case the facts summed up in the conclusion are not among those stated in the premises. They are different facts, as when one sees that the tide rises *m* times and concludes that it will rise the next time. These are the only inferences which increase our real knowledge, however useful the others may be.

In any problem in probabilities, we have given the relative frequency of certain events, and we perceive that in these facts the relative frequency of another event is given in a hidden way. This being stated makes the solution. This is therefore mere explicative reasoning, and is evidently entirely inadequate to the representation of synthetic reasoning, which goes out beyond the facts given in the premises. There is, therefore, a manifest impossibility in so tracing out any probability for a synthetic conclusion.

Most treatises on probability contain a very different doctrine.[12] They state, for example, that if one of the ancient denizens of the shores of the Mediterranean, who had never heard of tides, had gone to the bay of Biscay, and had there seen the tide rise, say m times, he could know that there was a probability equal to

$$\frac{m+1}{m+2}$$

that it would rise the next time. In a well-known work by Quetelet, much stress is laid on this, and it is made the foundation of a theory of inductive reasoning.[13]

But this solution betrays its origin if we apply it to the case in which the man has never seen the tide rise at all; that is, if we put $m = 0$. In this case, the probability that it will rise the next time comes out $^1/_2$, or, in other words, the solution involves the conceptualistic principle that there is an even chance of a totally unknown event. The manner in which it has been reached has been by considering a number of urns all containing the same number of balls, part white and part black. One urn contains all white balls, another one black and the rest white, a third two black and the rest white, and so on, one urn for each proportion, until an urn is reached containing only black balls. But the only possible reason for drawing any analogy between such an arrangement and that of Nature is the principle that alternatives of which we know nothing must be considered as equally probable. But this principle, is absurd. There is an indefinite variety of ways of enumerating the different possibilities, which, on the application of this principle, would give different results. If there be any way of enumerating the possibilities so as to make them all equal, it is not that from which this solution is derived, but is the following: Suppose we had an immense granary filled with black and white balls well mixed up; and suppose each urn were filled by taking a fixed number of balls from this granary quite at random. The relative number of white balls in the granary might be anything, say one in three. Then in one-third of the urns the first ball would be white, and in two-thirds black. In one-third of those urns of which the first ball was white, and also in one-third of those in which the first ball was black, the second ball would be white. In this way, we should have a distribution like that shown in the following table, where *w* stands for a white ball and *b* for a black one. The reader can, if he chooses, verify the table for himself.

wwww.					
wwwb.	wwbw.	wbww.	bwww.		
wwwb.	wwbw.	wbww.	bwww.		
wwbb.	wbwb.	bwwb.	wbbw.	bwbw.	bbww.
wwbb.	wbwb.	bwwb.	wbbw.	bwbw.	bbww.
wwbb.	wbwb.	bwwb.	wbbw.	bwbw.	bbww.
wwbb.	wbwb.	bwwb.	wbbw.	bwbw.	bbww.
wbbb.	bwbb.	bbwb.	bbbw.		
wbbb.	bwbb.	bbwb.	bbbw.		
wbbb.	bwbb.	bbwb.	bbbw.		
wbbb.	bwbb.	bbwb.	bbbw.		
wbbb.	bwbb.	bbwb.	bbbw.		
wbbb.	bwbb.	bbwb.	bbbw.		
wbbb.	bwbb.	bbwb.	bbbw.		
wbbb.	bwbb.	bbwb.	bbbw.		
bbbb.					
bbbb.					
bbbb.					
bbbb.					
bbbb.					
bbbb.					
bbbb.					
bbbb.					
bbbb.					
bbbb.					
bbbb.					
bbbb.					
bbbb.					
bbbb.					
bbbb.					
bbbb.					

In the second group, where there is one b, there are two sets just alike; in the third there are 4, in the fourth 8, and in the fifth 16, doubling every time. This is because we have supposed twice as many black balls in the granary as white ones; had we supposed 10 times as many, instead of

1, 2, 4, 8, 16

sets we should have had

1, 10, 100, 1000, 10000

sets; on the other hand, had the numbers of black and white balls in the granary been even, there would have been but one set in each group. Now suppose two balls were drawn from one of these urns and were found to be both white, what would be the probability of the next one being white? If the two drawn out were the first two put into the urns, and the next to be drawn out were the third put in, then the probability of this third being white would be the same whatever the colors of the first two, for it has been supposed that just the same proportion of urns has the third ball white among those which have the first two *white-white, white-black, black-white,* and *black-black.* Thus, in this case, the chance of the third ball being white would be the same whatever the first two were. But, by inspecting the table, the reader can see that in each group all orders of the balls occur with equal frequency, so that it makes no difference whether they are drawn out in the order they were put in or not. Hence the colors of the balls already drawn have no influence on the probability of any other being white or black.

Now, if there be any way of enumerating the possibilities of Nature so as to make the equally probable, it is clearly one which should make one arrangement or combination of the elements of Nature as probable as another, that is, a distribution like we have supposed, and it, therefore, appears that the assumption that any such thing can be done, leads simply to the conclusion that reasoning from past to future experience is absolutely worthless. In fact, the moment that you assume that the chances in favor of that of which we are totally ignorant are even, the problem about the tides does not differ, in any arithmetical particular, from the case in which a penny (known to be equally likely to come up heads and tails) should turn up heads m times successively. In short, it would be to assume that Nature is a pure chaos, or chance combination of independent elements, in which reasoning from one fact to another would be impossible; and since, as we shall hereafter see, there is no judgment of pure observation without reasoning, it would be to suppose all human cognition illusory and no real knowledge possible. It would be to suppose that if we have found the order of Nature more or less regular in the past, this has been by a pure run of luck which we may expect is now at an end. Now, it may be we have no scintilla of proof to the contrary, but reason is unnecessary in reference to that belief which is of all the most settled, which nobody doubts or can doubt, and which he who should deny would stultify himself in so doing.

The relative probability of this or that arrangement of Nature is something which we should have a right to talk about if universes were as plenty as blackberries, if we could put a quantity of them in a bag, shake them well up, draw out a sample, and examine them to see what proportion of them had one arrangement and what proportion another. But, even in that case, a higher universe would contain us, in regard to whose arrangements the conception of probability could have no applicability.

IV.

We have examined the problem proposed by the conceptualists, which, translated into clear language, is this: Given a synthetic conclusion; required to know out of all possible states of things how many will accord, to any assigned extent, with this conclusion; and we have found that it is only an absurd attempt to reduce synthetic to analytic reason, and that no definite solution is possible.

But there is another problem in connection with this subject. It is this: Given a certain state of things, required to know what proportion of all synthetic inferences relating to it will be true within a given degree of approximation. Now, there is no difficulty about this problem (except for its mathematical complication); it has been much studied, and the answer is perfectly well known. And is not this, after all, what we want to know much rather than the other? Why should we want to know the probability that the fact will accord with our conclusion? That implies that we are interested in all possible worlds, and not merely the one in which we find ourselves placed. Why is it not much more to the purpose to know the probability that our conclusion will accord with the fact? One of these questions is the first above stated and the other the second, and I ask the reader whether, if people, instead of using the word probability without any clear apprehension of their own meaning, had always spoken of relative frequency, they could have failed to see that what they wanted was not to follow along the synthetic procedure with an analytic one, in order to find the probability of the conclusion; but, on the contrary, to begin with the fact at which the synthetic inference aims, and follow back to the facts it uses for premises in order to see the probability of their being such as will yield the truth.

As we cannot have an urn with an infinite number of balls to represent the inexhaustibleness of Nature, let us suppose one with a finite number, each ball being thrown back into the urn after being drawn out, so that there is no exhaustion of them. Suppose one ball out of three is white and the rest black, and that four balls are drawn. Then the table on page 140 represents the relative frequency of the different ways in which these balls might be drawn. It will be seen that if we should judge by these four balls of the proportion in the urn, 32 times out of 81 we should find it $1/4$, and 24 times out of 81 we should find it $1/2$, the truth being $1/3$. To extend this table to high numbers would be great labor, but the mathematicians have found some ingenious ways of reckoning what the numbers would be. It is found that, if the true proportion of white balls is *p,* and *s* balls are drawn, then the error of the proportion obtained by the induction will be—

half the time within	$0.477\sqrt{\frac{2p(1-p)}{s}}$
9 times out of 10 within	$1.163\sqrt{\frac{2p(1-p)}{s}}$
99 times out of 100 within	$1.821\sqrt{\frac{2p(1-p)}{s}}$
999 times out of 1,000 within	$2.328\sqrt{\frac{2p(1-p)}{s}}$
9,999 times out of 10,000 within	$2.751\sqrt{\frac{2p(1-p)}{s}}$
9,999,999,999 times out of 10,000,000,000 within	$4.77\sqrt{\frac{2p(1-p)}{s}}$

The use of this may be illustrated by an example. By the census of 1870, it appears that the proportion of males among native white children under one year old was 0.5082, while among colored children of the same age the proportion was only 0.4977.[14] The difference between these is 0.0105, or about one in a 100. Can this be attributed to chance, or would the difference always exist among a great number of white and colored children under like circumstances? Here p may be taken at $^1/_2$; hence $2p(1-p)$ is also $^1/_2$. The number of white children counted was near 1,000,000; hence the fraction whose square-root is to be taken is about $\frac{1}{2{,}000{,}000}$. The root is about $\frac{1}{1400}$, and this multiplied by 0.477 gives about 0.0003 as the probable error in the ratio of males among the whites as obtained from the induction. The number of black children was about 150,000, which gives 0.0008 for the probable error. We see that the actual discrepancy is ten times the sum of these, and such a result would happen, according to our table, only once out of 10,000,000,000 censuses, in the long run.

It may be remarked that when the real value of the probability sought inductively is either very large or very small, the reasoning is more secure. Thus, suppose there were in reality one white ball in 100 in a certain urn, and we were to judge of the number 100 drawings. The probability of drawing no white ball would be $\frac{366}{1000}$; that of drawing one white ball would be $\frac{370}{1000}$; that of drawing two would be $\frac{185}{1000}$; that of drawing three would be $\frac{61}{1000}$; that of drawing four would be $\frac{15}{1000}$; that of drawing five would be only

$\frac{3}{1000}$, etc. Thus we should be tolerably certain of not being in error by more than one ball in 100.

It appears, then, that in one sense we can, and in another we cannot, determine the probability of synthetic inference. When I reason in this way:

> Ninety-nine Cretans in a hundred are liars;
> But Epimenides is a Cretan;
> Therefore, Epimenides is a liar:—

I know that reasoning similar to that would carry truth 99 times in 100. But when I reason in the opposite direction:

> Minos, Sarpedon, Rhadamanthus, Deucalion, and Epimenides, are all the Cretans I can think of;
> But these were all atrocious liars,
> Therefore, pretty much all Cretans must have been liars;

I do not in the least know how often such reasoning would carry me right. On the other hand, what I do know is that some definite proportion of Cretans must have been liars, and that this proportion can be probably approximated to by an induction from five or six instances. Even in the worst case for the probability of such an inference, that in which about half the Cretans are liars, the ratio so obtained would probably not be in error by more than $^1/_6$. So much I know; but, then, in the present case the inference is that pretty much all Cretans are liars, and whether there may not be a special improbability in that I do not know.

V.

Late in the last century, Immanuel Kant asked the question, "How are synthetical judgments *a priori* possible?"[15] By synthetical judgments he meant such as assert positive fact and are not mere affairs of arrangement; in short, judgments of the kind which synthetical reasoning produces, and which analytic reasoning cannot yield. By *a priori* judgments he meant such as that all outward objects are in space, every event has a cause, etc., propositions which according to him can never be inferred from experience. Not so much by his answer to this question as by the mere asking of it, the current philosophy of that time was shattered and destroyed, and a new epoch in its history was begun. But before asking *that* question he ought to have asked the more general one, "How are any synthetical judgments at all possible?" How is it that a man can observe one fact and straightway pronounce judgment concerning another different fact not involved in the first? Such reasoning, as we have seen, has, at least in the usual sense of the phrase, no definite probability; how, then, can it add to our knowledge? This is a strange paradox; the Abbé Gratry says it is a miracle, and that every true induction is an immediate inspiration from on high.*[16] I respect this explanation far more than many a pedantic attempt to solve the question by some juggle with probabilities, with the forms of syllogism, or what not. I respect it because it shows an appreciation of the depth of the problem, because it assigns an adequate cause, and because it is intimately connected—as the true account should be—with a general philosophy of the universe. At the same time, I do not accept this explanation, because an explanation should tell *how* a thing is done, and to assert a perpetual miracle seems to be an abandonment of all hope of doing that, without sufficient justification.[17]

It will be interesting to see how the answer which Kant gave to his question about synthetical judgments *a priori* will appear if extended to the question of synthetical judgments in general. That answer is, that synthetical judgments *a priori* are possible because whatever is universally true is involved in the conditions of experience. Let us apply this to a general synthetical reasoning. I take from a bag a handful of beans; they are all purple, and I infer that all the beans in the bag are purple. How can I do that? Why, upon the principle that whatever is universally true of my experience (which is here the appearance of these different beans) is involved in the condition of experience. The condition of this special experience is that all these beans were taken from that bag. According to Kant's principle, then, whatever is found true of all the beans drawn from the bag must find its explanation in some peculiarity of the contents of the bag. This is a satisfactory statement of the principle of induction.

When we draw a deductive or analytic conclusion, our rule of inference is that facts of a certain general character are either invariably or in a certain proportion of cases accompanied by facts of another general character. Then our premise being a fact of the former class, we infer with certainty or with the appropriate degree of probability the existence of a fact of the second class. But the rule for synthetic inference is of a different kind. When we sample a bag of beans we do not in the least assume that the fact of some

*[Original Peirce footnote:] Logique. The same is true, according to him, of every performance of a differentiation, but not of integration. He does not tell us whether it is the supernatural assistance which makes the former process so much the easier.

beans being purple involves the necessity or even the probability of other beans being so. On the contrary, the conceptualistic method of treating probabilities, which really amounts simply to the deductive treatment of them, when rightly carried out leads to the result that a synthetic inference has just an even chance in its favor, or in other words is absolutely worthless. The color of one bean is entirely independent of that of another. But synthetic inference is founded upon a classification of facts, not according to their characters, but according to the manner of obtaining them. Its rule is, that a number of facts obtained in a given way will in general more or less resemble other facts obtained in the same way; or, *experiences whose conditions are the same will have the same general characters.*

In the former case, we know that premises precisely similar in form to those of the given ones will yield true conclusions, just once in a calculable number of times. In the latter case, we only know that premises obtained under circumstances similar to the given ones (though perhaps themselves very different) will yield true conclusions, at least once in a calculable number of times. We may express this by saying that in the case of analytic inference we know the probability of our conclusion (if the premises are true), but in the case of synthetic inferences we only know the degree of trustworthiness of our proceeding. As all knowledge comes from synthetic inference, we must equally infer that all human certainty consists merely in our knowing that the processes by which our knowledge has been derived are such as must generally have led to true conclusions.

Though a synthetic inference cannot by any means be reduced to deduction, yet that the rule of induction will hold good in the long run may be deduced from the principle that reality is only the object of the final opinion to which sufficient investigation would lead. That belief gradually tends to fix itself under the influence of inquiry is, indeed, one of the facts with which logic sets out.

NOTES

1. An extensive account of how probability ought to be defined, where Peirce also distinguishes it from plausibility and verisimilitude, is found in the August 1910 letter to Paul Carus, see Chapter 9. See also the note appended at the end of chapter 3.

2. This definition of mathematics—which suggests that, like logic, Peirce sees mathematics as a classificatory science—is quite different from how Peirce's father defined the subject—namely, as the science which draws necessary conclusions—which is the definition that Peirce would steadfastly hold onto later.

3. In the preface to his *Logic of Chance* (London: Macmillan, 1866), John Venn contrasts his own view, which he calls the material view of logic, with the formal or conceptual view of logic. With the material view he means a view of logic "which regards it as taking cognisance of laws of things and not the laws of our minds in thinking about things"; see also Peirce's review of Venn's book (W2, sel. 8).

4. In the conceptualistic view, as distinguished by Venn, logic, including the logic of chance, concerns a study of the laws of our minds in thinking about things. According to Venn, De Morgan and Boole are conceptualists (ibid).

5. Augustus De Morgan, *Formal Logic: or, the Calculus of Inference, Necessary and Probable* (London: Taylor and Walton, 1847); see esp. the preface. The view that De Morgan is a conceptualist is shared with Venn (see previous note), who writes about De Morgan that "in his scheme Probability is concerned with formal inferences in which the premises are entertained with a conviction short of absolute certainty" (ibid).

6. In the *Century Dictionary* Peirce writes under "logarithm" (CD:3503):

> In each system of logarithms, the logarithms corresponding to any geometrical progression of natural numbers are in arithmetical progression: that is, if each number of the series is obtained from the preceding one by multiplying a constant factor into this preceding one, the each logarithm may be obtained from the preceding one by adding a constant increment or subtracting a constant decrement.

7. Peirce gives the following rendition of this law in the *Century Dictionary* (CD:3376): "the law that as the physical force of excitation of a nerve increases geometrically the sensation increases arithmetically, so that the sensation is proportional to the logarithm of the excitation."

8. In the *Century Dictionary* Peirce defines "balance of probabilities" as (CD:425): "the excess of reasons for believing one of two alternatives over the reasons for believing the other. It is measured by the logarithm of the ratio of the chances in favor of a proposition to the chances against it."

9. De Morgan, *Formal Logic.*

10. The "loosely speaking" suggests that hypothesis, or abduction, is to be included here also. See chapter 6 below.

11. Note that Peirce does not say that all mathematics is analytic, but only that all mathematical demonstrations are analytic.

12. In "A Theory of Probable Inference" Peirce calls this "the doctrine of Inverse Probabilities"—the view that the past occurrence or nonoccurrence of an event affects the probability of its occurrence in the future—and gives the same example of Quetelet (W4:441).

13. Adolphe Quetelet (1796–1874), *Théorie des Probabilités* (Brussels: A. Jamar, 1853), 43–47 (Pt. 2, Ch. 1).

14. The 1870 census was the first to provide detailed information about the black population, just five years after the end of slavery, on December 6, 1865.

15. Immanuel Kant, *Critique of Pure Reason,* B19.

16. Auguste Joseph Alphonse Gratry (1805–1872), *Logique* 4th ed., 2 vols. (Paris: Douniol, 1858), introduction; see also bk. 1 ch.1, bk. 3, ch. 4, and bk. 4, ch. 7. Peirce makes the same observation in W4:444.

17. As Peirce put it elsewhere, such an assertion blocks the road of inquiry (EP2:48).

Chapter 5

THE ORDER OF NATURE

In "The Order of Nature" Peirce argued against the notion, expressed, for instance, by John Stuart Mill, that the uniformity of nature is the sole warrant for induction. Instead Peirce argued that induction should be explained by the doctrine of probability that he developed in the previous two articles. The article initiates several themes that return in Peirce's 1887–1888 Guess at the Riddle *and in the cosmological papers that he wrote for the* Monist *in the beginning of the 1890s. Peirce rejected the argument from design—which states that only a god can explain the fact that there is order in the universe—on the ground that a universe without order is a contradiction in terms.*

The text reproduced below is as it originally appeared in Popular Science Monthly *13 (1878): 203–17. It is only minimally edited and italic brackets surround text added by the editor. The article is listed as the twelfth essay in* Search for a Method *(R 1582:2), but is not included in* How to Reason. *R 874 contains an uncorrected typescript of "The Order of Nature," made most likely with* Search for a Method *in mind, but its title is missing. Besides typographical errors, the typescript agrees with the published paper (the typographical mistakes also reveal that Peirce was not the typist). A discussion of the history of the paper is found in the introduction.*

I.

Any proposition whatever concerning the order of Nature must touch more or less upon religion. In our day, belief, even in these matters, depends more and more upon the observation of facts. If a remarkable and universal orderliness be found in the universe, there must be some cause for this regularity, and science has to consider what hypotheses might account for the phenomenon. One way of accounting for it, certainly, would be to suppose that the world is ordered by a superior power. But if there is nothing in the universal subjection of phenomena to laws, nor in the character of those laws themselves (as being benevolent, beautiful, economical, etc.), which goes to prove the existence of a governor of the universe, it is hardly to be anticipated that any other sort of evidence will be found to weigh very much with minds emancipated from the tyranny of tradition.

Nevertheless, it cannot truly be said that even an absolutely negative decision of that question could altogether destroy religion, inasmuch as there are faiths in which, however much they differ from our own, we recognize those essential characters which make them worthy to be called religions, and which, nevertheless, do not postulate an actually existing Deity. That one, for instance, which has had the most numerous and by no means the least intelligent following of any on earth, teaches that Divinity in his highest perfection is wrapped away from the world in a state of profound and eternal sleep, which really does not differ from non-existence, whether it be called by that name or not. No candid mind who has followed the writings of M. Vacherot can well deny that his religion is as earnest as can be.[1] He worships the Perfect, the Supreme Ideal; but he conceives that the very notion of the Ideal is repugnant to its real existence. In fact, M. Vacherot finds it agreeable to his reason to assert that non-existence is an essential character of the perfect, just as St. Anselm and Descartes found it agreeable to theirs to assert the extreme opposite.[2] I confess that there is one respect in which either of these positions seems to me more congruous with the religious attitude than that of a theology which stands upon evidences; for as soon as the Deity presents himself to either Anselm or Vacherot, and manifests his glorious attributes, whether it be in a vision of the night or day, either of them recognizes his adorable God, and sinks upon his knees at once; whereas the theologian of evidences will first demand that the divine apparition shall identify himself, and only after having scrutinized his credentials and weighed the probabilities of his being found among the totality of existences, will he finally render his circumspect homage, thinking that no characters can be adorable but those which belong to a real thing.

If we could find out any general characteristic of the universe, any mannerism in the ways of Nature, any law everywhere applicable and universally valid, such a discovery would be of such singular assistance to us in all our future reasoning, that it would deserve a place almost at the head of the principles of logic. On the other hand, if it can be shown that there is nothing of the sort to find out, but that every discoverable regularity is of limited range, this again will be of logical importance. What sort of a conception we ought to have of the universe, how to think of the *ensemble* of things, is a fundamental problem in the theory of reasoning.

II.

It is the legitimate endeavor of scientific men now, as it was twenty-three hundred years ago, to account for the formation of the solar system and of the cluster of stars which forms the galaxy, by the fortuitous concourse of atoms. The greatest expounder of this theory, when asked how he could write an immense book on the system of the world without one mention of its author, replied, very logically, "Je n'avais pas besoin de cette hypothèse-là."[3] But, in truth, there is nothing atheistical in the theory, any more than there was in this answer. Matter is supposed to be composed of molecules which obey the laws of mechanics and exert certain attractions upon one another; and it is to these regularities (which there is no attempt to account for) that general arrangement of the solar system would be due, and not to hazard.

If any one has ever maintained that the universe is a pure throw of the dice, the theologians have abundantly refuted him. "How often," says Archbishop Tillotson, "might a man, after he had jumbled a set of letters in a bag, fling them out upon the ground before they would fall into an exact poem, yea, or so much as make a good discourse in prose! And may not a little book be as easily made by chance as this great volume of the world?"[4] The chance world here shown to be so different from that in which we live would be one in which there were no laws, the characters of different things being entirely independent; so that, should a sample of any kind of objects ever show a prevalent character, it could only be by accident, and no general proposition could ever be established. Whatever further conclusions we may come to in regard to the order of the universe, thus much may be regarded as solidly established, that the world is not a mere chance-medley.

But whether the world makes an exact poem or not, is another question. When we look up at the heavens at night, we readily perceive that the stars are not simply splashed on to the celestial vault; but there does not seem to be any precise system in their arrangement either. It will be worth our while, then, to inquire into the degree of orderliness in the universe; and, to begin, let us ask whether the world we live in is any more orderly than a purely chance-world would be.

Any uniformity, or law of Nature, may be stated in the form, "Every A is B;" as, every ray of light is a non-curved line, every body is accelerated toward the earth's centre, etc. This is the same as to say, "There does not exist any A which is not B;" there is no curved ray; there is no body not accelerated toward the earth; so that the uniformity consists in the non-occurrence in Nature of a certain combination of characters (in this case, the combination of being A with being non-B).* And, conversely, every case of the non-occurrence of a combination of characters would constitute a uniformity in Nature. Thus, suppose the quality A is never found in combination with the quality C: for example, suppose the quality of idiocy is never found in combination with that of having a well-developed brain. Then nothing of the sort A is of the sort C, or everything of the sort A is of the sort non-C (or say, every idiot has an ill-developed brain), which, being something universally true of the A's, is a uniformity in the world. Thus we see that, in a world where there were no uniformities, no logically possible combination of characters would be excluded, but every combination would exist in some object. But two objects not identical must differ in some of their characters, though it be only in the character of being in such-and-such a place. Hence, precisely the same combination of characters could not be found in two different objects; and, consequently, in a chance-world every combination involving either the positive or negative of every character would belong to just one thing. Thus, if there were but five simple characters in such a world,† we might denote them by A, B, C, D, E, and their negatives by a, b, c, d, e; and then, as there would be 2^5 or 32 different combinations of these characters, completely determinate in reference to each of them, that world would have just 32 objects in it, their characters being as in the following table:

*[Original Peirce footnote:] For the present purpose, the negative of a character is to be considered as much a character as the positive, for a uniformity may either be affirmative or negative. I do not say that no distinction can be drawn between positive and negative uniformities.

†[Original Peirce footnote:] There being 5 simple characters, with their negatives, they could be compounded in various ways so as to make 241 characters in all, without counting the characters *existence* and *non-existence*, which make up 243 or 3^5.

Table I.			
ABCDE	AbCDE	aBCDE	abCDE
ABCDe	AbCDe	aBCDe	abCDe
ABCdE	AbCdE	aBCdE	abCdE
ABCde	AbCde	aBCde	abCde
ABcDE	AbcDE	aBcDE	abcDE
ABcDe	AbcDe	aBcDe	abcDe
ABcdE	AbcdE	aBcdE	abcdE
ABcde	Abcde	aBcde	abcde

For example, if the five primary characters were *hard, sweet, fragrant, green, bright,* there would be one object which reunited all these qualities, one which was hard, sweet, fragrant, and green, but not bright; one which was hard, sweet, fragrant, and bright, but not green; one which was hard, sweet, and fragrant, but neither green nor bright; and so on through all the combinations.

This is what a thoroughly chance-world would be like, and certainly nothing could be imagined more systematic. When a quantity of letters are poured out of a bag, the appearance of disorder is due to the circumstance that the phenomena are only partly fortuitous. The laws of space are supposed, in that case, to be rigidly preserved, and there is also a certain amount of regularity in the formation of the letters. The result is that some elements are orderly and some are disorderly, which is precisely what we observe in the actual world. Tillotson, in the passage of which a part has been quoted, goes on to ask, "How long might 20,000 blind men, which should be sent out from the several remote parts of England, wander up and down before they would all meet upon Salisbury Plains, and fall into rank and file in the exact order of any army? And yet this is much more easy to be imagined than how the innumerable blind parts of matter should rendezvous themselves into a world."[5] This is very true, but in the actual world the *blind men* are, as far as we can see, *not* drawn up in any particular order at all. And, in short, while a certain amount of order exists in the world, it would seem that the world is not so orderly as it might be, and, for instance, not so much so as a world of pure chance would be.

But we can never get to the bottom of this question until we take account of a highly-important logical principle* which I now proceed to enounce. This principle is that any plurality or lot of objects whatever have some character in common (no matter how insignificant) which is peculiar to them and not shared by anything else. The word "character" here is taken in such

*[Original Peirce footnote:] This principle was, I believe, first stated by Mr. De Morgan.[6]

a sense as to include negative characters, such as incivility, inequality, etc., as well as their positives, civility, equality, etc. To prove the theorem, I will show what character any two things, A and B, have in common, not shared by anything else. The things, A and B, are each distinguished from all other things by the possession of certain characters which may be named A-ness and B-ness. Corresponding to these positive characters, are the negative characters un-A-ness which is possessed by everything except A, and un-B-ness, which is possessed by everything except B. These two characters are united in everything except A and B; and this union of characters un-A-ness and un-B-ness makes a compound character which may be termed A-B-lessness. This is not possessed by either A or B, but it is possessed by everything else. This character, like every other, has its corresponding negative un-A-B-lessness, and this last is the character possessed by both A and B, and by nothing else. It is obvious that what has thus been shown true of two things is, *mutatis mutandis,* true of any number of things. Q. E. D.

In any world whatever, then, there must be a character peculiar to each possible group of objects. If, as a matter of nomenclature, characters peculiar to the same group be regarded as only different aspects of the same character, then we may say that there will be precisely one character for each possible group of objects. Thus, suppose a world to contain five things, $\alpha, \beta, \gamma, \delta, \varepsilon$. Then it will have a separate character for each of the 31 groups (with *non-existence* making up 32 or 2^5) shown in the following table:

TABLE II.

	$\alpha\beta$	$\alpha\beta\gamma$	$\alpha\beta\gamma\delta$	$\alpha\beta\gamma\delta\varepsilon$
α	$\alpha\gamma$	$\alpha\beta\delta$	$\alpha\beta\gamma\varepsilon$	
β	$\alpha\delta$	$\alpha\beta\varepsilon$	$\alpha\beta\delta\varepsilon$	
γ	$\alpha\varepsilon$	$\alpha\gamma\delta$	$\alpha\gamma\delta\varepsilon$	
δ	$\beta\gamma$	$\alpha\gamma\varepsilon$	$\beta\gamma\delta\varepsilon$	
ε	$\beta\delta$	$\alpha\delta\varepsilon$		
	$\beta\varepsilon$	$\beta\gamma\delta$		
	$\gamma\delta$	$\beta\gamma\varepsilon$		
	$\gamma\varepsilon$	$\beta\delta\varepsilon$		
	$\delta\varepsilon$	$\gamma\delta\varepsilon$		

This shows that a contradiction is involved in the very idea of a chance-world, for in a world of 32 things, instead of there being only 3^5 or 243 characters, as we have seen that the notion of a chance-world requires, there would, in fact, be no less than 2^{32}, or 4,294,967,296 characters, which would not be all independent, but would have all possible relations with one another.

We further see that so long as we regard characters abstractly, without regard to their relative importance, etc., there is no possibility of a more or less degree of orderliness in the world, the whole system of relationship between the different characters being given by mere logic; that is, being implied in those facts which are tacitly admitted as soon as we admit that there is any such thing as reasoning.

In order to descend from this abstract point of view, it is requisite to consider the characters of things as relative to the perceptions and active powers of living beings. Instead, then, of attempting to imagine a world in which there should be no uniformities, let us suppose one in which none of the uniformities should have reference to characters interesting or important to us. In the first place, there would be nothing to puzzle us in such a world. The small number of qualities which would directly meet the senses would be the ones which would afford the key to everything which could possibly interest us. The whole universe would have such an air of system and perfect regularity that there would be nothing to ask. In the next place, no action of ours, and no event of Nature, would have important consequences in such a world. We should be perfectly free from all responsibility, and there would be nothing to do but to enjoy or suffer whatever happened to come along. Thus there would be nothing to stimulate or develop either the mind or the will, and we consequently should neither act nor think. We should have no memory, because that depends on a law of our organization. Even if we had any senses, we should be situated toward such a world precisely as inanimate objects are toward the present one, provided we suppose that these objects have an absolutely transitory and instantaneous consciousness without memory—a supposition which is a mere mode of speech, for that would be no consciousness at all. We may, therefore, say that a world of chance is simply our actual world viewed from the standpoint of an animal at the very vanishing-point of intelligence. The actual world is almost a chance-medley to the mind of a polyp. The interest which the uniformities of Nature have for an animal measures his place in the scale of intelligence.

Thus, nothing can be made out from the orderliness of Nature in regard to the existence of a God, unless it be maintained that the existence of a finite mind proves the existence of an infinite one.

III.

In the last of these papers we examined the nature of inductive or synthetic reasoning. We found it to be a process of sampling. A number of specimens of a class are taken, not by selection within that class, but at random. These specimens will agree in a great number of respects. If, now, it were likely that a second lot would agree with the first in the majority of these respects, we might base on this consideration an inference in regard to any one of these characters. But such an inference would neither be of the nature of induction, nor would it (except in special cases) be valid, because the vast majority of points of agreement in the first sample drawn would generally be entirely accidental, as well as insignificant. To illustrate this, I take the ages at death of the first five poets given in Wheeler's *Biographical Dictionary.*[7] They are:

Aagard, 48.
Abeille, 70.
Abulola, 84.
Abunowas, 48.
Accords, 45.

These five ages have the following characters in common:

1. The difference of the two digits composing the number, divided by three, leaves a remainder of *one.*
2. The first digit raised to the power indicated by the second, and divided by three, leaves a remainder of *one.*
3. The sum of the prime factors of each age, including one, is divisible by three.

It is easy to see that the number of accidental agreements of this sort would be quite endless.[8] But suppose that, instead of considering a character because of its prevalence in the sample, we designate a character before taking the sample, selecting it for its importance, obviousness, or other point of interest. Then two considerable samples drawn at random are extremely likely to agree approximately in regard to the proportion of occurrences of a character so chosen. *The inference that a previously designated character has nearly the same frequency of occurrence in the whole of a class that it has in a sample drawn at random out of that class is induction.* If the character be not previously designated, then a sample in which it is found to be prevalent can only serve to suggest that it *may be* prevalent in the whole class. We may consider this surmise as an inference if we please—an inference of possibility; but a second sample must be drawn to test the question of whether the character actually is prevalent. Instead of designating beforehand a single character in reference to which we will examine a sample, we may designate two, and use the same sample to determine the relative frequencies of both. This will be making two inductive inferences at once; and, of course, we are less certain that both will yield correct conclusions than we should be that either separately would do so. What is true of two characters is true of any limited number. Now, the number of characters which have any considerable interest for us in reference to any class of objects is more moderate than might be supposed. As we shall be sure to examine any sample with reference to these characters, they may be regarded not exactly as predesignated, but as predetermined (which amounts to the same thing); and we may infer that the sample represents the class in all these respects if we please, remembering only that this is not so secure an inference as if the particular quality to be looked for had been fixed upon beforehand.

The demonstration of this theory of induction rests upon principles and follows methods which are accepted by all those who display in other matters the particular knowledge and force of mind which qualify them to judge of this. The theory itself, however, quite unaccountably seems never to have occurred to any of the writers who have undertaken to explain synthetic reasoning. The most widely-spread opinion in the matter is one which was much promoted by Mr. John Stuart Mill—namely, that induction depends for its validity upon the uniformity of Nature—that is, on the principle that what happens once will, under a sufficient degree of similarity of circumstances, happen again as often as the same circumstances recur.[9] The application is this: The fact that different things belong to the same class constitutes the similarity of circumstances, and the induction is good, provided this similarity is "sufficient." What happens once is, that a number of these things are found to have a certain character; what may be expected, then, to happen again as often as the circumstances recur consists in this, that all things belonging to the same class should have the same character.

This analysis of induction has, I venture to think, various imperfections, to some of which it may be useful to call attention. In the first place, when I put my hand in a bag and draw out a handful of beans, and, finding three-quarters of them black, infer that about three-quarters of all in the bag are black, my inference is obviously of the same kind as if I had found any larger proportion, or the whole, of the sample black, and had assumed that it represented in that respect the rest of the contents of the bag. But the analysis in question hardly seems adapted to the explanation of this *proportionate* induction, where the conclusion, instead of being that a certain event uniformly happens under certain circumstances, is precisely that it does not uniformly occur, but only happens in a certain proportion of cases. It is true that the whole sample may be regarded as a single object, and the inference may be brought under the formula proposed by considering the conclusion to be that any similar sample will show a similar proportion among its constituents. But this is to treat the induction as if it rested on a single instance, which gives a very false idea of its probability.

In the second place, if the uniformity of Nature were the sole warrant of induction, we should have no right to draw one in regard to a character whose constancy we knew nothing about. Accordingly, Mr. Mill says that, though none but white swans were know to Europeans for thousands of years, yet the inference that all swans were white was "not a good induction," because it was not known that color was a usual generic character (it, in fact, not being so by any means).[10] But it is mathematically demonstrable that an inductive inference may have as high a degree of probability as you please independent of any antecedent knowledge of the constancy of the character inferred. Before it was known that color is not usually a character of *genera,* there was certainly a considerable probability that all swans were white. But the further study of the *genera* of animals led to the induction of their non-uniformity in regard to color. A deductive application of this general proposition would have gone far to overcome the probability of the universal whiteness of swans before the black species was discovered. When we do know anything in regard to the general constancy or inconstancy of a character, the application of that general knowledge to the particular class to which any induction relates, though it serves to increase or diminish the force of the induction, is, like every application of general knowledge to particular cases, deductive in its nature and not inductive.

In the third place, to say that inductions are true because similar events happen in similar circumstances—or, what is the same thing, because objects similar in some respects are likely to be similar in others—is to overlook those conditions which really are essential to the validity of inductions. When we take all the characters into account, any pair of objects resemble one another in just as many particulars as any other pair. If we limit ourselves to such characters as have for us any importance, interest, or obviousness, then a synthetic conclusion may be drawn, but only on condition that the specimens by which we judge have been taken at random from the class in regard to which we are to form a judgment, and not selected as belonging to any sub-class. The induction only has its full force when the character concerned has been designated before examining the sample. These are the essentials of induction, and they are not recognized in attributing the validity of induction to the uniformity of Nature. The explanation of induction by the doctrine of probabilities, given in the last of these papers, is not a mere metaphysical formula, but is one from which all the rules of synthetic reasoning can be deduced systematically and with mathematical cogency. But the account of the matter by a principle of Nature, even if it were in other respects satisfactory, presents the fatal disadvantage of leaving us quite as much afloat as before in regard to the proper method of induction. It does not surprise me, therefore, that those who adopt this theory have given erroneous rules for the conduct of reasoning, nor that the greater number of examples put forward by Mr. Mill in his first edition, as models of what inductions should be, proved in the light of further scientific progress so particularly unfortunate that they had to be replaced by others in later editions.[11] One would have supposed that Mr. Mill might have based an induction on *this* circumstance, especially as it is his avowed principle that, if the conclusion of an induction turns out false, it cannot have been a good induction. Nevertheless, neither he nor any of his scholars seem to have been led to suspect, in the least, the perfect solidity of the framework which he devised for securely supporting the mind in its passage from the known to the unknown, although at its first trial it did not answer quite so well as had been expected.

IV.

When we have drawn any statistical induction—such, for instance, as that one-half of all births are of male children—it is always possible to discover, by investigation sufficiently prolonged, a class of which the same predicate may be affirmed universally; to find out, for instance, *what sort of* births are of male children. The truth of this principle follows immediately from the theorem that there is a character peculiar to every possible group of objects. The form in which the principle is usually stated is, that *every event must have a cause.*

But, though there exists a cause for every event, and that of a kind which is capable of being discovered, yet if there be nothing to guide us to the discovery; if we have to hunt among all the events in the world without any scent; if, for instance, the sex of a child might equally be supposed to depend on the configuration of the planets, on what was going on at the antipodes, or on anything else—then the discovery would have no chance of ever getting made.

That we ever do discover the precise causes of things, that any induction whatever is absolutely without exception, is what we have no right to assume. On the contrary, it is any easy corollary, from the theorem just referred to, that every empirical rule has an exception. But there are certain of our inductions which present an approach to universality so extraordinary that, even if we are to suppose that they are not strictly universal truths, we cannot possibly think that they have been reached merely by accident. The most remarkable laws of this kind are those of *time* and *space.* With reference to space, Bishop Berkeley first showed, in a very conclusive manner, that it was not a thing *seen,* but a thing *inferred.*[12] Berkeley chiefly insists on the impossibility of directly seeing the third dimension of space, since the retina of the eye is a surface. But, in point of fact, the retina is not even a surface; it is a conglomeration of nerve-needles directed toward the light and having only their extreme points sensitive, these points lying at considerable distances from one another compared with their areas. Now, of these points, certainly the excitation of no one singly can produce the perception of a surface, and consequently not the aggregate of all the sensations can amount to this. But certain relations subsist between the excitations of different nerve-points, and these constitute the premises upon which the hypothesis of space is founded, and from which it is inferred. That space is not immediately perceived is now universally admitted; and a mediate cognition is what is called an inference, and is subject to the criticism of logic. But what are we to say to the fact of every chicken as soon as it is hatched solving a problem whose data are of a complexity sufficient to try the greatest mathematical powers? It would be insane to deny that the tendency to light upon the conception of space is inborn in the mind of the chicken and of every animal. The same thing is equally true of time. That time is not directly perceived is evident, since no lapse of time is present, and we only perceive what is present. That, not having the idea of time, we should ever be able to perceive the flow in our sensations without some particular aptitude for it, will probably also be admitted. The idea of force—at least, in its rudiments—is another conception so early arrived at, and found in animals so low in the scale of intelligence, that it must be supposed innate. But the innateness of an idea admits of degree, for it consists in the tendency of that idea to present itself to the mind. Some ideas, like that of space, do so present themselves irresistibly at the very dawn of intelligence, and take possession of the mind on small provocation, while of other conceptions we are prepossessed, indeed, but not so strongly, down a scale which is greatly extended. The tendency to personify every thing, and to attribute human characters to it, may be said to be innate; but it is a tendency which is very soon overcome by civilized man in regard to the greater part of the objects about him. Take such a conception as that of gravitation varying inversely as the square of the distance.[13] It is a very simple law. But to say that it is simple is merely to say that it is one which the mind is particularly adapted to apprehend with facility. Suppose the idea of a quantity multiplied into another had been no more easy to the mind than that of a quantity raised to the power indicated by itself—should we ever have discovered the law of the solar system?

It seems incontestable, therefore, that the mind of man is strongly adapted to the comprehension of the world; at least, so far as this goes, that certain conceptions, highly important for such a comprehension, naturally arise in his mind; and, without such a tendency, the mind could never have had any development at all.[14]

How are we to explain this adaptation? The great utility and indispensableness of the conceptions of time, space, and force, even to the lowest intelligence, are such as to suggest that they are the results of natural selection. Without something like geometrical, kinetical, and mechanical conceptions, no animal could seize his food or do anything which might be necessary for the preservation of the species. He might, it is true, be provided with an instinct which would generally have the same effect; that is to say, he might have conceptions different from those of time, space, and force, but which coincided with them in regard to the ordinary cases of the animal's experience. But, as that animal would have an immense advantage in the struggle for life whose mechanical conceptions did not break down in a novel situation (such as development must bring about), there would be a constant selection in favor of more and more correct ideas of these matters. Thus would be attained the knowledge of that fundamental law upon which all science rolls; namely, that forces depend upon relations of time, space, and mass. When this idea was once sufficiently clear, it would require no more than a comprehensible degree of genius to discover the exact nature of these relations. Such an hypothesis naturally suggests itself, but it must be admitted that it does not seem sufficient to account for the extraordinary accuracy with which these conceptions apply to the phenomena of Nature, and it is probable that there is some secret here which remains to be discovered.

V.

Some important questions of logic depend upon whether we are to consider the material universe as of limited extent and finite age, or quite boundless in space and in time. In the former case, it is conceivable that a general plan or design embracing the whole universe should be discovered, and it would be proper to be on the alert for some traces of such a unity. In the latter case, since the proportion of the world of which we can have any experience is less than the smallest assignable fraction, it follows that we never could discover any *pattern* in the universe except a repeating one; any design embracing the whole would be beyond our powers to discern, and beyond the united powers of all intellects during all time. Now, what is absolutely incapable of being known is, as we have seen in a former paper, not real at all. An absolutely incognizable existence is a nonsensical phrase.[15] If, therefore, the universe is infinite, the attempt to find in it any design embracing it as a whole is futile, and involves a false way of looking at the subject. If the universe never had any beginning, and if in space world stretches beyond world without limit, there is no *whole* of material things, and consequently no general character to the universe, and no need or possibility of any governor for it. But if there was a time before which absolutely no matter existed, if there are certain absolute bounds to the region of things outside of which there is a mere void, then we naturally seek for an explanation of it, and, since we cannot look for it among material things, the hypothesis of a great disembodied animal, the creator and governor of the world, is natural enough.[16]

The actual state of the evidence as to the limitation of the universe is as follows: As to time, we find on our earth a constant progress of development since the planet was a red-hot ball; the solar system seems to have resulted from the condensation of a nebula, and the process appears to be still going on.[17] We sometimes see stars (presumably with systems of worlds) destroyed and apparently resolved back into the nebulous condition, but we have no evidence of any existence of the world previous to the nebulous stage from which it seems to have been evolved. All this rather favors the idea of a beginning than otherwise. As for limits in space, we cannot be sure that we see anything outside of the system of the milky-way. Minds of theological predilections have therefore no need of distorting the facts to reconcile them with their views.

But the only scientific presumption is, that the unknown parts of space and time are like the known parts, occupied; that, as we see cycles of life and death in all development which we can trace out to the end, the same holds good in regard to solar systems; that as enormous distances lie between the different planets of our solar system, relatively to their diameters, and as still more enormous distances lie between our system relatively to its diameter and other systems, so it may be supposed that other galactic clusters exist so remote from ours as not to be recognized as such with certainty. I do not say that these are strong inductions; I only say that they are the presumptions which, in our ignorance of the facts, should be preferred to hypotheses which involve conceptions of things and occurrences totally different in their character from any of which we have had any experience, such as disembodied spirits, the creation of matter, infringements of the laws of mechanics, etc.

The universe ought to be presumed too vast to have any character. When it is claimed that the arrangements of Nature are benevolent, or just, or wise, or of any other peculiar kind, we ought to be prejudiced against such opinions, as being the offspring of an ill-founded notion of the finitude of the world. And examination has hitherto shown that such beneficences, justice, etc., are of a most limited kind—limited in degree and limited in range.

In like manner, if any one claims to have discovered a plan in the structure of organized beings, or a scheme in their classification, or a regular arrangement among natural objects, or a system of proportionality in the human form, or an order of development, or a correspondence between conjunctions of the planets and human events, or a significance in numbers, or a key to dreams, the first thing we have to ask is whether such relations are susceptible of explanation on mechanical principles, and if not they should be looked upon with disfavor as having already a strong presumption against them; and examination has generally exploded all such theories.

There are minds to whom every prejudice, every presumption, seems unfair. It is easy to say what minds these are. They are those who never have known what it is to draw a well-grounded induction, and who imagine that other people's knowledge is as nebulous as their own. That all science rolls upon presumption (not of a formal but of a real kind) is no argument with them, because they cannot imagine that there is anything solid in human knowledge. These are the people who waste their time and money upon perpetual motions and other such rubbish.

But there are better minds who take up mystical theories (by which I mean all those which have no possibility of being mechanically explained). These are persons who are strongly prejudiced in favor of such theories. We all have natural tendencies to believe in such things; our education often strengthens this tendency; and the result is, that to many minds nothing seems so antecedently probable as a theory of this kind. Such persons find evidence enough in favor of their views, and in the absence of any recognized logic of induction they cannot be driven from their belief.

But to the mind of a physicist there ought to be a strong presumption against every mystical theory; and therefore it seems to me that those scientific men who have sought to make out that science was not hostile to theology have not been so clear-sighted as their opponents.

It would be extravagant to say that science can at present disprove religion; but it does seem to me that the spirit of science is hostile to any religion except such a one as that of M. Vacherot. Our appointed teachers inform us that Buddhism is a miserable and atheistical faith, shorn of the most glorious and needful attributes of a religion; that its priests can be of no use to agriculture by praying for rain, nor to war by commanding the sun to stand still.[18] We also hear the remonstrances of those who warn us that to shake the general belief in the living God would be to shake the general morals, public and private. This, too, must be admitted; such a revolution of thought could no more be accomplished without waste and desolation than a plantation of trees could be transferred to new ground, however wholesome in itself, without all of them languishing for a time, and many of them dying. Nor is it, by-the-way, a thing to be presumed that a man would have taken part in a movement having a possible atheistical issue without having taken serious and adequate counsel in regard to that responsibility. But, let the consequences of such a belief be as dire as they may, one thing is certain: that the state of the facts, whatever it may be, will surely get found out, and no human prudence can long arrest the triumphal car of truth—no, not if the discovery were such as to drive every individual of our race to suicide!

But it would be folly to suppose that any metaphysical theory in regard to the mode of being of the perfect is to destroy that aspiration toward the perfect which constitutes the essence of religion. It is true that, if the priests of any particular form of religion succeed in making it generally believed that religion cannot exist without the acceptance of certain formulas, or if they succeed in so interweaving certain dogmas with the popular religion that the people can see no essential analogy between a religion which accepts these points of faith and one which rejects them, the result may very well be to render those who cannot believe these things irreligious. Nor can we ever hope that any body of priests should consider themselves more teachers of religion in general than of the particular system of theology advocated by their own party. But no man need be excluded from participation in the common feelings, nor from so much of the public expression of them as is open to all the laity, by the unphilosophical narrowness of those who guard the mysteries of worship.[19] Am I to be prevented from joining in that common joy at the revelation of enlightened principles of religion, which we celebrate at Easter and Christmas, because I think that certain scientific, logical, and metaphysical ideas which have been mixed up with these principles are untenable? No; to do so would be to estimate those errors as of more consequence than the truth—an opinion which few would admit. People who do not believe what are really the fundamental principles of Christianity are rare to find, and all but these few ought to feel at home in the churches.

NOTES

1. Étienne Vacherot (1809–1897), French philosopher. See his two-volume *La métaphysique et la science, ou principes de métaphysique positive* (Paris:

Librairie de F. Chamerot, 1858), the first volume of which Peirce checked out of the Harvard library on January 30, 1869, and *La Religion* (Paris: Librairie Chamerot et Lauwereyns, 1869), bk. 2, ch. 5, esp. 302ff.

2. See Anselm's *Proslogion* and Descartes's fifth *Meditation.* Much later, in his "Neglected Argument for the Reality of God" (EP2:434–50, 1908), when arguing for the *reality* of God while denying his *existence,* Peirce would make a move not unlike the one he ascribes to Vacherot. See also Peirce's account of the a priori method in "The Fixation of Belief," section 5.

3. Pierre-Simon, Marquis de Laplace (1749–1827), allegedly made this comment in reply to Napoleon's question why his *Exposition du Système du Monde* (1796) contained not a single reference to God.

4. John Tillotson (1630–1694). See *The Works of John Tillotson, Late Archbishop of Canterbury* (London: Printed by J. F. Dove for Richard Priestly, 1820), 1:346. Also in more popular venues, such as Robert Chambers's two-volume *Cyclopedia of English Literature* (Boston: Gould, Kendall, and Lincoln, 1847), 1:434–37; quotation on 1:436.

5. Tillotson, *The Works of John Tillotson,* I:347.

6. August De Morgan, *Formal Logic* (London, Taylor and Walton, 1847), 39, and "On the Syllogism no. V, and on Various Points of the Onymatic System," *Transactions of the Cambridge Philosophical Society* 10 (1864): 428–87, see esp. 456 and 461–62.

7. *A Brief Biographical Dictionary.* Compiled and arranged by Charles Hole, with additions and corrections by William A. Wheeler (New York: Hurd and Huntington, 1866). The same example is found in "A Theory of Probable Inference" (W4:435).

8. In "A Theory of Probable Inference" Peirce draws the following lesson from the example (W4:435): "Yet there is not the smallest reason to believe that the next poet's age would possess these characters."

9. John Stuart Mill, *A System of Logic, Ratiocinative and Inductive: Being a Connected View of the Principles of Evidence, and of Scientific Investigation,* 6th ed. (London: Longmans, 1865). See esp. III.iii.1.

10. *A System of Logic* III.iii.2–3. Peirce had discussed Mill's *Logic* before in his fourth Lowell Lecture of 1866 (W1, item 44), including Mill's justification of induction in terms of the principle of the uniformity of nature (W1:415–23).

11. *The Collected Works of John Stuart Mill, Volume VII – A System of Logic Ratiocinative and Inductive* [1843], edited by John M. Robson (Toronto: University of Toronto Press, 1974) contains a complete account of the alterations of Mill in subsequent editions of the *Logic.*

12. George Berkeley, *An Essay Towards a New Theory of Vision* (1709), §§2–46.

13. This is Isaac Newton's law of universal gravitation, which, in Newton's view, was derived from empirical observations by induction (*Principia Mathematica* [1687], Book 3, General Scholium).

14. Rather than conjecturing that reason somehow exists independently of the universe, and that the universe is subsequently shaped in its image, Peirce later

argues that the reasonableness of the universe is a natural product of evolution itself. That is to say, without any preset plan or external supervision, the universe taken by itself in its evolutionary development becomes more and more guided by what we call reason. Since we too emerge within that universe, Peirce maintains that this reason not only governs the outer world we call nature, but also the inner world we call our mind. And because the reason, or reasonableness, that we call ours reflects the reasonableness of the world, we are able use our reason to understand the world (CP 5.604, 1903). Peirce speaks in this context also of *concrete reasonableness* (CP 2.34, 1902).

15. Application of the pragmatic maxim shows the concept of an "absolutely incognizable existence" to be without meaning. See "How to Make Our Ideas Clear," sect. 2. Earlier, in "Some Consequences of Four Incapacities," Peirce argues that we can have no conception of the absolutely incognizable (W2:240).

16. For later discussions, see "Design and Chance" (W4, sel. 79), "A Guess at the Riddle" (W6, sel. 22–28), and the *Monist* metaphysical papers of the early 1890s (W8, sel. 22–30).

17. The nebular hypothesis was proposed by Emanuel Swedenborg in 1734 and was further developed by Kant in 1755 and by Laplace in 1796. Peirce's father, Benjamin Peirce, discusses the hypothesis extensively in *Ideality in the Physical Sciences* (Boston: Little, Brown & Co., 1881).

18. On several occasions Peirce refers to John Tyndall's prayer test, mistakenly believing it to be a proposal to empirically test the efficacy of praying for rain. Tyndall, a physician, instead proposed a method for testing the efficacy of praying in sick wards. See his "Prayer for the Sick: Hints toward a Serious Attempt to Estimate Its Value," *London Contemporary Review* 20 (July 1872): 205–10. The command for the sun to stand still to help in battle is found in *Joshua* 10:12–13. Peirce later uses this to cast doubt on the a priori claim that the laws of nature must be absolutely exact (e.g. W6:204).

19. Peirce's opposition to fixed beliefs, or dogmas, in religion is a recurrent theme in his work, especially in the early 1890s when he got involved with the Open Court Publishing Company, which considered it its mission to reconcile science and religion. See, for instance, "*[*Morality and Church Creed*]*" (W8, sel. 38), "What Is the Religion of Science?" (which Paul Carus retitled "The Marriage of Religion and Science"; W9, sel. 50), and "What is Christian Faith?" (W9, sel. 52).

Chapter 6

DEDUCTION, INDUCTION, AND HYPOTHESIS

In this sixth and final article that appeared in the Illustrations series, Peirce discussed the three kinds of reasoning that can be extracted from the general form of the syllogism, and called them: deduction, induction, and hypothesis. Peirce devoted special attention to the third (which elsewhere he also refers to as retroduction and abduction), using examples from the history of science to show how this kind of reasoning differs from the other two, as traditionally only deduction and induction were distinguished.

The text is as it originally appeared in Popular Science Monthly *13 (August 1878): 470–82. It was written in April and May of 1878. The text is only minimally edited and italic brackets surround any text added by the editor. The article is listed as the thirteenth essay in* Search for a Method *(R 1582:2) and is not included in* How to Reason. *The paper was to be the second essay in Peirce's 1910* Essays on Reasoning*—the first essay combining "The Fixation of Belief" and "How to Make Our Ideas Clear." R 672 (1910), entitled, "On the Essence of Reasoning and of its Chief Varieties" contains the aborted beginning of this second essay. (The surviving material for the first essay is found in R 334 and is here reproduced in footnotes to "The Fixation of Belief" and "How to Make Our ideas Clear.") Unfortunately, Peirce never got very far when revising the sixth article for republication. Of the 1910 attempt only six sheets survive on which Peirce gives some examples that show that at least some animals possess* *the faculty of reason—a view that he thinks conforms to his doctrine of synechism. A more detailed discussion of the history of the sixth paper is found in the introduction.*

I.

The chief business of the logician is to classify arguments; for all testing clearly depends on classification.[1,2] The classes of the logicians are defined by certain typical forms called syllogisms. For example, the syllogism called *Barbara* is as follows:[3]

S is M; M is P:
Hence, S is P.

Or, to put words for letters—

Enoch and Elijah were men; all men die:
Hence, Enoch and Elijah must have died.

The "is P" of the logicians stands for any verb, active or neuter. It is capable of strict proof (with which, however, I will not trouble the reader) that all arguments whatever can be put into this form; but only under the condition that the *is* shall mean "*is* for the purposes of the argument" or "is represented by." Thus, an induction will appear in this form something like this:

These beans are two-thirds white;
But, the beans in this bag are (represented by) these beans;
∴ The beans in the bag are two-thirds white.

But, because all inference may be reduced in some way to *Barbara,* it does not follow that this is the most appropriate form in which to represent every kind of inference. On the contrary, to show the distinctive characters of different sorts of inference, they must clearly be exhibited in different forms peculiar to each. *Barbara* particularly typifies deductive reasoning; and so long as the *is* is taken literally, no inductive reasoning can be put into this form. *Barbara* is, in fact, nothing but the application of a rule. The so-called major premise lays down this rule; as, for example, *All men are mortal.* The other or minor premise states a case under the rule; as, *Enoch was a man.* The conclusion applies the rule to the case and states the result: *Enoch is mortal.* All deduction is of this character; it is merely the application of general rules to particular cases. Sometimes this is not very evident, as in the following:

All quadrangles are figures,
But no triangle is a quadrangle;
Therefore, some figures are not triangles.

But here the reasoning is really this:

Rule.—Every quadrangle is other than a triangle.
Case.—Some figures are quadrangles.
Result.—Some figures are not triangles.

Inductive or synthetic reasoning, being something more than the mere application of a general rule to a particular case, can never be reduced to this form.

If, from a bag of beans of which we know that $2/3$ are white, we take one at random, it is a deductive inference that this bean is probably white, the probability being $2/3$. We have, in effect, the following syllogism:

Rule.—The beans in this bag are $2/3$ white.
Case.—This bean has been drawn in such a way that in the long run the relative number of white beans so drawn would be equal to the relative number in the bag.
Result.—This bean has been drawn in such a way that in the long run it would turn out white $2/3$ of the time.

If instead of drawing one bean we draw a handful at random and conclude that about $2/3$ of the handful are probably white, the reasoning is of the same sort. If, however, not knowing what proportion of white beans there are in the bag, we draw a handful at random and, finding $2/3$ of the beans in the handful white, conclude that about $2/3$ of those in the bag are white, we are rowing up the current of deductive sequence, and are concluding a rule from the observation of a result in a certain case. This is particularly clear when all the handful turn out one color. The induction then is:

These beans were in this bag.
These beans are white.
∴ All the beans in the bag were white.

Which is but an inversion of the deductive syllogism.

Rule.—All the beans in the bag were white.
Case.—These beans were in the bag.
Result.—These beans are white.

So that induction is the inference of the *rule* from the *case* and *result.*

But this is not the only way of inverting a deductive syllogism so as to produce a synthetic inference. Suppose I enter a room and there find a number of bags, containing different kinds of beans. On the table there is a handful of white beans; and, after some searching, I find one of the bags contains white beans only. I at once infer as a probability, or as a fair guess, that this handful was taken out of that bag.[5] This sort of inference is called *making an hypothesis.* It is the inference of a *case* from a *rule* and *result.* We have, then—

DEDUCTION.

Rule.—All the beans from this bag are white.
Case.—These beans are from this bag.
∴ *Result.*—These beans are white.

INDUCTION.

Case.—These beans are from this bag.
Result.—These beans are white.
∴ *Rule.*—All the beans from this bag are white.

HYPOTHESIS.

Rule.—All the beans from this bag are white.
Result.—These beans are white.
∴ *Case.*—These beans are from this bag.

We, accordingly, classify all inference as follows:

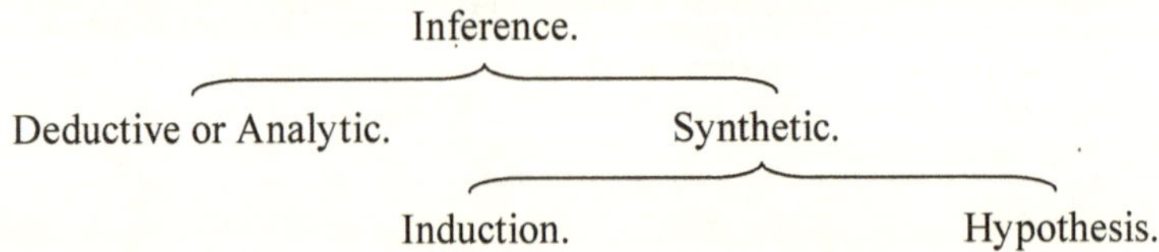

Induction is where we generalize from a number of cases of which something is true, and infer that the same thing is true of a whole class. Or, where we find a certain thing to be true of a certain proportion of cases and infer that it is true of the same proportion of the whole class. Hypothesis is where we find some very curious circumstance, which would be explained by the supposition that it was a case of a certain general rule, and thereupon adopt that supposition.[6] Or, where we find that in certain respects two objects have a strong resemblance, and infer that they resemble one another strongly in other respects.

I once landed at a seaport in a Turkish province; and, as I was walking up to the house which I was to visit, I met a man upon horseback, surrounded by four horsemen holding a canopy over his head. As the governor of the province was the only personage I could think of who would be so greatly honored, I inferred that this was he. This was an hypothesis.[7]

Fossils are found; say, remains like those of fishes, but far in the interior of the country. To explain the phenomenon, we suppose the sea once washed over this land. This is another hypothesis.

Numberless documents and monuments refer to a conqueror called Napoleon Bonaparte. Though we have not seen the man, yet we cannot explain what we have seen, namely, all these documents and monuments, without supposing that he really existed. Hypothesis again.[8]

As a general rule, hypothesis is a weak kind of argument. It often inclines our judgment so slightly toward its conclusion that we cannot say that we believe the latter to be true; we only surmise that it may be so. But there is no difference except one of degree between such an inference and that by which we are led to believe that we remember the occurrences of yesterday from our feeling as if we did so.

II.

Besides the way just pointed out of inverting a deductive syllogism to produce an induction or hypothesis, there is another. If from the truth of a certain premise the truth of a certain conclusion would necessarily follow, then from the falsity of the conclusion the falsity of the premise would follow. Thus, take the following syllogism in *Barbara:*

Rule.—All men are mortal.
Case.—Enoch and Elijah were men.
∴ *Result.*—Enoch and Elijah were mortal.

Now, a person who denies this result may admit the rule, and, in that case, he must deny the case. Thus:

Denial of Result.—Enoch and Elijah were not mortal.
Rule.—All men are mortal.
∴ *Denial of Case.*—Enoch and Elijah were not men.

This kind of syllogism is called *Baroco,* which is the typical mood of the second figure. On the other hand, the person who denies the result may admit the case, and in that case he must deny the rule. Thus:

Denial of the Result.—Enoch and Elijah were not mortal.
Case.—Enoch and Elijah were men.
∴ *Denial of the Rule.*—Some men are not mortal.

This kind of syllogism is called *Bocardo,* which is the typical mood of the third figure.

Baroco and *Bocardo* are, of course, deductive syllogisms; but of a very peculiar kind. They are called by logicians indirect moods, because they need some transformation to appear as the application of a rule to a particular case. But if, instead of setting out as we have here done with a necessary deduction in *Barbara,* we take a probable deduction of similar form, the indirect moods which we shall obtain will be—

Corresponding to *Baroco,* an hypothesis;
and, Corresponding to *Bocardo,* an induction.

For example, let us begin with this probable deduction in *Barbara:*

Rule.—Most of the beans in this bag are white.
Case.—This handful of beans are from this bag.
∴ *Result.*—Probably, most of this handful of beans are white.

Now, deny the result, but accept the rule:

Denial of Result.—Few beans of this handful are white.
Rule.—Most beans in this bag are white.
∴ *Denial of Case.*—Probably, these beans were taken from another bag.

This is an hypothetical inference. Next, deny the result, but accept the case:

Denial of Result.—Few beans of this handful are white.
Case.—These beans came from this bag.
∴ *Denial of Rule.*—Probably, few beans in the bag are white.

This is an induction.

The relation thus exhibited between synthetic and deductive reasoning is not without its importance.[8] When we adopt a certain hypothesis, it is not alone because it will explain the observed facts, but also because the contrary hypothesis would probably lead to results contrary to those observed. So, when we make an induction, it is drawn not only because it explains the distribution of characters in the sample, but also because a different rule would probably have led to the sample being other than it is.

But the advantage of this way of considering the subject might easily be overrated. An induction is really the inference of a rule, and to consider it as the denial of a rule is an artificial conception, only admissible because, when statistical or proportional propositions are considered as rules, the denial of a rule is itself a rule. So, an hypothesis is really a subsumption of a case under a class and not the denial of it, except for this, that to deny a subsumption under one class is to admit a subsumption under another.

Bocardo may be considered as an induction, so timid as to lose its amplificative character entirely. Enoch and Elijah are specimens of a certain kind of men. All that kind of men are shown by these instances to be immortal. But instead of boldly concluding that all very pious men, or all men favorites of the Almighty, etc., are immortal, we refrain from specifying the description of men, and rest in the merely explicative inference that *some* men are immortal. So *Baroco* might be considered as a very timid hypothesis. Enoch and Elijah are not mortal. Now, we might boldly suppose them to be gods or something of that sort, but instead of that we limit ourselves to the inference that they are of *some* nature different from that of man.

But, after all, there is an immense difference between the relation of *Baroco* and *Bocardo* to *Barbara* and that of Induction and Hypothesis to Deduction. *Baroco* and *Bocardo* are based upon the fact that if the truth of a conclusion necessarily follows from the truth of a premise, then the falsity of the premise follows from the falsity of the conclusion. This is always true. It is different when the inference is only probable. It by no means follows that, because the truth of a certain premise would render the truth of a conclusion probable, therefore the falsity of the conclusion renders the falsity of the premise probable. At least, this is only true, as we have seen in a former paper, when the word probable is used in one sense in the antecedent and in another in the consequent.[10]

III.

A certain anonymous writing is upon a torn piece of paper. It is suspected that the author is a certain person. His desk, to which only he has had access, is searched, and in it is found a piece of paper, the torn edge of which exactly fits, in all its irregularities, that of the paper in question. It is a fair hypothetic inference that the suspected man was actually the author. The ground of this inference evidently is that two torn pieces of paper are extremely unlikely to fit together by accident. Therefore, of a great number of inferences of this sort, but a very small proportion would be deceptive. The analogy of hypothesis with induction is so strong that some logicians have confounded them. Hypothesis has been called an induction of characters.[11] A number of characters belonging to a certain class are found in a certain object; whence it is inferred that all the characters of that class belong to the object in question. This certainly involves the same principle as induction; yet in a modified form. In the first place, characters are not susceptible of simple enumeration like objects; in the next place, characters run in categories. When we make an hypothesis like that about the piece of paper, we only examine a single line of characters, or perhaps two or three, and we take no specimen at all of others. If the hypothesis were nothing but an induction, all that we should be justified in concluding, in the example above, would be that the two pieces of paper which matched in such irregularities as have been examined would be found to match in other, say slighter, irregularities. The inference from the shape of the paper to its ownership is precisely what distinguishes hypothesis from induction, and makes it a bolder and more perilous step.

The same warnings that have been given against imagining that induction rests upon the uniformity of Nature might be repeated in regard to hypothesis.[12] Here, as there, such a theory not only utterly fails to account for the validity of the inference, but it also gives rise to methods of conducting it which are absolutely vicious. There are, no doubt, certain uniformities in Nature, the knowledge of which will fortify an hypothesis very much. For example, we suppose that iron, titanium, and other metals exist in the sun, because we find in the solar spectrum many lines coincident in position with those which these metals would produce; and this hypothesis is greatly strengthened by our knowledge of the remarkable distinctiveness of the particular line of characters observed. But such a fortification of hypothesis is of a deductive kind, and hypothesis may still be probable when such reënforcement is wanting.

There is no greater nor more frequent mistake in practical logic than to suppose that things which resemble one another strongly in some respects are any the more likely for that to be alike in others. That this is absolutely false, admits of rigid demonstration; but, inasmuch as the reasoning is somewhat severe and complicated (requiring, like all such reasoning, the use of A, B, C, etc., to set it forth), the reader would probably find it distasteful, and I omit it. An example, however, may illustrate the proposition: The comparative mythologists occupy themselves with finding points of resemblance between solar phenomena and the careers of the heroes of all sorts of traditional stories; and upon the basis of such resemblances they infer that these heroes are impersonations of the sun. If there be anything more in their reasonings, it has never been made clear to me. An ingenious logician, to show how futile all that is, wrote a little book, in which he pretended to prove, in the same manner, that Napoleon Bonaparte is only an impersonation of the sun. It was really wonderful to see how many points of resemblance he made out.[13] The truth is, that any two things resemble one another just as strongly as any two others, if recondite resemblances are admitted. But, in order that the process of making an hypothesis should lead to a probable result, the following rules must be followed:

1. The hypothesis should be distinctly put as a question, before making the observations which are to test its truth. In other words, we must try to see what the result of predictions from the hypothesis will be.

2. The respect in regard to which the resemblances are noted must be taken at random. We must not take a particular kind of predictions for which the hypothesis is known to be good.

3. The failures as well as the successes of the predictions must be honestly noted. The whole proceeding must be fair and unbiased.

Some persons fancy that bias and counter-bias are favorable to the extraction of truth—that hot and partisan debate is the way to investigate. This is the theory of our atrocious legal procedure. But Logic puts its heel upon this suggestion. It irrefragably demonstrates that knowledge can only be furthered by the real desire for it, and that the methods of obstinacy, of authority, and every mode of trying to reach a foregone conclusion, are absolutely of no value.[14] These things are proved. The reader is at liberty to think so or not as long as the proof is not set forth, or as long as he refrains from examining it. Just so, he can preserve, if he likes, his freedom of opinion in regard to the propositions of geometry; only, in that case, if he takes a fancy to read Euclid, he will do well to skip whatever he finds with A, B, C, etc., for, if he reads attentively that disagreeable matter, the freedom of his opinion about geometry may unhappily be lost forever.

How many people there are who are incapable of putting to their own consciences this question, "Do I want to know how the fact stands, or not?"

The rules which have thus far been laid down for induction and hypothesis are such as are absolutely essential. There are many other maxims expressing particular contrivances for making synthetic inferences strong, which are extremely valuable and should not be neglected. Such are, for example, Mr. Mill's four methods.[15] Nevertheless, in the total neglect of these, inductions are hypotheses may and sometimes do attain the greatest force.

IV.

Classifications in all cases perfectly satisfactory hardly exist. Even in regard to the great distinction between explicative and ampliative inferences, examples could be found which seem to lie upon the border between the two classes, and to partake in some respects of the characters of either. The same thing is true of the distinction between induction and hypothesis. In the main, it is broad and decided. By induction, we conclude that facts, similar to observed facts, are true in cases not examined. By hypothesis, we conclude the existence of a fact quite different from anything observed, from which, according to known laws, something observed would necessarily result. The former, is reasoning from particulars to the general law; the latter, from effect to cause. The former classifies, the latter explains. It is only in some special cases that there can be more than a momentary doubt to which category a given inference belongs. One exception is where we observe, not facts similar under similar circumstances, but facts different under different circumstances—the difference of the former having, however, a definite relation to the difference of the latter. Such inferences, which are really inductions, sometimes present nevertheless some indubitable resemblances to hypotheses.

Knowing that water expands by heat, we make a number of observations of the volume of a constant mass of water at different temperatures. The scrutiny of a few of these suggests a form of algebraical formula which will approximately express the relation of the volume to the temperature. It may be, for instance, that v being the relative volume, and t the temperature, the few observations examined indicate a relation of the form—

$$v = 1 + at + bt^2 + ct^3$$

Upon examining observations at other temperatures taken at random, this idea is confirmed; and we draw the inductive conclusion that all observations within the limits of temperature from which we have drawn our observations could equally be so satisfied. Having once ascertained that such a formula is possible, it is a mere affair of arithmetic to find the values of a, b, and c, which will make the formula satisfy the observations best. This is what physicists call an *empirical formula,* because it rests upon mere induction, and is not explained by any hypothesis.[16]

Such formulæ, though very useful as means of describing in general terms the results of observations, do not take any high rank among scientific discoveries. The induction which they embody, that expansion by heat (or whatever other phenomenon is referred to) takes place in a perfectly gradual manner without sudden leaps or innumerable fluctuations, although really important, attracts no attention, because it is what we naturally anticipate. But the defects of such expressions are very serious. In the first place, as long as the observations are subject to error, as all observations are, the formula cannot be expected to satisfy the observations exactly. But the discrepancies cannot be due solely to the errors of the observations, but must be partly owing to the error of the formula which has been deduced from erroneous observations. Moreover, we have no right to suppose that the real facts, if they could be had free from error, could be expressed by such a formula at all. They might, perhaps, be expressed by a similar formula with an infinite number of terms; but of what use would that be to us, since it would require an infinite number of coefficients to be written down? When one quantity varies with another, if the corresponding values are exactly known, it is a mere matter of mathematical ingenuity to find some way of expressing their relation in a simple manner. If one quantity is of one kind—say, a specific gravity—and the other of another kind—say, a temperature—we do not desire to find an expression for their relation which is wholly free from numerical constants, since if it were free from them when, say, specific gravity as compared with water, and temperature as expressed by the centigrade thermometer, were in question, numbers would have to be introduced when the scales of measurement were changed. We may, however, and do desire to find formulas expressing the relations of physical phenomena which shall contain no more arbitrary numbers than changes in the scales of measurement might require.

When a formula of this kind is discovered, it is no longer called an empirical formula, but a law of Nature; and is sooner or later made the basis of an hypothesis which is to explain it. These simple formulæ are not usually, if ever, exactly true, but they are none the less important for that; and the great triumph of the hypothesis comes when it explains not only the formula, but also the deviations from the formula. In the current language of the physicists, an hypothesis of this importance is called a theory, while the term hypothesis is restricted to suggestions which have little evidence in their favor. There is some justice in the contempt which clings to the word hypothesis. To think that we can strike out of our own minds a true preconception of how Nature acts, is a vain fancy. As Lord Bacon well says: "The subtlety of Nature far exceeds the subtlety of sense and intellect: so that these fine meditations, and speculations, and reasonings of men are a sort of insanity, only there is no one at hand to remark it."[16] The successful theories are not pure guesses, but are guided by reasons.

The kinetical theory of gases is a good example of this. This theory is intended to explain certain simple formulæ, the chief of which is called the law of Boyle.[17] It is, that if air or any other gas be placed in a cylinder with a piston, and if its volume be measured under the pressure of the atmosphere, say fifteen pounds on the square inch, and if then another fifteen pounds per square inch be placed on the piston, the gas will be compressed to one-half its bulk, and in similar inverse ratio for other pressures. The hypothesis which has been adopted to account for this law is that the molecules of a gas are small, solid particles at great distances from each other (relatively to their dimensions), and moving with great velocity, without sensible attractions or repulsions, until they happen to approach one another very closely. Admit this, and it follows that when a gas is under pressure what prevents it from collapsing is not the incompressibility of the separate molecules, which are under no pressure at all, since they do not touch, but the pounding of the molecules against the piston. The more the piston falls, and the more the gas is compressed, the nearer together the molecules will be; the greater number there will be at any moment within a given distance of the piston, the shorter the distance which any one will go before its course is changed by the influence of another, the greater number of new courses of each in a given time, and the oftener each, within a given distance of the piston, will strike it. This explains Boyle's law. The law is not exact; but the hypothesis does not lead us to it exactly. For, in the first place, if the molecules are large, they will strike each other oftener when their mean distances are diminished, and will consequently strike the piston oftener, and will produce more pressure upon it. On the other hand, if the molecules have an attraction for one another, they will remain for a sensible time within one another's influence, and consequently they will not strike the wall so often as they otherwise would, and the pressure will be less increased by compression.

When the kinetical theory of gases was first proposed by Daniel Bernoulli, in 1738, it rested only on the law of Boyle, and was therefore pure hypothesis.[18] It was accordingly quite naturally and deservedly neglected.[19] But, at present, the theory presents quite another aspect; for, not to speak of the considerable number of observed facts of different kinds with which it has been brought into relation, it is supported by the mechanical theory of heat.[20] That bringing together bodies which attract one another, or separating bodies which repel one another, when sensible motion is not produced nor destroyed, is always accompanied by the evolution of heat, is little more than an induction. Now, it has been shown by experiment that, when a gas is allowed to expand without doing work, a very small amount of heat disappears. This proves that the particles of the gas attract one another slightly, and but very slightly. It follows that, when a gas is under pressure, what prevents it from collapsing is not any repulsion between the particles, since there is none. Now, there are only two modes of force known to us, force of position or attractions and repulsions, and force of motion. Since, therefore, it is not the force of position which gives a gas its expansive force, it must be the force of motion. In this point of view, the kinetical theory of gases appears as a deduction from the mechanical theory of heat. It is to be observed, however, that it supposes the same law of mechanics (that there are only those two modes of force) which holds in regard to bodies such as we can see and examine, to hold also for what are very different, the molecules of bodies. Such a supposition has but a slender support from induction. Our belief in it is greatly strengthened by its connection with the law of Boyle, and it is, therefore, to be considered as an hypothetical inference. Yet it must be admitted that the kinetical theory of gases would deserve little credence if it had not been connected with the principles of mechanics.

The great difference between induction and hypothesis is, that the former infers the existence of phenomena such as we have observed in cases which are similar, while hypothesis supposes something of a different kind from what we have directly observed, and frequently something which it would be impossible for us to observe directly. Accordingly, when we stretch an induction quite beyond the limits of our observation, the inference partakes of the nature of hypothesis. It would be absurd to say that we have no inductive warrant for a generalization extending a little beyond the limits of experience, and there is no line to be drawn beyond which we cannot push our inference; only it becomes weaker the further it is pushed. Yet, if an induction be pushed very far, we cannot give it much credence unless we find that such an extension explains some fact which we can and do observe. Here, then, we have a kind of mixture of induction and hypothesis supporting one another; and of this kind are most of the theories of physics.

V.

That synthetic inferences may be divided into induction and hypothesis in the manner here proposed,[*21] admits of no question. The utility and value of the distinction are to be tested by their applications.

Induction is, plainly, a much stronger kind of inference than hypothesis; and this is the first reason for distinguishing between them. Hypotheses are sometimes regarded as provisional resorts, which in the progress of science are to be replaced by inductions. But this is a false view of the subject. Hypothetic reasoning infers very frequently a fact not capable of direct observation. It is an hypothesis that Napoleon Bonaparte once existed. How

---

*[Original Peirce footnote:] This division was first made in a course of lectures by the author before the Lowell Institute, Boston, in 1866, and was printed in the *Proceedings of the American Academy of Arts and Sciences*, for April 9, 1867.[22]

is that hypothesis ever to be replaced by an induction? It may be said that from the premise that such facts as we have observed are as they would be if Napoleon existed, we are to infer by induction that *all* facts that are hereafter to be observed will be of the same character. There is no doubt that every hypothetic inference may be distorted into the appearance of an induction in this way. But the essence of an induction is that it infers from one set of facts another set of similar facts, whereas hypothesis infers from facts of one kind to facts of another. Now, the facts which serve as grounds for our belief in the historic reality of Napoleon are not by any means necessarily the only kind of facts which are explained by his existence. It may be that, at the time of his career, events were being recorded in some way not now dreamed of, that some ingenious creature on a neighboring planet was photographing the earth, and that these pictures on a sufficiently large scale may some time come into our possession, or that some mirror upon a distant star will, when the light reaches it, reflect the whole story back to earth. Never mind how improbable these suppositions are; everything which happens is infinitely improbable. I am not saying that *these* things are likely to occur, but that *some* effect of Napoleon's existence which now seems impossible is certain nevertheless to be brought about. The hypothesis asserts that such facts, when they do occur, will be of a nature to confirm, and not to refute, the existence of the man. We have, in the impossibility of inductively inferring hypothetical conclusions, a second reason for distinguishing between the two kinds of inference.

A third merit of the distinction is, that it is associated with an important psychological or rather physiological difference in the mode of apprehending facts. Induction infers a rule. Now, the belief of a rule is a habit. That a habit is a rule active in us, is evident. That every belief is of the nature of a habit, in so far as it is of a general character, has been shown in the earlier papers of this series. Induction, therefore, is the logical formula which expresses the physiological process of formation of a habit. Hypothesis substitutes, for a complicated tangle of predicates attached to one subject, a single conception. Now, there is a peculiar sensation belonging to the act of thinking that each of these predicates inheres in the subject. In hypothetic inference this complicated feeling so produced is replaced by a single feeling of greater intensity, that belonging to the act of thinking the hypothetic conclusion. Now, when our nervous system is excited in a complicated way, there being a relation between the elements of the excitation, the result is a single harmonious disturbance which I call an emotion. Thus, the various sounds made by the instruments of an orchestra strike upon the ear, and the result is a peculiar musical emotion, quite distinct from the sounds themselves. This emotion is essentially the same thing as an hypothetic inference, and every hypothetic inference involves the formation of such an emotion.[22] We may say, therefore, that hypothesis produces the *sensuous* element of thought, and induction the *habitual* element. As for deduction, which adds nothing to the premises, but only out of the various facts represented in the premises selects one and brings the attention down to it, this may be considered as the logical formula for paying attention, which is the *volitional* element of thought, and corresponds to nervous discharge in the sphere of physiology.

Another merit of the distinction between induction and hypothesis is, that it leads to a very natural classification of the sciences and of the minds which prosecute them.[24] What must separate different kinds of scientific men more than anything else are the differences of their *techniques*. We cannot expect men who work with books chiefly to have much in common with men whose lives are passed in laboratories. But, after differences of this kind, the next most important are differences in the modes of reasoning. Of the natural sciences, we have, first, the classificatory sciences, which are purely inductive—systematic botany and zoölogy, mineralogy, and chemistry. Then, we have sciences of theory, as above explained—astronomy, pure physics, etc. Then, we have sciences of hypothesis—geology, biology, etc.

There are many other advantages of the distinction in question which I shall leave the reader to find out by experience. If he will only take the custom of considering whether a given inference belongs to one or other of the two forms of synthetic inference given on page 170, I can promise him that he will find his advantage in it, in various ways.

NOTES

1. Compare Peirce's comments on chemistry as a classificatory science in the opening paragraph of "The Doctrine of Chances" (Chapter 3 above).

2. See also the opening paragraph of "The Doctrine of Chances," with notes.

3. Barbara belongs to a set of mnemonic names for the valid forms of syllogism established by the early thirteenth century. Syllogisms with names that begin with B, including Baroco and Bocardo (see below), reduce to Barbara.

4. In "On the Natural Classification of Arguments" of 1867 Peirce shows that "the general formula of all argument" is a syllogism in Barbara (W2:27–28). For a discussion of Peirce's argument, see Wilfrid Hodges, "The Scope and Limits of Logic," in Dale Jacquette, ed., *Philosophy of Logic* (Amsterdam: Elsevier, 2007), 44–46.

5. In 1902 Peirce observes (CP 2.102):

> I was too much taken up in considering syllogistic forms and the doctrine of logical extension and comprehension ... As long as I held that opinion, my conceptions of Abduction necessarily confused two different kinds of reasoning. When after repeated attempts, I finally succeeded in clearing the matter up, the fact shone out that probability proper had nothing to do with the validity of Abduction, unless in a doubly indirect manner.

6. This form of argument Peirce also terms *abduction* and *retroduction.* Some (e.g. Gilbert Harman, Peter Lipton) identify it with so-called "inference to the best explanation." See, however, Daniel G. Campos, "On the Distinction Between Peirce's Abduction and Lipton's Inference to the Best Explanation," *Synthese* 180.3 (2011): 419–42. For a more detailed discussion of abduction, see EP2, sel. 16 "Pragmatism as the Logic of Abduction" (1903). See also chapter 8, where Peirce discusses practical and scientific retroduction.

7. Peirce describes this encounter in a fictional tale titled "Embroidered Thessaly" (W8:299), a tale that was inspired by his own travels in the area. See W8:455–56 for historical details.

8. For a more detailed account of this, see Peirce's 1901 "On the Logic of Drawing History from Ancient Documents, Especially from Testimonies" (HP-PLS 2:705–62; partially included in EP2 as sel. 8). The three examples—which are inferences from effect to cause, thus involving a reasoning backwards in time—exemplify why Peirce later changed the name for this type of inference into retroduction.

9. See Peirce's discussion of analytic versus synthetic reasoning in "The Probability of Induction," sect. V.

10. See "The Principle of Induction," sect. I.

11. Peirce maintains the same view in "A Theory of Probable Inference" (W4:419). After the turn of the century Peirce rejects this view, classifying "induction of characters" as a type of induction.

12. Peirce argues against this way of justifying induction in "The Order of Nature," sect. III (chapter 5 above).

13. In 1827 the French physicist Jean-Baptiste Pérès (1752–1840), publishes a satirical pamphlet, *Comme quoi Napoléon n'a jamais existé ou Grand Erratum* (1827), in which he shows that the same sort of arguments that purportedly prove the Scriptures to be merely mythical in origin, can be used to show that Napoleon never existed but is only an allegorical figure—a personification of the sun. See Albert Sonnenfeld, "Napoleon as Sun Myth," *Yale French Studies* 26 (1960): 32–36.

14. These are the first methods of fixing belief (see "The Fixation of Belief," sect. V). The sentiment returns in Peirce's First Rule of Reason, "that in order to learn you must desire to learn, and in so desiring not be satisfied with what you already incline to think," with the corollary that follows from it: "Do not block the way of inquiry" (RLT:178).

15. These are the methods of agreement, of difference, of residues, and of concomitant variation. See *Logic,* vol. 1, Bk. 3, Ch. 8.

16. As Peirce explains in the *Century Dictionary* (CD:3375): "A mere empirical formula which satisfies a series of observations sufficiently, but would not hold in extreme cases, is not considered as a law."

17. Francis Bacon, *Novum Organum: Or True Suggestions for the Interpretation of Nature* (London: Longman, 1858), Bk. 1, Aphorism 10.

18. In the *Century Dictionary* Peirce defines this law as follows (CD:3376): "at any given temperature the volume of a given mass of gas varies inversely as the pressure which it bears."

19. See Daniel Bernoulli's *Hydrodynamica, sive de viribus et motibus fluidorum commentarii* (Argentorati: J.R Dulseckeri, 1738). In this work, which marks the origin of the modern kinetic theory of gasses, Bernoulli derives Boyle's law by calculating the force exerted on a movable piston by the impacts of *n* particles moving with speed *v* in a closed space of total volume V.

20. Bernoulli conjectured that a gas consists of a large multitude of minute particles that move rapidly in a random pattern. This view conflicted with the opinion generally held at the time, most prominently by Newton that the particles stay more or less in place, repelling each other from a distance. Newton further showed that if this repulsion followed the inverse-square law, the product of pressure and volume (PV) would be constant.

21. Peirce is here referring to the then recent work of Rudolf Clausius, James Clerk Maxwell, and Ludwig Boltzmann, which culminates in the Maxwell-Boltzmann distribution. See esp. Maxwell's "Illustrations of the Dynamical Theory of Gases," *Philosophical Magazine* fourth series 19.124 (1860): 19–32, 20.130 (1860): 21–37.

22. What remains of the 1866 Lowell lectures is found in W1 sel. 41–51. The reference is most directly to item 51, a privately printed pamphlet designed to accompany the lectures. The 1867 paper is "On the Natural Classification of Arguments" (W2, sel. 3).

23. Later, Peirce argues that perception operates in accordance with the logic of hypothetic inference, or abduction—albeit that it is instinctive rather than voluntary, and unconscious rather than conscious. That is to say, in perception considered as abduction only the conclusion enters the realm of our deliberative understanding. Peirce calls such a conclusion a *perceptual judgment*, and whenever we reflect on what we did perceive (the *percept* as he calls it), we take the perceptual judgment, such as "this line is longer than that one," "the glass in the top window is broken, or "I hear a train in the distance," as testimony of what we did perceive (CP 7.643, 1903). For a discussion, see e.g. Susan Haack "How the Critical Common-sensist Sees Things." *Histoire, Épistémologie, Langage* 16.1 (1994): 9–33, and Sandra B. Rosenthal "Peirce's Pragmatic Account of Perception: Issues and Implications," in Cheryl Misak, ed., *The Cambridge Companion to Peirce* (Cambridge: Cambridge University Press, 2004).

24. Critical of the many attempts of giving an a priori classification of the sciences, Peirce calls for an empirical classification based on what is actually taking place in terms of living scientific activity. See esp. "On Science and Natural Classes" of 1902 (EP2, sel. 9), and "An Outline Classification of the Sciences" of 1903 (EP2, sel. 18). See also Beverly Kent, *Logic and the Classification of the Sciences* (Montreal: McGill-Queen's University Press, 1987).

Chapter 7

ESSAYS TOWARD THE INTERPRETATION OF OUR THOUGHTS

[My Pragmatism,]

[Set forth in Two Chapters.]*

The first manuscript sheet suggests that the idea of adding a preface emerged at this point. Initially, Peirce aimed to launch directly into "The Fixation of Belief," as he did in R 620:53–56, but then he crossed out the title and chapter designation, and began with a preface instead. The preface is set off in square brackets. Peirce also indicated that the titles "Essays Toward the Interpretation of our Thoughts" and "My Pragmatism" are only provisional titles to the volume and the essay.

The essay begins with a rather longish historical account of the Metaphysical Club and its cast of characters, as well as Peirce's encounter with Appleton, the publisher of Popular Science Monthly. *When revising "The Fixation of Belief," of which two versions are reproduced below, Peirce spends much time explaining the importance of good reasoning and that*

*A four-page earlier branch covering the same terrain survives as R 620:53–56. Rather than calling it a preface, Peirce opens that branch with "[Chapter I.] The Fixation of Belief," which is preceded by the Bible verse "By their Fruits ye shall know them" (Matt 7:16).

the ability to reason well is a trait that can only be acquired with great difficulty. Peirce further separates logic from psychology, arguing that logic is a normative science and that good reasoning requires self-control. Peirce next discusses, and criticizes, the views of John Stuart Mill, which lands him into a discussion of the nominalism-realism debate with references to the Scotists and the Ockhamists. In Version II Peirce adds a more extensive discussion of probability, discussing, among others, the views of Laplace.

The first run consists of sheets Peirce numbered 1–18 (R 620:2–19). The sheets, which contain a date and time stamp as well, were composed April 6–12, 1909. Several discarded branches of various lengths follow them. Two of these branches are here included. The first is a branch that Peirce began on May 12, 1909. The sheets are numbered 19–39 (R 620:20–44) and 42–51 (R 620:43–52). Pages 40–41 have not been recovered. Peirce discarded this branch twelve days later, on May 24. On that day he began working on the second branch included below, numbering the sheets 21–42 (R 621:2–37), 43–50 (R 623:2–10), 51–82 (R 625:2–34). This second branch breaks off with a sheet he dated June 24. Peirce seemingly stops working on this project, picking up the pieces again on August 1 with a brand new attempt to write a preface. This preface survives in R 631 and R 632. This later attempt, which is not included in this edition, is aborted on August 26 and does not reach beyond the attempt of writing a preface. On September 4 Peirce begins again anew. Much of that attempt, which carries to October 27, is reproduced below in chapter 8. Where significant insight might be gained into the thought process that went into the final wording of the text, Peirce's alterations are marked in footnotes. Interline additions that are not accompanied by a deletion of text are not normally identified in notes. Alterations are given only where they are thought to be significant and related to the central topic of the chapter.

[PREFACE.[*] The main part of this Essay,—the characterization of Belief and of Doubt, the argument as to the effective[†] aim of inquiry,[‡] the description of four methods directed toward that aim, with the criticisms of them, the discussion of the proper function of thinking, and the consequent maxim for attaining clear concepts,—reproduces almost verbatim a paper I read,—it must have been in 1872,—to a group of young men who used, at the time, to

[*] [PREFACE] *after deleted:* [CHAPTER I.] The Fixation of B

[†] effective] *R 620:53 reads* practical

[‡] as to the ... inquiry,] *appears in R 620:53 above deleted:* that the aim of Reasoning is really nothing but the fixation of belief,

meet once a fortnight in Cambridge, Mass., under the name of 'The Metaphysical Club,'—a name chosen to alienate such as it would alienate.[1] Its constitution was likewise effective, consisting in a single clause forbidding any action by the Club as a collective body. This saved it from wasting the only intrinsically precious element in the world as many societies waste it, in the idle frivolity they call 'business.' Besides, since without action there could be no officers and, in particular, no secretary nor any acknowledged record of debate, any gentlemen who might be desirous of distinguishing themselves or of, as it were, taking out patents to protect their rights in such philosophical ideas as they might invent, would be provided with a sufficient motive for holding their peace, and abandoning the Delta (or playground) of discussion to those who only sought to draw as near to the truth as they could. It proved quite the most successfully organized body of students for genuine educative efficiency, in contradistinction to saw-dust-stuffing, that ever I had the good fortune to be placed in;—a model worth a few pages of description. The member whom we all of us acknowledged as the ripest student of philosophy among us,—or rather, all but two of our number did so,—was Chauncey Wright, then 42 years old, who had been a student of mathematics under my father, and I believe under Jeffries Wyman of comparative anatomy.[2] I am sure he was still a student of Dr. Asa Gray in botany.[3] He was a bachelor and eked out a small competence by computing for the *American Ephemeris*.[4] When I had first known him, in 1857 or 8 (he was graduated A.B. in 1852,) he was a follower of Hamilton in philosophy; but he had been greatly impressed by the *Philosophie Positive* of Comte,[5] and later had adopted with great ardour and studied in utmost detail the philosophy and logic of J.S. Mill. Later, but before that conversion was quite complete, he had become a thorough Darwinian. I do not know how promptly he accepted the doctrine 'Origin of Species,' because from before its appearance until the following summer I was in a remote wilderness,[6] and merely heard of the appearance of such a work that had created an immense sensation. I did not at all realize that the greatest mental awakening since Newton and Leibniz had begun. Against Spencer[7] and any belief in general evolution Wright's Comtism had armed him.[8] He used to liken the general eonic tendencies of the Universe to "weather," an utterance that, by the way, did not sound unlike Spencer himself, in his first manner.[9] But at the time I made Wright's acquaintance he seldom missed an opportunity to disparage Spencer, and made even the truly remarkable *Principles of Psychology* an object of derision. Although Wright did accept the *Origin of Species*, he did not at first realize at all the breadth of its meaning. He accepted it as disclosing no more than it professed to prove (or render probable,)* the origin of *species*, but not necessarily the origin of broader classes. He emphatically protested against its being understood as foreshadowing any general cosmogony. It was not until Darwin's other books, especially the one on Climbing Plants and the *Descent of Man*,[10] that something of the true significance of the new light was acknowledged by him; and even then he restricted the principle to biology. When Mill's *Examination of Hamilton* came out in the spring of 1865,[11] I put the volume into my portmanteau and betook myself alone to a sea-side hotel, long before the season had begun to open, in order that I might study it in solitude; and it influenced me decidedly, and helped me to clear up my opinions. But my mother's milk in philosophy had been the *Critik der reinen Vernunft*;[12] and by that time, I not only knew both editions almost by rote, and had pondered deeply every point, but I had compared them with the writings of other great philosophers, Spinoza, Leibniz, and English philosophical classics, and had devoted several years to the study of scholasticism, from which study I had risen more or less a scholastic realist though not yet the more than scholastic realist which riper studies have converted me. I did not then go further than the stand-point of Scotus,—a half-baked, apologetic, diffident realism that I have long since outgrown.[13] It was far enough, however, to save me from Mill's individualistic nominalism,—to enable me, indeed, to see that that doctrine is nothing but Metaphysics, in the most objectionable sense, a hypothesis not only quite without support, but in conflict with the indispensable postulate of all philosophy, all science, and all reasoning; namely, the Postulate that perceptual experience is intelligible. Wright, on the contrary, regarded the doctrine as the most certain truth in the world; and the effect upon him of the *Examination of Hamilton* was to complete the demolition of what little remained of his early Hamiltonianism. As to that he and I agreed pretty well; but we must have fought out nearly a thousand close disputations regular set-tos concerning the philosophy of Mill, perfectly dispassionate of course, before the Metaphysical Club had been started. These discussions had been of the greatest service to me; for he was ten years older than I, was well trained both in science (as I was, too,) and in philosophy, and an extremely exact thinker, as thinkers go. In the course of those years, my Kantism got whittled down to small dimensions. It was little more than a wire,—an iron wire, however. On the other hand, I found great pabulum in Berkeley, and good nutriment, too, in Spinoza, strange as that may sound to one who knows only his *Ethica*,—and has penetrated no more than the rind of that most misinterpreted of books.[14]

**Parenthetical remark added interline.*

Wright had no great gift for expression, but he insisted upon exactitude in the use of words, which I lacked. He wrote a generally intelligible style, workmanlike and meeting the expectations of readers of the Quarterlies. He did not think in words, but in diagrams and arrays, like a mathematician, and in recollections of real things modified by imagination, like a man of physical science. I may mention that after his death (as a bachelor) in 1875, a volume of his writings was published with a memoir, under the editorship* of Charles Elliot Norton.[15]

Another member of our 'Metaphysical Club' who directly influenced the present essay, as I am not aware that Wright did, was a profound lawyer, a most genial man, and a great admirer of Alexander Bain.[16] Nicholas St John Green was his name. His exact age I cannot tell, but he was graduated the year before Wright, that is to say, in 1851.[17] He made no pretensions as a student of philosophy, but I heartily admired the style in which his powerful and matured good sense got to the kernels of the different questions. In those days, the courts often made the decisions of living men's real rights and interests to hang upon the emptiest scholastic distinctions to a degree hardly credible to those whose recollections do not go back so far; and Green's interpretation of the true Law of Contributory Negligence, which looks straight at the facts of human life, still seems to me admirable.[18] It was surely sound law if the law is to be defined as the perfection of human reason. Green's interpretation freed it from nonsensical distinctions between immediate and remote, direct and indirect causation; and it made a deeper and more lasting impression on me than it would have done but for my own efforts to separate the wheat from the chaff of the scholastic doctrines. His exposition of the working of the Criminal Law, with which he was intimately acquainted in consequence—well, in consequence, before aught else, of his being a veritably humane and human person, human and humanely human, humane and humanly humane; and then, specifically, in consequence of his having been, before the war, the partner and, as Dickens would say, the jackal, of that very able and notorious personage, Benjamin F. Butler, Esq., who, at one time, desirous of pushing some position of the Supreme Court of Massachusetts to absurdity, made it known that he would defend all criminal cases of a certain class without fees.[19] In this way, Mr. Green had become well acquainted with criminals, and with the working of the criminal law, which he regarded as so ill-adapted to any rational purpose that it would be hardly an exaggeration to pronounce it insane. In the first place, he thought that dread of legal punishment, as

* editorship] *after deleted:* distinguished

distinguished from dread of conviction, acted scarcely at all to deter persons from crime. He knew that it had not the slightest effect upon the criminal classes, further than to prevent a burglar from a wholly unnecessary murder; and when a man previously respectable is convicted of crime, the legal punishment may well be welcomed by him as putting him out of the way of suffering the far heavier social punishment. The only crimes from which dread of legal punishment might be expected to be deterrent he thought were those in which an individual of one class of society victimizes a member of another with which his own regards itself as being at war; such as blacks against whites, Hebrews against Gentiles, tenants against landlords, operatives against mill-owners, and the like; and in those cases (which were much rarer than now,) the known difficulty of conviction must minimize that dread. At any rate, the deterrent effect must be small. And how much does this little cost? He assured me that he had often found himself defending a man who had been previously convicted twenty or thirty times, and each time at the cost of the whole machinery of a court for some days. He said it would be far cheaper to maintain such a man all his life in what to him would be luxury,—not to speak of the losses of his victims, which were really far in access of any practical appraisal of them. He would warm up upon this topic, and exclaim, "I ask you this question,—this problem if you will: Why do we not treat the professional criminals, at least, as insane, and instead of punishing them, sequester them for life amid elevating surroundings, make them work, and prevent them from propagating their bread of insanity? It would cost the state no more, convictions would be easily obtained, and the vast losses of individuals would be altogether saved.[20] The professional criminal *is* insane; for the only practical definition of insanity is incapability to regulate one's conduct and affairs with ordinary sense. Now it would be a self-contradiction to suppose any man to be capable of acting contrary to his resultant impulse at the time; and it would be in flagrant conflict with everybody's personal experience, to suppose there is any man on earth that possesses the power, under every hurricane of temptation, of awakening in himself whatever impulse may picture itself before his imagination as fine, regardless of his settled habits. Such a capacity would be proof of a superhuman sanity far removed indeed from the poor thief's real weakness. Mr. Respectability knows this well enough; that is to say, he knows that our Criminal Law fails to protect society, and ought somehow to be reformed; but he himself purposes no Quixotic waste of time in endeavours to rationalize Society. Besides, since he does not expect himself to be robbed, he sees no reason for forgoing the gratification of his wrath against the criminal class. Yet he knows well enough that the thief's legal punishments are little more than a flea-bite to the thief himself, that it is only his innocent children and his women-folks, for whom he don't care much, upon whom the crushing misery falls; so that Mr. Respectability's calm acquiescence in the evil and impotent rage at the criminal is a bit of insanity on *his* part that, reciprocating with the counter-insanity of the professional thief, serves to keep Society steady in maintaining the Law in its venerable* barbarity, while paying its enormous annual tribute to preserve the criminality of the criminal classes." I have given this to convey some idea of the style of Green's reasoning, drawing what little I could from recollection and the rest from imagination undoubtedly more or less tinged by subconscious memories. I am sure I have heard him say something more or less like what I have set down, and I have done my best to imitate,—not at all the lucidity of his diction, but the structure and colors of his thought, in an imaginary restoration of his discourse, with a view to enabling the reader to figure to himself how Green's vein of reason would naturally have deflected mine into the channel of this essay. Meantime it must not be supposed that Green preached social generalities from the house-tops. It was only to cool and discreet heads that he would unfold such thoughts; and he limited his own endeavours to such items of reformed law as he thought he might induce the Court to accept. Undoubtedly he did contribute to the improvement of the Common Law as defined in Massachusetts. He died in 1876. Two other members of our little club have passed away. One of them had already published an extensive philosophical treatise which has reached a second edition. It was John Fiske, the author of *Outlines of Cosmic Philosophy*.[21] The other, Francis Ellingwood Abbot, only attended one meeting.[22] Perhaps he profited by that one, however. For I that evening argued in favour of scholastic realism, and endeavoured to show that it was identical with the belief of all modern physicists. For the doctrine of the scholastic realists was that *some*,—not *all*,—generals have an "objective" truth. Now, the main body of physicists hold it to be a real truth that every particle of matter has a component acceleration toward every other equal to a certain fixed quantity multiplied by the unchanging mass of that other,† and divided by the square of its distance from that other, and the physicist holds, not only that this happens to be the fact now, but that it will be so, unless some great revolution in the whole constitution of the universe should occur, and further, that it *would be so* under all circumstances that *might* occur, even if they do not. For he is continually

* venerable] *above deleted:* criminal

† other,] *before deleted:* (which is constant in all its reactions,)

reasoning on that basis, especially in disproving hypotheses.* Moreover, physicists hold most decidedly that this always would be the case, whatever men might think, and even if there were no men nor other living beings at all;—an opinion flatly contrary to that of the nominalistic metaphysicians, who make the laws of nature to be mere fabrications of the human mind. It was nearly half a generation later, in 1885, when, Abbot, in a book entitled *Scientific Theism*,[23] came out with the same opinion I had so often urged; and since this opinion is very unusual among philosophists, it seems to me quite possible that, although he had, no doubt, quite forgotten that meeting, seeds of the doctrine were that night sown in his mind, to blossom many years later. I must confess that I had, about that time, repeatedly argued to the same effect in print.[24] But had Abbot's attention been called to such a contention in print, he would have mentioned it in his own publication of the same opinion.

All the members of our little club have been more or less marked for their philosophical breadth ever since that time.† We may, at any rate, be permitted to exult that we had among us the foremost psychologist of the present time, Professor William James, member of the Institute of France, of the Royal Prussian Academy, of the Royal Danish Academy, and of the first of such bodies, the "Academia dei Lyncei,"[25] as well as Mr. Justice Holmes of the Supreme Court of the United States and formerly Chief Justice of Massachusetts, of whose profundity as a philosophical jurist I may not presume to say a word.[26]

Some years later, I was so fortunate as to meet, as a fellow passenger on a steamer to Liverpool, Mr. William H. Appleton, the eminent publisher, especially of books of scientific philosophy; who then engaged me, unknown as I was, to write half-a dozen articles for the *Popular Science Monthly* on the Logic of Science.[27] They were written and duly appeared under the inappropriate title (whom to propitiate I have forgotten) of *Illustrations of the Logic of Science*. My Metaphysical Club paper on Pragmatism almost without other alteration than amplifications and padding, served for my first two articles. One set of amplifications in the second article occupied considerable space and were highly important.‡ Namely, so far as my pragmatism is a doctrine, it is doctrine that the significance of any intellectual thought consists in the particular manner in which it

---

* *might* ... hypotheses.] *interline above deleted:* might occur and that the constancy of the mass of every particle is equally universal.

† time] *before deleted:* , "present company alone excepted,"

‡On a draft sheet dated April 10 (R 620:61–62), Peirce writes about the essay he presented to the Metaphysical Club:

tends, and will tend, to regulate the thinker's conduct. Be careful, honest reader, not to misunderstand this, as I myself misunderstood it, at a time when the subject had been absent from my attention for a number of years, when, in Professor Baldwin's *Dictionary of Psychology and Philosophy,* I represented myself as having formerly supposed that the significance of a concept lies in a *doing.*[28] I never said that, but that that significance lies in the *form,* not essentially of any actual doing, but of the doing so far as it shall be truly determined by the concept. Very well; my second magazine article, now, at last, is explicitly presented, as it was always intended, as the second chapter of an essay amplifying the paper read to the Club with illustrations of the maxim of clearness that had not been needed at first, on account of my having put forth sundry such interpretations at previous meetings, before I had undertaken to formulate their common motive. One of these illustrations plainly shows, I think, that, at the time I wrote the magazine article, I did not fully comprehend the "maxim of pragmatism," in spite of my having enunciated it correctly enough. In the present edition, I have inserted between heavy square brackets my present views on the subject. Owing to the tenets of Professor James which I think have tended toward this error of my own, I now call the theory of the "maxim" *Pragmaticism,* and leave my old term *Pragmatism* in the sense James has given to it, to denote the general tendency of thought in respect to which James, Schiller, Vailati, the writers in *Leonardo,* and I agree, in opposition to most metaphysicians.[29] Another amplification of the original paper, first introduced in the magazine article lies in its first section, where the pragmatic interpretation is treated as the highest of three grades of clearness of apprehension. I still think this to be correct, though I should now subdivide the second grade into two, and the third grade into three. I prefer, however, not to load down the present essay with complications, however important they be. I have, in this edition, omitted certain passages which to a good many readers would seem like padding,—and in a certain sense,

---

the essay was, so far as I know, the earliest attempt to formulate pragmatism, which, as a *practice*, had been, I think, best illustrated by Berkeley, especially in his two works on vision. The *word* "pragmatism," which I had invented for the *practice* that I had incessantly preached in the club, and *also* for my *formulation* of it (which I still believe will eventually be found accurately to define the essence of its vitality)—that word had been used in the paper read to the club; but I had struck it out in going to press. I did not even insert it in making up the list of philosophical terms for the *Century Dictionary.* For no great notice had been taken of the matter until James brought forward pragmatism, and in doing so, had substituted for my logical definition a broadly philosophical definition, easy to comprehend, although, according my persistent way of thinking, the broadly philosophical definition of concepts capable of definition from a logical point of view are shallow just in proportion as they are broad.

indeed, so they were, albeit they had served a definite purpose in aid of the conveyance of my own apprehension of the idea before my mind.[30]

[version I]

[CHAPTER I.] THE FIXATION OF BELIEF.

Few people care to study logic, because everybody conceives himself proficient enough in the art of reasoning, already. But this satisfaction, I notice, is limited to each one's own reasoning, and does not extend to that of other men.*†

---

*This is the opening of the "Fixation of Belief" as it originally appeared in *Popular Science Monthly.* The opening square bracket following it suggests that Peirce is still considering republishing the original article interspersed with additions in square backets.

†In a rejected branch, dated April 13 (R 620: 64–66), Peirce began the following insertion:

> [There are, indeed, a few of us who certainly do not conceive ourselves to be proficient enough already in logic, inasmuch as we are prosecuting research into some of its problems on the tenterhooks of doubt as to what our results will be. Each of us, after anxious consideration as to what he can hope to accomplish takes up a problem, and after acquainting himself, as far as possible, with all that has hitherto been discovered that bears upon it, plans, under the surest lights he can, a line of research in which every step taken must be taken, as in any other science, by attaining some new experience. Mathematics, for example, is just as much an experiential science as any other, although mathematical experiences are not experiences of external perception but are experiences of a certain class of imaginations connected by reasonings more or less as the experiences of any other science are connected. The logician's experiences though quite different from those of the mathematician, nevertheless resemble his in being experiences of imagination. The problem that has chiefly engrossed me for a number of years, might conceivably be completely solved by novel experiences of about 59 thousand kinds. But as I have no hope of having such a number of new experiences, I have been working toward a solution by another process by which less than a hundred novel experiences promises to furnish me with a decided clue; and so in case I shall be enabled to work steadily upon the matter, in two or in three years more I shall have a method of thinking ready for presentation to intelligent persons who have not occupied themselves with abstract thought by which they will be able,—not, indeed, to think about their daily business any better than they do at present, but rationally to satisfy themselves about some of the mysteries of life,—not in detail, yet in a manner that will be of practical advantage to them.

In a rejected branch, dated April 13 (R 620: 68), Peirce began the following insertion:

> [That "everybody" will, indeed, heartily endorse this observation with no trace of any egotistical gobbling of it as if it were meant for himself alone. On the contrary, he wishes others to have the benefit of it, especially those who obstinately disagree with him, and if he awards any share of it to himself, it will be with such a touching humility of pen or lip that even he himself must acknowledge it to be beautiful. *[...]*

[But why say it? It is a platitude that everybody that everybody preaches and nobody takes to heart. I use the word "everybody" here, as "everybody" uses it, but not as including the few conscientious reasoners who have rendered the scientific reasoning of today incomparably superior to that of fifty years ago, and who, in every age since Gerbert was believed by "everybody" to have sold his soul to the devil because he possessed the art now known as "vulgar arithmetic," have brought about a gradually increasing accuracy of thought.[31] It chanced[*] that just as I had copied for this edition the first two sentences of this chapter, and was pausing to make up my mind whether or no to add myself to the number of unheeded preachers of a truism, my eyes fell upon an open page of a recent periodical, and I was startled out of my reverie at finding myself reading the following words:

> Among men there is no habit more inveterate than the persuasion of each individual that he personally is immune from slips of reasoning. All around him during every day of his life he takes notice how badly other people reason, without ever saying to himself that probably he is[32†]

[*] chanced] happened

[†]In a rejected branch Peirce continues here as follows (R 620:69–72, April 13):

> he is like other people in the same regard." Thereupon he goes on to express a "misgiving" as to the non-Euclidean geometry. He means that those who have accepted that system of geometrical postulates as not involving any contradiction have not sufficiently taken into account the possibility of their having slipped somewhere in their reasoning. Considering that those persons count among their number *all* the eminent mathematicians now living, without a dissenting voice among them, though it is abstractly possible that they are all in the wrong, just as they may be in thinking that twice three amounts to six, and considering also that during the last fifty years or so, the whole subject has been pretty well threshed out, that possibility would seem to have no great probability. It does not enter into my purpose to criticize the writer's reasoning. He seems not to be understandingly acquainted either with Cayley's celebrated memoir or with Klein's two luminous papers in the *Mathematische Annalen* of 1872 or about that time; but his attack is limited to two *expressions* in Lobatchewski's little book, which is as elementary as possible and is certainly correct in substance.[33] Lobatchewski's fundamental hypothesis is that given any real straight line and any real point not in that line in their common plane there are two straight lines through that point and in that plane that are on the limit between all right lines through the same point and in the same plane that cut the given line at finite distances from the given point, on the one hand, and on the other, all straight lines through the same point and in the same ~~line~~ plane that cut the ~~first~~ given line at no points at ~~finite~~ real distances from the given point. These two lines through the given point he speaks of as "parallel" *to the given line towards that side* of the perpendicular let fall from the given point upon that line on which the parallel is nearer to the given line than is the given point. One of his theorems is rather loosely expressed thus: "Two straight lines which are parallel to a third are also parallel to one another"; but of course he means *towards the same side*; for otherwise, two lines that pass through a given point parallel to a given line would be, according to this theorem, parallel to each other, which, since they both pass through that point, would

There came the end of the page. "What!" I mentally exclaimed, "Are exceptions to my generalization so plenty that I am to find them everywhere? This writer must surely be an exception; for who would utter a warning in so studied a style only to disregard it himself?" However, on examining the context, I found that the admonition was addressed to a group of some hundred of the most careful reasoners in the world; the case being quite parallel to that of an American who should accuse all the inhabitants of the British Empire of false reckoning for saying "A pint of cold water weighs a pound and a quarter," instead of "A pint's a pound all the world round," the latter being in point of fact false. What we ought to say is "When the barometer at thirty inches and thirty hundredths is found Uncle Sam's peculiar pint of boiling water weights mighty near an avoirdupois pound. So the writer not only subscribed to my assertion but illustrated the truth of it most obligingly in his own person; as I believe just for the *[sake]* of so giving it a new emphasis.[34]

In the course of these essays I shall have occasion more than once or twice myself to contest some common opinion; and although I have done my very best, in multiplied reëxaminations of each subject from every point of view that I could discover during many years, to search out any errors, yet after all I can do no more than lay my reasonings before my reader and appeal to his judgment.

The attainment of sound reasoning is a discipline closely analogous to the attainment of moral virtue. Each of us, I hope, has been duly instructed as to how the latter is to be achieved; yet for the sake of showing its parallelism with the attainment of sound ratiocination, the reader will pardon my repeating[*] the method of it. Namely, it beings with running over in one's mind how one has conducted oneself on each occasion, and in asking oneself whether one's actual conduct was in all respects such as was best calculated to effect the achievement of one's heart's desire. The first step toward this is to recognize and get a lively image of what one's heart's desire really is; for it is just there that men oftenest go wrong. One must recollect that he will make a dire mistake if after having attained what he had labored for during

---

be too absurd to be supposed to be according to his meaning, even if his demonstration were not there to show clearly that he meant parallel towards the same side. Not but a line parallel to another is so throughout its whole length, but toward the side of its parallelism it approaches that other. It is pretty difficult to imagine how the writer could fail to see what Lobatchewski meant, since he *proves* it very clearly, and since that which the writer takes him to mean is so glaringly absurd; but anybody who looks into the literature of circle-squaring soon learns that the contentious type of mind has a wonderful faculty of throwing dust into its own eyes.

[*] repeating] *above deleted:* rehearsing

many a long year, one should find it mere dust and ashes in one's mouth. One's very realest desire is not whatever one may happen to desire, but is what he *would* desire upon sufficient information and a sufficiently detailed imagination. Woe to him if he never actually comes to desire it but only, too late, regrets that he didn't. After settling with himself what his real desire must be, he has next to review his recent conduct and find in what respects and why it was not calculated to bring about the desired result, and how otherwise he should have acted. He is thereupon to consider what must be his rule of action in the immediate future, in order to advance most swiftly to his true end. Having discovered what this rule should be, he next has to *resolve,* i.e., sincerely promise himself that his conduct shall conform to it. But Hell is paved with good resolutions; and therefore to this promise must be attached good security, or, in other words the *resolve,* which is composed of that plastic stuff that dreams are made of, must be baked into the hard brickbat of a real *determination* of the habit-machinery of his organism, which shall have force to govern his actions. A determination is a *virtual* habit. *Nota bene* that the word "virtual" is one of those excellently definite English words that are derived from the lingo of medieval scholasticism, a part of our speech that, now when so many men, coming home after years spent in Germany, have somewhat lost their mastery of their mother tongue,* are apt to get misunderstood. The English *"virtual"* and the German *"virtuel"* differ widely in their meanings. *Virtuel* seems to be a tolerable synonym of *möglich,* possible. But "virtual," followed by any common noun, say 'N,' makes an appellative phrase which denotes anything which, while it is not an 'N,' has, nevertheless, the characteristic behaviour and properties of an 'N.' For example, in the Eighth Book of *Paradise Lost,* Adam inquires of Raphael,

> Love not the heav'nly Spirits? And how thir Love
> Express they? By looks onely? Or do they mix
> Irradiance, *virtual* or immediate touch?

By "virtual touch," Milton's Adam meant something that *was not* touch, but brought all the delight that touch can bring. So a determination *is not* a habit, since it does not result from repeated performances, on the same sort of occasion, of the sort of action that it will cause to be again performed on the same sort of occasion; but it works all the effects of habit, and is, therefore, strictly speaking, a "virtual habit." The truth is that an oft-repeated performance in the imagination on the same sort of imagined

* mother tongue] *above deleted:* vernacular

occasion of one and the same sort of action will create a real disposition to a real performance of an action of the same sort on the same sort of real occasion. The effectiveness of this disposition appears to be not only as great relatively to that of a habit induced by real performances on real occasions, as the proportional vividness of the imaginations that induce it are to the vividness of the perceptions that would induce the habit; but in truth to be far greater. The effectiveness of the virtual habit relatively to that of a real habit, is, I say, unquestionably far greater than in proportion to the vividness of the imaginations that induce the former relatively to the vividness of the perceptions of the experiences that induce the latter. There must, therefore, I venture to think, be a sort of self-hypnotizing effect when we strain, in some obscure way, to influence our future behaviour by calling up as vividly as we can the image of a given sort of stimulus[*] and that of our responding to it in the desired way. For we seem to command our organism or our soul or whatever it may be, as if we said to it: "You will act *thus:* do you hear? THUS! THUS!! **THUS!!!**" It is strikingly analogous learning one's part in a drama by saying it over repeatedly, with a peculiar sort of commanding effort. The vividness of the imaginations seem to be a factor of the effectiveness of the induced disposition. The intensity of the effort is a second factor and the number of repetitions a third.[†] This I believe to be the method of moral self-control.[‡] I am much more confident that what I have described is precisely parallel in all its features to the method of intellectual self-control; now self-control is certainly the principal ingredient[§] of efficiency, *alias* virtue, whether moral or intellectual. I shall leave the psychological problems involved in these questions to the psychologists. Not but that I may have made systematic experiments of my own. But the psychological problems appear to me to be as difficult as they are important and I shall prefer to learn what I can about them from those who devote their lives to such questions. There remains an entirely independent and previous investigation of a purely logic kind. In this chapter, however, I shall confine myself to the first question of the series; namely, What is that that we aim at when we reason?

I may as well, however, say a word about the indifference to logic which we find not on the part of "everybody,"—since "everybody" is more or less

---

[*] stimulus] *above deleted:* occasion

[†] third.] *period added by editor after removal of:* , though I don't mean to claim any quantity *[*illeg.*]*

[‡] moral self-control.] *above deleted:* learning moral virtue] *above deleted:* moral discipline

[§] ingredient] *above deleted:* factor

a fool, but on the part of really superior minds. Such minds desire to understand the theory of every kind of doing in which they take part when any question of superiority between different methods of doing it is open; and that there is such question concerning reasoning, is obvious. For such minds, therefore, the theory of reasoning cannot but be a matter of lively concern;* and we find that in fact it is so. They are, indeed, indifferent to such logic as is set forth from professorial chairs; but that is because in that they discern mere will-o'-the-wisp gleams and no *lanterna pedibus,* no light to guide their researches.† The occupant of a chair of logic in an institution that had before his day been an efficient agency in the advance of science, having once brought out a system of distribution and nomenclature for the different kinds of processes of the mind, a system that was marked with slapdash haste and with the most extraordinary lack of reflexion upon the real functions of the different processes and of their relations to the purposes of thinking, I remarked to him that I feared that the young men whose minds should be engrossed with the study of his system would not turn out particularly efficient reasoners. He drew himself up to his full professorial height, and with the utmost disdain replied, "I am not occupied with teaching my students to reason: I am teaching them Science." Now in order to appreciate the plight of contemporary university instruction in logic as it is illustrated by this little anecdote just suppose a parallel remark to have been made by a professor of some science that is in a healthy condition;—say chemistry. Suppose I had remarked to a professor of chemistry, that I feared a course of work proposed by him for his students would not make them good analysts. Chemical analysis no more constitutes chemistry than reasoning constitutes logic;—rather less so. But what should we say to a professor of chemistry who professed to teach his students "The science of chemistry," without caring whether they made good analyses or not? Should we not say that his students might as well occupy themselves with committing a handbook of the science to heart, and that a man who was not a good analyst could not advance the science of chemistry? It is much more true that a young man who cannot discriminate good reasoning from bad, can be of no use to the world as a logician. My little anecdote

* lively concern;] *before deleled:* And if they seem not to be so, it is because they cannot find in the logic-books any light that does not seem like that of a will-o'-the-wisp.

† their researches.] *before deleled:* I once remarked to the author of a system of nomenclature for the processes of the mind, a system that was marked by the most complete ignorance of the essential functions of those processes in the development of truth

(which I have modified in an inessential particular for the sake of preventing identification and avoiding personalities) well illustrates the idle futility of the logic that the textbooks of our age. They hardly touch the real science, and as far as they do so, they teach false doctrine, without even serious attempts to defend it. I am referring to the entire scientific age, that era that I reckon from A.D. 1543, the year of publication of the *De Revolutionibus,* as its initial epoch.[*35] The logic of this age has been infantile and contemptible in comparison with its actual reasonings. Some books have been produced that have contributed some materials for a worthy logic. The First Book of Francis Bacon's *Novum Organum* is an eloquent and wholesome homily upon the proper frame of mind of a true investigator.[36] The Second Book is occupied with the questions of the true logic of induction. It unconsciously follows the Stoics in pronouncing the Method of Exclusions to be that true logic;[37] and without throwing any real light upon the problem, it does unwittingly take one step toward the solution of it, as far as that method could solve it, by rendering it clear that the Method of Exclusions must be excluded from any list of its possible solutions. In the XIXth century, Whewell's *Novum Organum Renovatum,*[38] and other books did really contribute some truths of high value to the subject, although they were inevitably vague, by specifying some of the conditions of successful research of the higher kinds and by showing the relation of these to the doctrines of Kant's *Critik der reinen Vernunft.* Whewell had the

---

[*]In a discarded attempt to compose an introduction, dated March 29, Peirce provides the following aborted description of what he means by "science" (R 618:3–4, 1909):

> The word *science,* with the corresponding Latin and Greek words, *scientia* and ἐπιστήμη, has been used at different times in three principal senses. As long as Aristotle reigned, science was defined as "knowledge through principles," which was as much as to say, syllogistically demonstrated knowledge. But this manifestly conflicts with modern notions, since, in the first place, it excludes all the physical, natural, psychological, historical, and other sciences which for us constitute science *par excellence,* and in the second place makes the conclusion of a syllogism science, while its at least as well-known premises may not be science. Therefore, this use of the word is to be utterly rejected. Coleridge in his introduction to the *Encyclopaedia Metropolitana* defined science as systematized knowledge. But this places the emphasis upon a character of secondary importance and ignores the heart and vitalizing spirit of science. It makes science consist in the text-book, which every man of science more nearly regards as the mortuary urn that preserves the ashes of what formerly lived as science. What are the leading characters which animates the true scientific man in whose head and heart geuine science lives? The first is the effective passion to find out the very truth of some subject. The second is that this truth does not animate the scientist because of its mere personal interest or for any other ulterior reason, but simply for its pure interest in itself as truth. In saying this I do not in the least exclude practical science, such as a science of the steam-engine, and the *[...]*

advantage of a real and high investigator's acquaintance with science, on the one hand, and that of a powerful speculative understanding on the other. At the time of the appearance of the first of these works of Whewell, the *History of the Inductive Sciences,* in 1837,[39] John Stuart Mill, a young man born in 1806 may 20, of quite unusual intellectual force,—partly inheritance, partly family traditions,—had for seven years be preparing to write a treatise of Logic. This treatise was ready in the April of 1842, and was, I believe, actually published in that year.[40] It was, all things considered, the most remarkable treatise on logic that the scientific era had ever produced; so that it is well worth my while* and that of my reader that I should make a just appraisal of it. In the first place, what was John Mill, and what were his qualifications for so very great an undertaking? He was a young man of great natural and precocious powers of intellect, and, like the majority of such young men, of remarkable docility. Seeing that he was so docile, in estimating his affirmative qualifications, we must begin by considering the affirmative qualifications of his guides. Now these were two of the greatest dialecticians of history, his father, James Mill, and Jeremy Bentham. All this prepares us to find two of the three intellectual qualities which we do find in John Stuart Mill; namely, his subtilty and his consecution of thought. The third of his qualities he certainly neither inherited nor obtained by copying his two great models. It was peculiarly his own, and in several particulars overcame his consistency. This quality was a remarkable candour, which sometimes obliged him to recognize truths that did not very well hang together with his primal positions. But subtilty, consecution, and candour by no means suffice to make a great reasoner.† He needs a flock of other qualifications that John Mill did not possess in any particularly high degree, such as a keen sense for the novel; the chess-player's power of grasping together in their relations a vast mass of items without losing sight of any of them; a sort of intellectual music in his soul by which he recognizes and creates symmetries, parallels and other relationships of forms; the mathematician's generalizing faculty, which was exemplified in the introduction of the decimal point into numerical notation, (though it would have been more perfect if it had been a mark upon the

---

* my while] *before deleted:* to say just how much it is worth.

†In a discarded branch dated May 17–19, (R 620:91–98) Peirce writes a comma instead of a period and continues thus: "as we shall see if we note what are the characteristics of the reasonings of the great masters in the different kinds of sciences. In mathematics, we find that by the perceptions of the analogies and symmetries of forms the great masters attain generalizations." The sentence is followed by an unfinished example of how Leibniz's solution to solving linear equations with multiple variables (or unknowns) inspired the theory of determinants.

place of the zero power of the base, instead of between that and the place of the *minus*-one-power,) and appears to still better advantage in the introduction of negative and fractional exponents, Harriott's plan of transposing all the terms of an equation to one side,[41] in the invention of determinants, imaginary quantities and their representation by points on a sphere, domains of rationality, */and/* Riemann's surfaces.* For one man to sit down and write a logic at all adequate to modern thought perhaps surpasses human powers.[42] Certainly, a logician should bring to his task, the ripe fruit of a lifetime's religious devotion; and he should have an inside acquaintance with the chief sciences of reasoning, since they are */the/* field of action of modern ratiocination; and by an inside acquaintance, I do not mean what is now taught in every university but a training in so dealing with the higher problems of science as to merit recognition as a peer in some main respects of the first scientific minds of the age. Mill possessed such a acquaintance with no science except Political Economy; and even that he did not treat as Marshall and others do, and as Cournot had already begun to do, but rather in the literary way of the prescientific ages.[43] His work unquestionably ranks relatively as very high; but instead of being the fruit of a life's immolation, it was produced by a young man of 36 who was engrossed with the duties of an office salaried at £600, who had been for many years contributing an elaborate article to every number of some Quarterly Review, as well as writing constantly for the newspapers. It is significant that every one of the examples of scientific reasoning that Mill selected in his first edition as preëminently worthy of admiration are now recognized as unsound; while he passed by without notice others that have stood the test of time. But by far the greatest flaw in his work,—a flaw which if it could have been escaped would have left the treatise, with all its other faults, a wholesome *medicina mentis,* is that from cover to cover it is narrowed to being an expression of a particular theory of metaphysics,—as completely unfounded as systems of metaphysics have usually been. This one is, indeed, a more dangerous poison than any other system of metaphysics for the reason that those who have been inoculated with are unable to recognize that there is any metaphysics in it; they are under the delusion that it is a pure formulation of experience. That this renders it dangerous, (supposing I am right in saying that it really is no better than a metaphysical conjecture) is plain enough; and yet even this is not the worst of it. The worst of it is that it is in its essence inimical to all reasoning. The abhorrent doctrine is that *universals,* that is, whatever is general, such as *species* and *genera*, the

*Riemann's surfaces.] *above deleted:* and many other mathematical inventions.

chemical elements and other kinds, the laws of nature, etc. are not *real,* but* *[two sheets missing]* that opinion by his father and his father's intimate friend Jeremey Bentham. Now his father and he, as well as almost all other militant nominalists acknowledge William of Ockham as the "ver-erabilis inceptor" of the Nominalistic opinion and frequently quote his logic as if it were the perfect truth. So we may as well turn to the explicit statements of that immortal, though very pestilential, book.† I shall argue this upon another page, contending myself here with utmost brevity the difference between the two metaphysical opinions, and indicating how it can have any practical importance. Amid the multiplicity of studies that importune the man who desires to take some part in the development of our civilization, any attempt‡ to delve into the huge tomes of the medieval monks would be so manifestly unwise that it may be well to inform the reader that the controversy between the followers of Scotus and of Ockham in the XIVth and XVth centuries, had as little to do with the famous disputes between realists and nominalists in the XIth and XIIth centuries, as the thermodynamics, of Clausius, Rankine, and Kelvin has to do with Francis Bacon's discussion of heat in Book II of the *Novum Organum.*[44] They breathe altogether different intellectual atmospheres, owing to the works of Aristotle having been translated into Latin in the early XIIIth century. There may even be among my readers well-disciplined minds who might confuse§ that great logician and Minorite Franciscan[45] friar Scotus to whom I refer, who died late in the year 1308, and whose proper name seems to have been John Duns; in modern spelling Dunce;** and by "a dunce" was originally understood a follower of Scotus, an enemy of the literary *renais-sance,* in the toils of whose argumentation it was next to impossible to avoid

* but] *most likely followed by something like:* figments of the imagination.

† pestilential, book.] *before deleted:* I will quote from the best edition of it, that of 1438. It is a pretty large book of 126 leaves of folio closely printed in the most abbreviated style: a person not versed in medieval abbreviations would find many pages on which he could not read a single sentence.

‡ any] *above partially deleted:* the folly of attempting

§It seems Peirce got carried away while writing this sentence and forgot that he began by referring to a confusion. Possibly, Peirce had in mind the rather common confusion of Johannes Duns Scotus with Johannes Scotus Eriugena, a ninth-century Neoplatonist.

** Dunce;] *before deleted:* just as pens, defens, ons have become pence, defence, once, etc. | *before deleted:* whose family ~~may have given~~ very likely gave its name to the Merse town of Duns, or Dunse, though there is some reason to believe that he himself was born in Northumbrian Dunstance. His later Irish followers

being entrapped. But as time went on, from vexation at such plaguey disputants, mingled with not a little dread of them, men gradually passed to contempt of them; and so the common noun took on its modern meaning.*

As for William of Ockham, he came from a parish of that name, whose fine church and tower date from the logician's time. Although he was educated at Merton College, where the Minorites were influential he became a secular priest and from the many ecclesiastical preferments that he early received, as well as from a slight tone in his writings that I fancy I detect, I venture to guess that his family was somewhat superior. He was certainly a man of thorough earnestness; since otherwise he would not have resigned his rich benefices in order to become a mendicant friar. It is plain that he took the same view of Nominalism that Mill did, and that, indeed most philosophers since the scientific era began have done. That is to say, he did not look upon it as a metaphysical opinion but, on the contrary, as a sort of positivism, or refusal to believe in anything one does not experience; and it must be acknowledged that such is its character. Only it cannot, from the nature of thing, stop at *un*belief but is perforce pushed on to *dis*belief; and to be a consistent nominalist one would have to go on to absolute skepticism,—Pyrrhonism, which, as everybody knows is as impossible as for a knitting-needle to stand up steady and motionless upon a table. All this I shall thoroughly and plainly argue, when the proper time comes; and I fully expect to carry my intelligent reader along with me,—the full argument would carry us too far afield just now. Naturally my reader does not

---

*Original Peirce footnote:

The tomb of Scotus in the Minoriten-Kirche in Cologne originally bore the inscription:

> Scotia me genuit, Anglia me suscepit,
> Gallia me docuit, Colonia me tenet,

and taking this in connexion with his being called *'Scotus,'* with the name Duns, and with the positiveness with which antiquarians of the town of Duns in Merse assert that that was his native place, I infer, though with very little confidence, that there exist records of a family of that name there. This imparts an appearance of truth to the statement that there is, or once was, in the library of the Minorite's Merton College in Oxford a MS of the *"Opus Oxoniense"* of Scotus, the colophon of which reads, *"Explicit lectura Doctoris Subtilis, in Universitate Oxoniensi, super libros Sententiarum, soilicet Doctoris Ioannis Dons, nati in quadam villa de Emylden, vocatur Dunstane in comitatu Northumbriae,"* etc. The whole story may, of course, be made of whole cloth; but if there was such a colophon, its scribe at least knew enough of that part of Northumbria to know that there is a parish there of Embledon, near the ancient castle of Dunstaneburgh, which, as far as I know, may have been built by the Duns family of Merse (if there was such a family, which is not far from pure conjecture,) and have been the birth-place of the great Scotus. It is true that *Dun* is properly a place name; but there is no more reason for a family being named the "Hills" than the "Duns," the meaning of the two words being nearly the same.

at present believe that I shall convince him of the truth of my position; and I cannot blame him. But I, who have carefully studied the question in all its ramifications for nearer fifty years than forty, am confident that every intelligent man will soon be won over to my opinion. In this place I shall merely set out the main argument, without taking up any of the objections which will occur to the reader. But strict justice will be meted out later.

In order to understand what Nominalism* is, it is requisite, first, to examine the meaning of the philosophical words *'real,'* and *'universals,'* and *'substance.'* 'Real,' Latin *realis,* as a term of law, in such phrases as 'real property' goes back to the early Middle Ages. But that word, as far as I have been able to make out, has no connection whatever with our vernacular word 'real,' beyond their identity in outward form, and their common derivation from the many-sensed Latin *res.* I have met with this latter adjective, in its Latin form, *realis,* in one of those translations of Aristotle that were executed in the early years of the XIIIth century; and I think I have found it in Aquinas. But he uses it so little that it does not appear at all in the *Thomas Lexicon* of Dr. Schütz;[46] and some respectable writers attribute the invention of it to Duns Scotus. Its common use in later scholasticism, and through that in the vernacular of English, French, Provençal, Spanish, Portuguese, Italian, German, Dutch, and doubtless other modern tongues is entirely due to him; and it is not true that it was ever used in the great dispute of the realists and nominalists of the XIIth century. The word did not exist at that time, unless the Law-term be regarded as the same word.

---

*A discarded branch, dated May 22, includes the following discussion of Roscellin (R 620:100–2):

> Nominalism is the metaphysical doctrine that "Universals" are of the nature of words and are made by the persons in whose thought or language they occur. It would be simpler to define it as the doctrine that Universals are figments and not real; and this would agree with all varieties of nominalism which have been rife in modern times or in those of Duns and Ockham. But, singularly enough, this definition would not apply to Roscellin's opinion, although Roscellin has always been regarded as the most extreme of all nominalists. We only know what his opinion was by the account of Archbishop Canterbury, who would probably have burned the poor, harmless fellow at the stake if he could have laid hands upon him. Anselm's statement of Roscellin's opinion, however, agrees so well with what we know to have been the views of some obscurer men of his time, the middle of the XIth century, which was not long after the midnight of the Dark Ages in Northern Europe, that this statement is quite credible that he held universals to be *"flatus vocis,"* the breath of the voice. Now this breath is certainly a real thing; it is the air that escapes from the lips in uttering a word, and was always regarded as real. As to the credibility of Anselm's report, it must be remembered that at that time it was the usual belief that the soul, "anima," is breath or air; and from that point of view it was not at all impossible for minds in the prevalent state of culture of the time to suppose that a Universal, since Universals seem to be the most spiritual part of discourse, is so much of the soul as escapes in uttering the corresponding word.

The sense in* which it ought to be employed in a scientific philosophical nomenclature must follow the usage of Scotus, which substantially agrees with that which it bears in the English vernacular. For Scotus as for ordinary speech, every thing, event, or circumstance that is not a figment, made up by some person or persons, whether intentionally or unconsciously, like the battle of the Centaurs and Lapithæ, perpetual motion (the dreamed of contrivance doing work gratis), Aladdin and his wonderful lamp, the music of the spheres, etc., is *real*.†

[version II]‡

[CHAPTER I.] THE FIXATION OF BELIEF.

Few people care to study logic, because everybody conceives himself proficient enough in the art of reasoning, already. But this satisfaction, I notice, is limited to each one's own reasoning, and does not extend to that of other men.

[But why say it? It is a platitude that everybody preaches and nobody takes to heart. I use the word "everybody" here, as "everybody" uses it, but not as including the few conscientious reasoners who have rendered the scientific reasoning of today incomparably superior to that of fifty years ago, and who, in every age since Gerbert was believed by "everybody" to have sold his soul to the devil because he possessed the art now known as "vulgar arithmetic," have brought about a gradually increasing accuracy of thought. It chanced that just as I had copied for this edition the first two sentences of this chapter, and was pausing to make up my mind whether or no to add myself to the number of unheeded preachers of a truism, my eyes fell upon an open page of a recent periodical, and I was startled out of my reverie at finding myself reading the following words:

> Among men there is no habit more inveterate than the persuasion of each individual that he personally is immune from slips of reasoning. All around him during every day of his life he takes notice how badly other people reason, without ever saying to himself that probably he is[47]

---

* sense in] *above deleted:* meaning

†This concludes Version I. Although the text ends in a complete sentence and does so while filling up the entire sheet, this first version must be considered incomplete, and Peirce most likely never finished it. Even assuming that he concluded his examination of the word "real," he is yet to begin a discussion of the terms "universal," and "substance," and even if he had finished that, he would at best only have completed the first bracketed insertion.

‡The text below begins with an exact reproduction of the transcription made for Version I—albeit with the editorial notes removed—up unto the phrase "There came the end of the page."

There came the end of the page; but it is enough. It read as if I had already fallen upon an exception to my general rule. Yet it was quite possibly no exception. For supposing, for the sake of argument, that the writer was about to urge an opinion contrary to that of a whole body of as accurate reasoners as the world contains, and an opinion so slenderly supported as to rest, let us say, merely upon his fancying he had come upon an error in a work by one of the weaker of those reasoners, this fancy being due to his own failure to follow a very simple proof, and to his so not understanding what the reasoner whom he misunderstood really meant to assert. Then, were that the case, do you suppose that he would have meant, himself, to take to heart the warning of his preaching, or do you suppose that he would have illustrated the truth of his (or perhaps her)[48] jeremiad—it sounds rather feminine,—by calling upon that whole body of admirably clear-sighted reasoners to mend their ways? This question may, and must, go unanswered; but under the latter supposition it would be a pretty striking example of the truth of my first sentence; all the more so should the writer in the periodical have turned out to be a writer on logic.[49]

I must say that the indifference of "everybody" to the science of logic is pretty thoroughly excused by the state of this science since the scientific era began. During this glorious age, just one truly capital contribution has been made to logic; just one, at all to compare with the splendid discoveries that have been made in all the physical and several of the psychical sciences. That one was a contribution that held within it the fertile germs of ideas which were destined as soon as they were developed to render logic, in vitally important respects, perhaps I may venture to say were destined, if developed in time, to render logic directly and indirectly, *in the main,* adequate to represent and criticize the reasoning of this same scientific era that seems now to be so swiftly drawing to its end. The contribution I refer to was the Doctrine of Chances. Yet I suppose that, even now, treatises on logic are in process of getting written which are not intended so much as to contain even a strict definition of mathematical Probability, unless it be one which falls into an obvious and a particularly vicious circle.[50] In another Essay in this volume it is my intention to discuss the relation of this doctrine (or "theory," in the mathematical sense,) to the explanation of the force of arguments.* Now and then, during the two generations and more

*Original Peirce footnote:

In this place I will just say that the straight-forward application of the doctrine of chances explains little in logic; but the proof that it cannot explain certain arguments serves to indicate how they are to be explained; and after these have been explained, still further problems stand out more clearly and definitely than before, and thus their solution is facilitated.

last past, works of high importance for logic have appeared, sufficient to impart joyful confidence in the future. William Whewell, after illuminating, himself, a wonderful range of sciences, dynamics, crystallography, optics, the tides, chemistry, the mathematics of political economy,* and the germinal principle of Gothic architecture, to all of which subjects he brought enduring improvements, and thus being splendidly equipped with one half of the panoply of a teacher of logic, in 1837 commenced a series of works upon the logic of science.[51] They were marked by an inside acquaintance with science; both in general and in particular; and by a genius of a very high order. Indeed, these works flew so high above the heads of the common herd† logicians, that very few, I fear, have profited‡ greatly by them. They still remain, however, though defective in vital points, the best exposition we possess of the essential nature of research after truth.

In 1839 appeared the first volume of the *Cours de philosophie positive,* upon the missive of which its deliverer, Auguste Comte,[52] then in his forty-second year, had been working with assiduity, penetrating genius, and frigidity that was arctic to the scalding-point, for twenty years.§ It was a philosophy eminently "Pragmatistic" in its aspirations, covering as it did, politics, exact science and practical religion, while rejecting as chaff metaphysics, theology, and psychology.

---

*Original Peirce footnote:

> His contributions to this indispensable improvements on Economics may be found in the *Transactions of the Cambridge Philosophical Society,* and extend from 1829 to 1851. Cournot's first book, a very useful and stimulating one, notwithstanding a fallacy about a partial differential equation that is a sad disgrace to a professor of mathematics, appeared in 1838. The mathematical method must prevail in economics; and if ever we succeed in interpreting the communications of the inhabitants of Mars, I will risk a nickel that their very first dispatch will read: "O yes, we know all about your attempts to learn from us. Well learn this: you are the most stupid of all organized beings. You even allow creatures to have a voice or a vote in your government who hardly can solve a partial differential equation in its generality and have not mastered the higher reaches of the calculus of variations. Bring a diploma from a colony of bees before you presume to seek instruction from Martians!"

† the common herd] *above deleted:* almost all others

‡ that … have profited] *replaces partially deleted:* that I fear I am the only one who has profited

§Original Peirce footnote:

> That Comte has been, almost to this late day,—70 years after his meridian passage,—one of the most grossly abused of men, is proof enough that he is too great for the powers of estimation of ordinary men, just as is the case with Rousseau. Two typical clergymen, whose books are virtuous and worldly to the core, have accused Comte of plagiarism;—plagiarism of his three stages, or eras, of thought, the theological, the metaphysical, and the positive, and plagiarism of his classification of the sciences, according to which

In the main, and making due allowance for looseness both of expression and of thought, his contributions to logic were of the true metal of science, and have supported that touch-stone that all men of science recognize as decisive.[*] A most characteristic example, and one of no little importance, was his declaration that no hypothesis ought to be considered in science unless it be capable[†] of "verification." He never once states what he means

---

the sciences form a series from the most general to the most narrow, each borrowing principles from those that precede it, especially from the last of them, and borrowing facts from its juniors (those that follow it in order,) especially ~~from the first of them~~ each from the one next after it in the series. I do not know whether there is any truth in the accusation or not; for I have not searched all Comte's philosophical writings beginning with 1820, any more than I find evidence that his accusers have done. Nor do I care much. As to the "law" of the three stages, its truth is not so well established as is, for example, the law that a man's supreme desire, in the three great acts of the drama of life, is successively love, reputation, and wealth, for the reason that the drama of history is not yet played out; and besides, the three stages are mere moods, and as I do not infer from the drama of life that money is most desirable of all things, much less do I infer from the succession of moods in history that materialism is the truest philosophy. But as to Comte's classification of the sciences or rather of the laws of scientific evolution that are at the bottom of it, they appear to be well-supported by arguments both *a priori* and *a posteriori,* if they are held along with the eminently positivistic conviction that it is beyond human powers to know what *undiscovered* sciences and their relationships are to be. I suppose that Comte, like every man, got suggestions from other men, while an accurate statement of just what they were and whence derived was more than he could make, especially in a course of lectures. His whole life seems to bespeak scrupulous honour, and I should shrink from uttering any hint of a suspicion of plagiarism on his part, as an act of baseness on my own; and all the more so seeing that I am obliged to add that I think his earnest and devoted strivings after philosophic truth were mostly failures. French notions of honour are not exactly my own; but the conscientiousness and minute consistency with which they are carried into the lives of respectable Frenchmen, at all costs, commands my respect; and Auguste Comte appears to me worthy of his full share of it.

I would trust an average Presbyterian clergyman a long way, both because of his virtue and because of his worldliness; but not quite so far as I would a man like Comte.

[*] Footnote added by Peirce:

This was expressed in late Greek as follows: ἀπὸ τῶν καρπῶν αὐτῶν ἐπιγνώσεσθε αὐτούς· μήτι συλλέγουσιν ἀπὸ ἀκανθῶν σταφυλὰς ἢ ἀπὸ τριβόλων σῦκα; οὕτω πᾶν δένδρον ἀγαθὸν καρποὺς καλοὺς ποιεῖ, τὸ δὲ σαπρὸν δένδρον καρποὺς πονηροὺς ποιεῖ· οὐ δύναται δένδρον ἀγαθὸν καρποὺς πονηροὺς ἐνεγκεῶ, οὐδὲ δένδρον σαπρὸν καρποὺς καλοὺς ποιεῖν.... ἄραγε ἀπὸ καρπῶν αὐτῶν ἐπιγνώσεσθε αὐτούς.[53] That is to say, if you want to know whether a proposition be true or not, apply it. Force it to prophesy what will be the result of an experiment and having first taken down its answer, go on and actually try that experiment. For a veracious proposition absolutely *cannot* bring forth lying prophecy, and a mendacious proposition is unlikely to utter a prophecy that turns out to be otherwise than false, in proportion to the degree of its falsity. The best rules of logic hold good for the conduct of life.

[†] unless it be capable] *above deleted:* which ~~is~~ would not be susceptible

by "vérifiable," but the context would seem to show that he meant that the hypothesis should imply nothing that cannot be directly observed in case it* be true. This is certainly a rule that breaks down at once as soon as it is put to the test.† It would, for example, forbid us to suppose any historical fact to be true; since what is past and gone cannot be observed. If, however, what he meant, in his confused, though vigorous, manner of thinking, was that a hypothesis which neither suggests nor seems likely to suggest any experiments or observations which, in case it were false, would‡ refute it by conflicting with some§ consequence that could be drawn from it, is an idle thing that should not be allowed to occupy a scientific man's attention, then the maxim is a very sound one as well as one that was much needed at the time Comte first put it forth.

An event for me personally highly favorable to opportunities for watching the effects of Comte's utterances occurred in the very year his great work began to appear: namely, I was that year born, and as a consequence, am perhaps in a better situation than my reader to testify to the gradual but marked improvements in men's reasoning, as Comte's writings came to win wider respect, until well into the sixties. Yet the reader, even if he has skipped my footnotes, will not expect, for all the right honour I render him, to find me swallowing in one bolus all this singular prophet ever said, as a society of his English worshippers have done. Those *sociétaires,* however, command a modicum of my respect for recognizing, with Comte, a side of Truth to which other nominalists are mostly quite blind; I mean its spiritual, its religious side, which all truth has.[54] But I opine that a man of science ought to be esteemed and rated chiefly as a discoverer, if he be one, and not a mere observer; the latter's merit depending mostly on the largest errors that he is apt to make, no matter how extreme his accuracy at most times; while the former should be judged chiefly by his finest and most difficult discoveries, with little reference to his oversights. Particularly I would insist on this in a science like logic, that has not yet settled down to a quite mature condition. I cannot call to mind a single utterance of Comte's with which I am altogether satisfied, and the general drift of his philosophy, I hold to be mistaken. He did not contribute to any science of discovery. If he was a scientist at all, he was one of those whose function it is to arrange and codify the discoveries of others. Some of his predictions sound quite ridiculous, now. He was decidedly too inclined to place restrictions upon

---

* it] *before deleted:* should

†This is … the test.] *above deleted:* But this would be simply ridiculous.

‡, in case … would] *above deleted:* might

§ some] *before deleted:* necessary or probable

research, by pronouncing this or that to be forever beyond the possibility of discovery, thus seeking to bar certain avenues to discovery;—perhaps the most pernicious vice into which a teacher of the philosophy of science can fall. He declared, for example, that we never could have any knowledge of the chemical composition of the heavenly bodies, a sort of knowledge that began to come in abundance within a few years after this declaration.[55] And examples of his arrogance in forbidding certain lines of research could easily be multiplied. Nevertheless, I hold him in high honor, heightened by his gropings for a religion, utterly inadequate as was his creed.

In the spring of 1842 there appeared *A System of Logic, Ratiocinative and Inductive,* by John Stuart Mill, a man of 36, the third clerk in the Examiner's Office of India House,*—a highly responsible and laborious position,—and a constant writer for the quarterly reviews and for the newspapers.[56] He was a man of great subtlety of thought, candour, and power of work, as well as a master of the pen. The very title of his work made an epoch in logic, in recognizing scientific reasoning to be as deserving of attention as syllogism. But he had no inside acquaintance with any other science than political economy, and even to that he did not apply the mathematical methods that had been introduced by Whewell and Cournot. It would, indeed, be an impossible mistake to regard Mill as a scientific man. Thus ignorant of what he needed, almost above all, to know, he was yet as a quarterly reviewer, a professor of omniscience. In that capacity, he made it a prominent purpose of his treatise to "annihilate" Whewell, one of the greatest men of science then living. Now a genuine scientific man, pregnant with glorious truth, does not interest himself in "annihilating" individuals: he disdains such work, and rather entertains something like contempt for men who undertake to execute annihilation whether upon those who do not annihilate themselves or upon those who do. Such, at least, has been my experience of true *savants*. One of the most instructive points that Whewell had brought out,—though he pushed it somewhat too far,—was a parallel between the evolution of an inductive science and the development of cognition, as set forth in Kant's immortal *Critik der reinen Vernunft;* and Mill especially directed his attack against that parallel, although it appears from a casual remark in one of his later writings that even in later life he, a professed philosopher, had never so much as peeked between the covers of the *Critik*. When I came across that naïve confession, it acted upon me as if I had unwittingly taken a swallow of strong peppermint. Its coolness quite took my breath away. For my own first serious readings had been in German mainly, and had been followed by hard studies, first in Greek,

---

* ,— …,—] *above deleted:* at a salary of £1200

and then in scholastic folios; so that it was only pretty late that I became aware of how common it is for an English writer upon an abstruse subject to mention in his preface that he knows nothing at all of the literature of the subject upon which he undertakes to instruct his readers. Above all, a quarterly reviewer was nothing if he was not bold. We need not weep that the race is extinct. Nevertheless, Mill had some high qualifications for his task; and considering that to that task, whose proper execution would have called for at least one half the total energies of a full life-time, he gave only seven years,—no! what am I saying? *not* seven years, but so much leisure and energy for the work as was left for it and other miscellaneous occupations in seven years during which his industry in India House advanced him from a salary of £800 up to £1200, besides writing elaborate reviews quarterly, and newspaper articles almost weekly; considering all that, his performance was well-nigh prodigious, whatever we are to think of it from a scientific point of view. No man who has wooed a virgin Truth, and knows what labors She demands of her suitors will imagine for one moment that Mill succeeded in plucking the heart out of the sphinx name Logic in those seven years. Such supercilious methods of approach can win no more than they deserve. They may win the applause of Europe for many generations, if the applause of those generations happens to be cheap in that direction; but they will not win the heart of Truth. Mill's very Preface discloses to a perspicacious reader just how his work was put together. It begins thus:—

> This book makes no pretence of giving to the world a new theory [of logic—this is just what it does, however; and of course Mill knew it—] ... it is an attempt ... to embody and systematize ... the best ideas which have been either promulgated on its subject ... or conformed to by accurate thinkers in their scientific inquiries ... In the existing state of cultivation of the sciences, there would be a very strong presumption against any one who should imagine [a fault in logic, this; no matter what he is going to say] that he had effected a revolution in the theory of the investigation of truth.[57]

This would be shallow if it were sincere. It is not so, since Mill must have known that his theory of inference from particulars to particulars would constitute so far as it was accepted just such "a revolution in the theory of the investigation of truth"[58] as he pretends, for a rhetorical purpose, to decry. To suppose otherwise would be to place him lower in the scale of intelligence than would be consistent with the fabric of his works. Nor is there any presumption against his theory,—utterly erroneous as it is,—on that ground or on any other founded on the "burden of proof." Scientific men recognize a presumption against any man who announces a startling discovery while displaying gross ignorance or misunderstanding of what has already been published that bears on his conclusion. It would be ridiculous to put Mill in that category, though he was not very well read in logic; and with the exception mentioned no "burden of proof" can properly be used in science. It is merely an artificial device to prevent legal processes from going to needless prolongations; and in science discussions cannot and ought not to be abridged in that way. It is true that the phrase is sometimes used in such discussions; but it is usually nonsensical, and where it is not so is a misnomer.

As for the argument that logic must be nearly perfected, since science is so successfully prosecuted, it is easily disposed of by simply reviewing the facts of the case. To begin with, it is generally agreed that man is a rational animal; and Mill does not dispute it. Now that proposition means that man has an instinct for reasoning somewhat as birds have for flying and fish for swimming; and it is not supposed that either of those classes of animals are endowed with an understanding of dynamics.* If it be imagined that man's rationality consists in his inability to do anything successfully of which he does not understand the theory, the facts are contrary to that hypothesis, too. The reasoning of the special kind that this or that man of science has constantly to do is the chiefest part of his business, just as the special kinds of inference that are continually occurring in any other business are the chiefest part of that; and the performance of the particular line of acts of inference or judgment that are so continually recurring is learned in the business of science just as it is in any kind of business, not by reading rules about it nor by conscious reasoning about it, but by acquiring unconsciously or "subconsciously" (if you think that expression any more explanatory,) a habit of reasoning rightly, the agency of acquiring it being self-criticism, self-blame or self-praise, and energetic resolutions fortified by self-hypnotism, under the influence of constant experience and to the success or failure of his reasoning. Did Mr. Mill, I wonder, entertain the notion that medical men learn diagnosis out of books,—and, God save the mark, out of logic-books! True, there are in science, as in some other businesses, extremely important kinds of reasoning as to which, owing to their infrequency and the want of experiences as to their success or failure, no such training can occur; but in regard to them it cannot be said that science is so near the point of perfection that its reasoning could gain

* dynamics.] Peirce writes hydrodynamics instead, but as hydrodynamics does not apply to the flying of birds (which rather falls within the domain of aerodynamics) the more general science of dynamics is put in its place.

no advantage from a broader outlook. Mill's remark seems to be founded on the assumption that successful investigators are able to give the logical explanation of their own success. But during life-long intimacy with them, I have never known a single instance of such understanding. Very seldom have I seen any symptom of their understanding what such an explanation would consist in. They are quite content to rely upon* their common sense, or instinct for reasoning. One might expect this developed in them to a particularly high degree; but I cannot say that I have been able to observe that it is so. Every man of science conducts such investigation as he seems† to have within reach the best equipment for, external and internal; and it would seem to me a conjunction of events almost too improbable to occur twice in a millennium that the same man should have the requisite turns at once of curiosity and genius, together with the leisure and other facilities for any capital research into physics, and at the same time all the corresponding requisites for a much more unusual, indeed unprecedented discovery in a field at the very antipodes of science, where logic is. The most determinative character‡ of Mill's *Logic*, from which almost every other was generated, pervading all he ever wrote on philosophical topics and peeping out quite naively from the above extract from the preface to his ablest work, and imbibed from his mother's breast, was a touching confiding faith in the metaphysical teachings of his father and Jeremy Bentham. Those were what he meant when he spoke of "the best ideal promulgated by speculative minds." James Mill, the father's *Analysis of the Phenomena of the Human Mind*[59] is perhaps the clearest and most concise work on the subject, as well as being very striking and instructive; and Bentham's many writings are rendered less clear only by too much striving after clearness, so that they read too much like legal indentures. They are even better worth reading than his intimate friend, the elder Mill's. But the clearness of both is mainly due to the fewness and familiarity of the elements of thought of which they build up their fundamental doctrine.

What is this doctrine? It is a blend of several ingredients. The first is that state of belief which some people manifest who are fond of the saying "Seeing is believing." That is to say, it is a refusal to acknowledge that anything is Real,—the reader and I will examine into the meaning of this notion after we shall have duly considered the proper method of doing so,

* rely upon] *above deleted:* exercize

† seems] *above deleted:* happens

‡ determinative character] *above deleted:* distinctive feature

and until then we cannot finally decide what ought to be so called,—except material things and their motions. If you ask them whether consciousness is not Real, they will reply, Why, the grey matter of a waking brain is conscious, just as it is grey and soft and warm; but these are only appearances: they are not Real. There is a sect of such philosophers who may add that the softness and the warmth of the grey matter, though only appearances, are appearances of a great Reality called the energy of the universe; the warmth being the manifestation of a part of the energy that is, for the time, located in single real atoms or electrons, and in that state is called "kinetic" energy, while the softness is a manifestation of a part of the energy that is located in pairs of atoms or electrons, and is called, while in that state "potential" energy, and the electrical investigator has put forth a hypothesis of inner connexions to account for the two modes of appearance of energy. This is to make reality the same as permanence, or what Kant calls *Subſiſtenz, oder das Daſeyn der Subſtanz.*[60] Mill and Bentham could not consistently subscribe to that doctrine. For if a body receives one calory of heat from a source, A, and then another calory from a source, B, and finally lost half a calory to a body, C, we have not the slightest reason to think that there would be any meaning whatever in asking how much the half-calory so gained by C would have come from A, how much from B, and how much from the previous supply. It is as if M should say to N "The law requires you to pay me a dollar." and M should reply "If the law requires me to pay you a dollar, which of my dollars does it require me to pay?" To make energy Real, without supposing Hertz's* inner connexions[61] (a supposition contrary to Comte's maxim,) is contrary to the doctrine of Mill and Bentham, according to which nothing is Real unless it be definitely individual, or, as the logicians say, unless it be a Designate (that is Designable) Individual, or something all whose parts are in themselves susceptible of identification, whether we can identify them or not. This doctrine is the Nominalism of Ockham. Nominalism in itself is the doctrine that nothing that is *general* is real; a *General* subject of an assertion being a subject that,† within certain limits the person addressed may choose at will; as "Any American born, thirty-five years of age, is eligible to the office of President of the United States," or "Any chemical element may be taken as the one whose atomic weight shall be 1, or as 16, or as *any* other number," where both the element and the number are general. But an *Indefinite* subject[62] is one

* Hertz's] *spelled by Peirce:* Herz's

† subject of ... that,] above deleted: proposition being one of whatever subject,

whose full identification the *[...]* from his childhood.* The doctrine is that nothing is real but individual things and individual facts; or, as Mill himself says in the third chapter of his book, Individual Feelings, Individual Minds, Individual Bodies, Individual Facts of Likeness and of Unlikeness between pairs of Individual Real Objects, and Individual Facts of Successions and Simultaneities in Time.[63] [But he cannot consistently have meant that likenesses and unlikenesses, successions and simultaneities are *themselves [*not*]* real but only that the facts which the mind regards under these *aspects* are so.[64] But Mill's candour was constantly landing him in inconsistencies and this is probably an instance of it.] Nominalism had been, it is true, the universal opinion of all modern philosophers, excepting, perhaps, Schelling; but those philosophers had been, with very few exceptions, vague literary writers; and the few who were not so took on that style of thinking whenever they approached this question. Never during the whole scientific era had this question been in any degree threshed out. Many men (Hobbes, Locke, Berkeley, Reid, Dugald Stewart, Karl Pearson, and many others) have expressed themselves with vehemence about it, or to use a short word have scolded about *[*it*]*; but if you examine what they write critically and without bias, you will find it has not the substance of a frayed out cloth. We have evidence of the truth of this right before us; for though J.S. Mill was looked upon, at the time and for long after, perhaps even today by the majority of readers, as having examined this question at least as carefully as any other writer, and he professes to be teaching the principles which have been "conformed to" by the men of science; and yet he does not enumerate among the realities of the Universes the laws of nature. On the contrary, he introduces the term "uniformities" in place of "law" in order to express his belief that the only realities in those laws are "resemblances" between different pairs of facts.[65] He does not perceive that while a known resemblance is strictly limited to the objects to which it is known to apply, the characteristic of a law and all that gives a law its high importance and utility is that, as soon as it is known, it informs us as to *future* and as yet unknown experiences, which it could not do if it did not express a *real habit* of the things to whose behavior it relates.† Now a habit

*Some text could not be recovered, and it is difficult to estimate how much material is missing. On 1 September 1909 Peirce replaced a now-missing single sheet with at least seven sheets of which only the first six were found. It is likely that Peirce never completed the revision he started on September 1. Six days later, on September 7, Peirce started all over again, and since he does not get all that far in this new attempt it is impossible to say whether he intended to incorporate any of the previously written material in it or not.

† the things … it relates.] *above deleted:* nature.

consists in the fact that man or a thing *would* usually behave in definite way upon *any* definite sort of occasion and is thus by definition general. Mill does not even recognize that mental or physiological habits are real, keenly as most of us have had to feel their power for good and for ill. If a gambler has left his stake on the red until it has doubled three or four times, he will either pocket it, thinking it time for the luck to turn, or he will reason, as Mill would have us do, from a mere resemblance, and let it lie; but a sound reasoner will know that precisely because a run of luck is a mere resemblance, nothing whatever can be inferred from it in regard to future events. Mill himself was most characteristically on the point of stumbling upon the essential difference between a resemblance and a law at the end of the third section of the third chapter of his third book,[66] where he remarks that as soon as a chemist has made sure of some new reaction with a single specimen of pure gold, he infers with entire confidence that every piece of gold in the universe would behave in the same way under the same circumstances.[67] "Now mark," he continues, "another case, and contrast it with this. Not all the instances which have been observed since the beginning of the world, in support of the general proposition that all crows are black, would be deemed a sufficient presumption of the truth of the proposition, to outweigh the testimony of one unexceptionable witness who should affirm that in some region of the earth not fully explored, he had caught and examined a crow, and had found it to be gray."[68]

Who, that were not under the domination of a strange bias of mind, could so emphasize the difference between the two cases mentioned, as Mill here does, and yet fail to perceive that the difference is that in the latter what we have observed what is very likely no more than a *resemblance,* while in the former innumerable *successful predictions* have afforded us ample warrant for believing that there is an underlying *property* or *habit* of gold, a *law* of its nature, which is *not* a mere resemblance between actualities, but *determines what will be.* That very bias of mind is a good example of the force and reality of such a general tendency. Mill, in the chapter I first quoted (the third of his treatise, after the introduction,) distinctly refuses to admit either *laws,* or *properties,* or *habits* to be real, and different* from mere resemblances. Indeed, the whole associationalist school, with the signal exception of its founder, the Rev. Mr. Gay,[69] were curiously given to slurring over the reality of habits, although an Association is nothing in the world but a mental habit. This Ockamistic Nominalism, or Individualism, which prevented Mill from discovering the true theory of reasoning (or at least, the main proposition of the true theory,) was due to a bias that had

* different] *above deleted:* distinct

more or less affected all the philosophers of the scientific era, not as a result of any discussion about it, but as a mere uncontrolled habit, handed down by tradition and prejudice from the *littérateurs* of the *renaissance*.) Those elegant imitations of Tully, did not, you may well believe, undertake any stringent investigation into this question; and my reading of the few among them whom one may compliment with the title of philosophers makes me pretty confident that they had neither the exactitude not the intellectual industry required for such a research. But as well as I have been able to make out as the result of unstinted application,* the universal prevalence of nominalism in modern† times came about as follows.

Early in the XIVth century, the marvelous keenness and subtlety of mind of John Dunce the Scot had rendered his metaphysical opinions, which were of a guardedly realistic tinge (i.e. acknowledging the reality of many general characters,) are dominant in the great universities of northern Europe. But his subtilty had made his metaphysics so very intricate and in some particulars so excessively difficult to understand; and after he died at an early age in 1308, his followers carried its complication very nearly perhaps to the point of utter unintelligibility. Efforts were consequently made by Durandus of St Pourçain and Petrus Aureolus[70] to simplify the Scotistic doctrine; and finally (perhaps as early as 1320) William of Ockham (a parish of Surrey) boldly came out that there are things called spheres on account of them resembling one another, but denying that there is any real spherity common to them.[71] Sphericity, hardness, and other properties are, he declared, mere grammatical forms. This substantially, and neglecting slight points of difference, was the very doctrine taught by James Mill to his son and other pupils and fully accepted by that son's other teacher, Jeremy Bentham; and they were all readers of Ockham. Upon Ockham's deliverance of his opinions, at once a great amount of disputation arose and continued for a century. For in those times "everybody" took logic *au sérieux,* and controversies were carried on *vivâ voce,* under a moderator enforcing a code of rules called *obligations,* which required the disputants to stick to the point, and answer each other's arguments explicitly and formally. The result of these disputations was that when the revival of learning came, shortly after 1453, it found the Scotists, or the Dunces, as the humanists called them, after their master, in complete possession of the great northern universities, where metaphysics and logic were chiefly cultivated. Of course, the Dunces of the universities, largely Franciscan friars, and ignorant of polite literature, were obscurantists; but the word "dunce" by no means in those days implied

---

* unstinted application,] *above deleted:* a good deal of industry

† modern] *above deleted:* our

stupidity. On the contrary, what it did distinctly imply as you may find by consulting the *Oxford Dictionary* or the *Century Dictionary,* was a skill in argumentation with which the humanists were utterly unable to cope. In their predicament, they got what help they could from the Ockhamists, who still formed a party on account of their maintaining a certain attitude toward the pope; and some of them, such as Eck,[72] were no mean logicians; and so the humanists, knowing nothing of metaphysics, naturally took up the Ockhamist position on that subject, and all the more readily since it was only by rhetorical arguments that it was susceptible of being defended. When the humanists got possession of the universities, politics aiding, the old masters lost their chairs, the system of disputation languished and became a mere formality, metaphysics ceased to pay close attention to their opponents' arguments, and the whole subject degenerated into unsupported talk, as now; but the standpoint of nominalism was maintained, both as a heritage and as easy ground for a rhetorician to defend.

That, as well as I have been able to trace the matter out, is the way in which it has come about that almost all modern philosophers have been led into nominalism, error though it demonstrably is. Many of them call their variety of nominalism, "conceptualism"; but it is essentially the same thing; and their not seeing that it is so is but another example of that loose and slapdash style of thinking that has made it possible for them to remain nominalists. Their calling their "conceptualism" a middle term between realism and nominalism is itself an example in the very matter to which nominalism relates. For while the question between nominalism and realism is, in its nature, susceptible of but two answers: yes and no; they make an idle and irrelevant point which had been thoroughly considered by all the great realists; and instead of drawing a valid distinction, as they suppose, only repeat the very same confusion of thought: which made them nominalists. The question was whether all properties, laws of nature, and predicates of more than an actually existent subject are, without exception, mere figments* or not.† The conceptualists seek to wedge in a third posi-

---

* figments] *before deleted:* and nothing more, or whether some of them are true whatever you and I may think about them

†Original Peirce footnote:

> It must not be imagined that any notable realist of the XIIIth or XIVth century took the ground that any "universal" was what we in English should call a "thing," as it seems that, in an earlier age, some realists and some nominalists, too, had done; though perhaps it is not quite certain that they did so, their writings being lost. Their very definition of a "universal" admits that it is of the same generic nature as a word, namely, is: "quod natum aptum est *praedicari* de pluribus." Neither was it their doctrine that any "universal" *itself* is real. They might, indeed, some of them, think so; but their realism did not

tion conflicting with the principle of excluded middle. They say, "Those universals are real, indeed; but they are only real thoughts." So much may be said of the philosopher's stone. To give that answer constitutes a man a nominalist. Are the laws of nature, and that property of gold by which it will yield the purple of Cassius no more real than the philosopher's stone? No, the Conceptualists admit that there is a difference; but they say that the laws of nature and the properties of chemical species are results of thinking. The great realists had brought out all the truth there is in that much more distinctly long before modern conceptualism appeared in the world. They showed that the general is not capable of full actualization in the world of action and reaction but is of the nature of what is thought, but that our thinking only apprehends and does not create thought, and that that thought may and does as much govern outward things as it does our own thinking. But those realists did not fall into any confusion between the real fact of having a dream and the illusory object dreamed. The conceptualist doctrine is an undisputed truism about *thinking,* while the question between nominalists and realists relates to *thoughts,* that is, to the objects which thinking enables us to know.[*]

---

consist in *that* opinion, but in holding that what the word *signifies,* in contradistinction to what it can be truly *said of,* is real. Anybody may happen to opine that "the" is a real English word; but that will not constitute him a realist. But if he thinks that, whether the word "hard" itself be real or not; the property, the character, the predicate, *hardness,* is not invented by men, as the word is, but is really and truly in the hard things and is one in them all as a description of habit, disposition, or behaviour, *then* he *is* a realist.

[*]In a discarded branch, dated June 7, Peirce continues at this point as follows (R 624:2–3, 8):

> This excursus upon nominalism has been so long and, at the same time, so unconvincing that I fear that my reader, if not I myself, may need reminding what the purpose of this introductory chapter was to be. My intention has been to consider~~, calmly and sympathetically,~~ how far the present indifference to logic is reasonable. To that end I am inquiring how much modern logic has done for the improvement of modern reasoning. Not that the highest recommendation of a science necessarily *lies in* the utility of its practical applications, but that the surest *sign* of health and vigour in anything that lives and grows, (and science, which consists in cooperative discovery by the best known methods, essentially must grow and must be alive,) lies in its fertility. It is the moral of the fig-tree, which expresses, by the way, the very essence of the method of science: the hypothesis which brings forth predictions that ripen into verifications is preserved, while the hypothesis that is barren of such good fruit is hewn down and cast into the fire. It is the breeder's selection.
>
> It is in accordance with that principle that I am considering the existent indifference to logic; since by far the surest way to fell a study, chop it up, and cast it into the stove is simply to neglect it. Refute it and show it up as dire fraud, imbecile nonsense, or visionary extravagance and your refutation may prove to be just the tonic to endue it with vigor. Witness mediumistic "manifestations," homœopathy, and phonetic spelling.

Perhaps the eye of some stanch admirer of Mill, such as abound among the clearest minds, may fall upon these pages with indignant disgust at my insistent depreciation of his hero. But he will quite mistake my attitude. I, too, greatly admire him. In philosophic subtilty, he was almost the rival of Leibniz, who was, excepting Kant, the greatest philosophic intellect among the moderns. I shall give an example of this presently. Of course, Mill had not one ray of the originality of Leibniz. Leibniz likewise began by being a strong nominalist; but this character wore off, little by little as he matured, and quite disappeared when he finally enunciated his "law of continuity." Nominalism is naturally highly attractive to a brilliant intellect that has not yet attained its meridian power in logic: the doctrine is so charmingly simple.* This simplicity would in verity be a considerable† argument in its favor, were there any room to doubt its falsity. But there will remain no such doubt for a mind that shall have well considered the simple objection to it, which may be put in a nutshell, thus: A very extensive experience shows that if a considerable number either of things or of facts are studied together on account of some marked peculiarity, P, which is common to each and all of them, and‡ if, upon examination, it be found that some

*In a discarded draft, Peirce writes about Mill (R 624:5):

> He is a difficult writer to refute on account of his remarkable candour, which causes him to admit propositions in one place which he almost directly denies in another; so that it is hard to make out what his position is and even a careful student of Mill will be inclined to exclaim at what he supposes to be my misrepresentations of his doctrines. Nevertheless, having often and often gone over the ground with one of Mill's subtlest disciple,[73] I am completely confident that I do not misrepresent him in the least. Mr. Jevons showed up several of these inconsistencies in some strong articles in the *Fortnightly Review*. For example, although he denies that there is any real Law in nature, (the consistent nominalistic opinion being that of Dr. Karl Pearson, Poincaré, and others that we manufacture those law, just as they say we manufacture God,) he nevertheless yielded to Whatley's unpardonable blunder that an induction as a syllogism whose major premiss expresses the great Law of the Uniformity of Nature. Which great Law thus absolutely proves the reality of all the other Laws of Nature. Then, as if this were not inconsistency enough, he goes on to say that this Law is proved by Induction; that is, by a syllogism of which it is itself the major premiss. He formulates this Law in Book III, Chap. iii, §1, thus: "What happens once, will, under a sufficient degree of similarity of circumstances, happen again, and not only again, but always." In this very essay, I shall show that this statement has absolutely no meaning; and in another essay in the volume shall show that nature is quite as un-uniform as any world could possibly be imagined to be, and shall examine some other attempted formulations of a general uniformity in nature, showing that it is "subjective," as the lingo is, that is, "is all in your I."

† considerable] *above deleted:* strong

‡ considerable number … them, and] *above deleted:* large collection of surprising facts are found having some unusual feature in common, that does not justify any surprise, since that feature was the very reason for their being considered together *[…]*

entirely different and surprising property, Q, belongs* to every one of the collection, and it is consequently suggested that perhaps, owing to some unknown cause, everything or nearly everything that has the peculiarity, P, possesses the property, Q; and if, in order to test that conjecture, a few more things or facts having the property, P, be collected, and it be found, upon examination, that every one of them, too, possesses the property, Q, then we find that we can predict, with much confidence, that the next thing or fact that shall come to our notice as possessing the peculiarity, P, will be found to possess the property, Q; and such prediction will very rarely fail. Now it is very true that all the modern text-books on Probabilities that ever I have seen pronounce this to be a natural consequence of the Doctrine of Chances. They tell us that, whatever whole number n may be, if the first n events of examining new P-objects result in their being all found to have the property Q, then the probability is $\frac{n+1}{n+2}$, (i.e. the odds are $n + 1$ against 1, that the next examination of a new P-object will have that same property Q; and I should not at all wonder if the greatest living astronomer, Professor Simon Newcomb, were to pronounce this correct.[74] The principle on which the result rests was first set forth by a certain Rev. Mr. Bayes,[75] otherwise unknown to science. It was taken up by Laplace in his work on the subject, which is as weak on the logical side as it is strong on the mathematical side; and no writer of a text-book has dared to contradict Laplace.[76] The eminent mathematician, Boole, did dispute this position, but Boole's own presentation of the subject was not as clearly well-founded on the logical side, as it should have been.[77] Curiously enough Mill himself, in the first edition of his *Logic,* contested Laplace's result with much rhetorical effect; but his mind proved wayside soil for the good seed; and in the second or third edition the chapter was replaced by a newly written one in which Mill surrendered to the mathematicians. In another essay in this volume I shall show, conclusively and clearly, that the books are entirely wrong. At present, I must contend myself with referring the reader to that other chapter, with making here a condensed and general argument to the contrary, and with pointing out a couple of highly suspicious consequences of the Bayes-Laplace doctrine.† One of these is that, according to this, if the

* belongs] *above deleted:* be found to belong to

†In a discarded branch Peirce writes the following (R 627:2–3; June 14):

> The reasoning of the Calculus of Probabilities, otherwise known as the "Doctrine of Chances," is purely mathematical. That is to say, given the three following definitions, all the rest is so. The first of these definitions is that the probability that any description of event, E, will happen if any other description of event, F, happens, is the proportion, out of the number of individual occurrences of the events of the kind, F, in the long run, of the number of such occurrences that will also be occurrences of the kind E. The

prediction that the first object, (whether thing or fact,) would be found to have the surprising property, Q, were made without any basis of previous observations whatsoever, the value of n would be zero, so that, according to the formula, the probability would be ½, (i.e. there would be an even chance) that the first P-object examined would have the property, Q, in spite of this property being, by hypothesis, of a very surprising kind. In the next place, if a novice whose knowledge were confined to the theory of probabilities were to enter the casino at Monte Carlo, and placing a franc on the red were to see it doubled four times running, which happens every day, he ought to conclude that he might safely bet 5 to 1 that it would be doubled again if he left it on the red. In other words, it expresses an approval of the gambling mania; for it is just that kind of expectation that really *disconnected* events will follow laws such as really connected events follow, that leads to the gambler's frenzy. One more example: I never noticed particularly the number of a lottery ticket that won the capital prize; but why would it not be a good scheme to look up the old records and find such a number, and then buy the ticket bearing that number? For here, since I should have $n = 1$, the odds would be 2 to 1 that that number would win the capital prize. Or, indeed, in case one should dream that a certain number won a prize of $100,000, why would not the odds be 2 to 1 that the ticket of that number, in the first lottery offering such a prize, would fetch it, seeing that there is nothing in the mathematical operations that produce the rule,—the four operations, addition, subtraction, multiplication, and division,—that distinguish sleeping experiences from waking ones? I wish it to be understood that the conclusive refutation of the whole part of the current doctrine of probabilities which assigns probabilities to inductive conclusions is wholly independent of such objections as the three I have just adduced. For I am aware that defenders of the formula may "crawl out" of the first two, at least. That is, while not squarely meeting the real arguments that underlie them, they can show that there would always be circumstances which would forbid the use of

purely logical operation of analyzing the meaning of the expression "the long run," an analysis which is performed in very slovenly ways by different mathematicians, and by none more than by Laplace, will be set forth in a separate essay of this volume. The reader will get a sufficient, if not very satisfactory, notion of the meaning, without any explanation here. Second, two kinds of events are said to be "mutually exclusive" if, and only if, no individual occurrence can be at once of each of those two kinds. Thirdly, two kinds of events are said to be mutually independent, if, and only if, the probability of either, if the other happens, is equal to its probability if that other does not happen. These definitions are somewhat confused; but I think they will be understood well enough for the purposes of this argument, which will be repeated more fully in the Essay on Probability.

Peirce continued with an example that remained unfinished.

the formula in the manner I have supposed. Those objections set forth real grounds of suspicion without being conclusive. But I will now give, in brief and abridged form, a more fundamental objection, as conclusive as it is simple.

Any sound argument, whatsoever, that is to say, any truth, A, simple or complex, which contains a sound reason for believing a proposition, B, must represent a fact or facts that are *in re,* and not merely in our minds, connected with the fact stated in B. This is self-evident. One fact cannot, of itself, afford any evidence of a totally disconnected fact; and in order that an argument from one fact should, of itself, warrant any belief or approach to belief in another fact, the real connection or concatenation that exists between the facts as they are *in re* must be stated in the premises. "Hold on," I hear the reader exclaim, "do you mean to say that an argument from analogy has no value whatever." Certainly, I do mean just that, understanding the word "analogy" in the sense of an agreement that does not amount to any real connection between the events, and excluding the *minus* values of arguments that lead to error. For instance, I do not admit that the fact that, at the last supper of Our Lord, there were thirteen persons present of whom one went out and hanged himself, is any real ground for the slightest uneasiness about sitting down to dinner with twelve other persons, unless it be supposed or surmised on some other grounds, that from that time forward a real curse was placed on such assemblages, or some other real cause and real circumstance really connecting the Lord's supper with the dinner-party in question. Nor can I easily believe that the most superstitious persons would be disturbed by finding themselves participating in a dinner of thirteen, unless they thought they had some other ground for some such surmises as I have mentioned. When I turn over in my mind instances of arguments from analogy that have come up in conversations or discussions in which I have taken part, I find that in those that have struck me as having a *plus* value, there has always been evidence of real connection between the objects compared; and if the reader's experience has been different, I suspect it is because Nominalistic habits of thinking have prevented his recognizing the reality of connections in many cases. A believer in Weissmann's doctrine would probably admit that forms that share the same germ-plasm, even if it be differently modified in them, are really connected: yet if he is to be a consistent Nominalist, he must not do so, since the portions of germ-plasm in different individuals are "numerically" different (to use the bad translation of a Greek phrase that passes current with logicians); and their unity is, after all, nothing but unity of behaviour.[78] The argument from analogy that Mars is inhabited is tolerably strong; but its strength is found mainly in the fact that the very same sunlight, with the very same constitution (except for a larger proportion of the medicinal blue,) that generates all the teeming life of our planet shines there as well. Yet not the identical waves that break on the earth break on Mars: but only waves of the same *properties;* and nominalists contend that it is our minds that make the properties. On the other hand, when people of a rationalistic turn of mind jump to the conclusion that some rule that it would have been advantageous to adopt when public habits now thoroughly engrained were just beginning to take shape, could still be advantageously set up as an object of endeavor now when it can only be brought about by first destroying firmly settled conventions, (say in respect to phonetic spelling, the base of numeration, manners of dress, familiar principles of law, and the like,) too slight to warrant the enormous expenditure that the adoption of their conclusion would entail.* It is an analogy, to borrow Grover Cleveland's phrase, between a theory and a condition, or say, between a fantasy† and an actual state of things.[79] I continue to insist that nothing is plainer than that if two things have, in their real being, nothing to do with each other, a fact about the one can supply no knowledge or approach to knowledge about the other; and further on, when I shall have laid the proper foundation for it, I shall demonstrate that the truth is as I say. At present I am only seeking to shoot a glimmer of truth into my reader's consciousness; because nothing is less convincing than an absolute demonstration when its conclusions do‡ violence to all one's habits of belief; just as nothing is more blinding than too bright a light suddenly shot into eyes that have long been in utter darkness.

How well do I remember how, in consequence of missing a railway connexion at a small junction in Virginia, where only four beds were available, of which three had been surrendered to the ladies, and the fourth to an older man and to an invalid, eleven of us young fellows sat down to make a jolly night of it at a game of brag, and how on our exchanging cards it turned out that all our names began with P: Pinckney, Poindexter, Parmalee, Palmer, Peterson, Pratt, Poare, Poole, Patterson, Palfrey, and I cannot think of the eleventh? And how, not long after dawn, the railway whistle was heard, and soon after a twelfth fellow was descried coming up the hill with his carpet-bag of the period, who could perforce be added to our company; and how I then remarked that, according to Laplace and the

---

* too slight ... would entail.] *above deleted:* not to amount to ~~any~~ a connection sufficient for a basis of reasoning.

† fantasy] *above deleted:* dream

‡ conclusions do] *Peirce wrote, with the concluding letter of "conclusions" seemingly added later:* conclusions does

treatises on the Doctrine of Chances, the odds were twelve to one that his name, too, would be found to begin with P, but that I would willingly bet one to ten that it would not, an offer that was promptly taken up by Sam Pratt? And how, when he came, I immediately fell into pretendedly interested conversation with him, and, handing him my card, begged the favor of his name? And how he told me it was Sa*l*mon (with the *l* pronounced,) whereat I winked at Pratt in triumph, while the new-comer was fumbling for his card, which, being produced read "Psalmen," to the eternal triumph of Nominalism? I ought to add that my recollecting almost all the names is not so very surprising, considering that I had invented the whole story not half a minute previously. (Which furnishes answers to my above interrogations.) I think, however, that it testifies, none the less, to the true valuation of those mathematicians' good sense. For the calculus of probabilities, they tell us, is "good sense systematized."

It is further to be taken into account, in estimating the value of that rule of the treatises on Probabilities, that the Doctrine of Chances is put forth as a mathematical "theory," as the mathematicians, adhering to the Greek, call such a body of doctrine; and as a mathematical theory, its reasoning is bound to be *mathematical reasoning.* Now there are two characteristics of mathematical reasoning that merit attention in the present examination of that Nominalistic rule. The first of these is that the premises, or "data," of a mathematical reasoning are, for the purpose of drawing the conclusion, mere "hypotheses," as Riemann and the better logicians among the mathematicians call them.[80] In order to show what that signifies, I call your attention to the fact that in ordinary, non-mathematical reasoning, if my story had been a real occurrence, and the twelfth man's name had really been "Psalmen," that event would have constituted a not inconsiderable confirmation of the rule of the treatises, while it utterly loses all argumentative value as soon as we learn that it is fictitious. With a mathematical reasoning, however, it is not so. If the data happen to be veridical, the conclusion is so, likewise. But whether the data are so or not, the conclusion remains their necessary accompaniment, so that, as for the reasoning itself, it remains just as conclusive in an imaginary world as in a real one, instead of losing all probative force in case the premises are fictitious. For that reason, my story was very pertinent as showing that the rule cannot have been correctly deduced by mathematical reasoning, seeing that it makes such a difference as it is fact or fiction. The point is rather a subtile one, which I devised the story to bring out. Regarding the rule of the books as a sort of law of nature, if the story had been true, we feel that the circumstance could be regarded as putting it to the test, and that, had it been so, the fact that the twelfth man's name turned out to begin with P, as only about one name out of fifteen ordinarily do, would have been a tolerably strong confirmation of the law, while when we learn that the story is fictitious the argument completely evaporates. But when we remember that there is no pretence that the rule is a law of nature, but that it is claimed as a mathematical result that the odds *[*must be*]*[*] 12 to 1, we see that if this be so, it must hold equally in fiction as in fact that such are the odds informing the twelfth name beginning[†] with P, reason moves us to declare that the rule, far from being a correct result of any codification of sound judgment,[‡] is mere nonsense.

The second characteristic of mathematical reasoning is that the state of things concluded must, to be a sound mathematical result, have been among the states of things actually expressed[§] in the data, though presumably in a hidden manner, the effect of the reasoner's reasoning being merely to draw the reasonee's attention to the circumstance that he had admitted it, in admitting those data (reasoner and reasonee being most usually alternating *rôles* of the same mind; the suggesting, proposing, appealing *rôle* and the *rôle* of apprehending, weighing, and adopting,—or else rejecting). This statement will probably stagger the reader at first. "What?" he may say "do you mean, for example that in stating that two couriers moving at velocities v_1 and v_2, are, the one at t_1 seconds after noon of a certain date at the point distant d_1 along a road, and the other at t_2 seconds after the same noon at a point distant d_2 along the same road, I am stating when and where they will be together?" No, I reply, I do not say you expressly *state* it, but I do say that you precisely *say* it, in that sense of the word "say" in which what you say is, not the words you *utter*, but the state of things you affirm, which is the *meaning* of your statement.[**] For by saying that the couriers advance at the velocities v_1 and v_2, you mean and say that they move at the *constant* velocities v_1 and v_2; and the meaning of saying that a courier moves at the uniform velocity v_1 or v_2 along a road is to affirm that, at *all instants* he is at a distance along that road further than he was at all previous instants by the products of the differences between the instants multiplied into his constant velocity v_0 or v_1; and in saying this of *all* instants you say it of *each* of them, and among them of the instant which is

[*] it is … the odds] *above deleted:* it must, as a mathematical result be just as much | *[*must be*]* *is an editorial insertion based on the deleted text.*

[†] Beginning] *above deleted:* should begin

[‡] sound judgment,] *above deleted:* good sense

[§] actually expressed] *above deleted:* premised

[**] *say* it … your statement.] *above deleted:* state it, if one pays attention to your meaning.

$$\frac{t_1v_1 - t_2v_2 - d_1 + d_2}{v_1 - v_2}$$

seconds after the noon of the date referred to; and thus you say that, at that instant, the first courier is at that instant at the distance

$$\frac{t_1v_1 - t_2v_2 - d_1 + d_2}{v_1 - v_2}v_1 - t_1v_1 + d_1$$

along the road, which, by the definitions of the arithmetical operations, means the distance

$$\frac{(t_1 - t_2)v_1v_2 + d_2v_1 - d_1v_2}{v_1 - v_2};$$

while you say that at that same instant the second courier is at the point at the distance

$$\frac{t_1v_1 - t_2v_2 - d_1 + d_2}{v_1 - v_2}v_2 - t_2v_2 + d_2$$

along the road, which means at the distance

$$\frac{(t_1 - t_2)v_1v_2 + d_2v_1 - d_1v_2}{v_1 - v_2},$$

that is to say at the very same point as that at which the first courier is situated at the same instant; and that is what it means to say that at the time

$$\frac{t_1v_1 - t_2v_2 - d_2 + d_1}{v_1 - v_2}$$

the two couriers will be at the same point, distant

$$\frac{(t_1 - t_2)v_1v_2 + d_2v_1 - d_1v_2}{v_1 - v_2}$$

along the road. This illustrates the important sense in which it is true that mathematical reasoning can conclude nothing that has not been actually said in its data, the only function of the reasoning being to direct attention to this point, which is partly said in one datum and partly in others. In other words, it only "syllogizes" or "says together" what had been said separately.

Now applying that to the pretension of the text-books on probabilities, to say only that on the n first occasions a certain kind of event is observed, is to say just nothing at all as to what will or can be observed on the $(n + 1)^{th}$ occasion; and consequently it is impossible that the mathematical reasoning that professes to deduce a certain probability that the same kind of event will happen on that $(n + 1)^{th}$ occasion should be correct,—unless, indeed, the statement that "the probability of the same kind of event's happening on that $(n + 1)^{th}$ occasion is $\frac{n+1}{n+2}$" be entirely meaningless. Since the books do not say in any definite way what it means, that may, not improbably, be the case; for the mathematicians are far more likely to talk nonsense than to be wrong in their algebraical transformations. It will occur to you, reader, that perhaps their reasoning is not of the kind that I call mathematical, notwithstanding its conclusion being numerical and being pronounced with the absolute confidence of a mathematical result. In another place in this volume I shall lay this pretended reasoning before you, and shall show what its fundamental viciousness is. Meantime, I will say to you that you will then find it to be, in truth, *not* mathematical, but, not to be so, only because it is incorrect; that it is intended to be mathematical; but, in fact, owing to a meaningless definition, it is no reasoning at all; having no sense at all. Meantime, I appeal to your own Understanding to confirm my simple position that no kind of sound reasoning can conclude from a series of facts another fact utterly disconnected from them.

I will fabricate an example to illustrate this. There was a gentleman of high intelligence and ample means who, having lived all his life in his native province of the Sze-chuen, but having formed a friendship with one of the Western Barbarians, was persuaded to accompany the latter on a visit to America. So, starting from Lui-po on Kin-sha, or Golden Sand River, which is simply an upper stretch of the Yang-tse-Kiang, they went down in native boats to Nanking, travelling thence to Shang-hi, and there had the good luck to find a comfortable American sailing-ship bound for New York. For the American, who could have passed an examination in the Five Classics with high honours, was determined that his Chinaman brother should form his conception of the Western World as he had formed his of the Celestial Empire, by his own reflections upon his own experiences. He merely taught the mandarin as much English during the voyage as would enable him to paddle his own canoe. Arrived in New York, the Chinaman was conducted in an automobile, with all articles of instant necessity to a fashionable and quiet hotel, and to a suite of three rooms and bath, by the American, who had then immediately to return to the pier to complete the Custom House formalities and to arrange for the delivery of the three servants and the luggage. The Chinaman, left alone, was reposing on the bed after his bath, when his attention was attracted to two small white buttons on the wall. By that time, the dusk was coming on, though he could examine the buttons perfectly. They looked as if they were meant to be pushed; so he pushed one of them. Just as he did so, the room was instantaneously illuminated, and looking up, the stranger saw that half a dozen lights were shining brilliantly in the chandelier. He could not imagine how this had been effected; for it did not occur to him that his pressing the button on the wall had anything to do with it. That action seemed to have remained without any effect at all: he had thought that perhaps it would cause some secret door to open; but it seemed that the apparatus was out of order,—unless, indeed, the other button had to be pushed as well. It did not seem possible, however, that the two had to be pushed at once; since it would have been so easy to arrange for a single button to do whatever the simultaneous pushing of two would do. The facts that the walls were decorated in false panels with projecting pilasters between them, and that the buttons were placed on a pilaster by the side of the door,—and possibly, likewise the sight of the transom-window over the door,—helped to fasten his conjecture to the idea of an opening door. He had already started to look into the other rooms to see whether a door might not have opened in a wall of one of them, when it suddenly struck him that the obvious explanation almost doubtless was that one button caused the door to open while the other made it shut itself, and that he had pressed the wrong one. Confident of this, he returned and pressed the second button, when almost instantly the lights went out leaving the room seemingly considerably darker than it had been at first. This time the simultaneity of the darkening and the pressure of the button struck him forcibly. "Can it be," he exclaimed within himself, "that the pressing upon these buttons are in any way *connected* with the kindling and the quenching of those lights. Perhaps it would be no more marvelous than the kindling and the quenching are in themselves. If such connexion there be, why may I not predict that as soon as I press the first button again, the lights will be shining?" Thereupon, he did press the first button, and of course the lights shone so instantly that he could hardly tell whether his pressing had caused their kindling, or their kindling had startled him into pressing the button. He shouted out the Chinese for 'Heurêka,' and would have repeated the performance of Archimedes, had there been room for it within the four corners of Chinese etiquette. He repeatedly pressed one button after the other, each time announcing aloud before he did so the effect he was about to produce. After that, he left the lights shining and contended himself with extinguishing them and lightning them up again at intervals of just one hour each. When the American came in he recounted to him his whole experience. He said, "You see, at first it did not occur to me that there could be any real connexion between the button-pressing and the sudden illumination; and, of course, if *[...]*[*]

NOTES

1. For a discussion of the Metaphysical Club and the presentation Peirce is referring to, see introduction.

2. Chauncey Wright (1830–1875). Much of his work is collected in *The Evolutionary Philosophy of Chauncey Wright* (3 vols., edited by Frank X. Ryan; Bristol: Thoemmes Press, 2000). See also Edward H. Madden, *Chauncey Wright and the Foundations of Pragmatism* (Seattle: University of Washington Press, 1963). Jeffries Wyman (1814–1874), American naturalist and anatomist, and Hersey Professor of Anatomy at Harvard (1847–1874).

3. Asa Gray (1810–1888) was generally considered the most important American botanist of the nineteenth century. He was Professor of Natural History at Harvard from 1842 to 1873.

4. Wright worked as a computer for the *American Ephemeris and Nautical Almanac* from its inception, in 1852, till 1872.

5. August Comte (1798–1857), *Cours de Philosophie Positive* (Paris: Bachelier, 1830–1842).

6. From the fall of 1859 till the late spring of 1860, Peirce was in Biloxi, Mississipi, as part of a US Coast Survey expedition.

7. Herbert Spencer (1820–1903), British philosopher, biologist, and sociologist. Though now largely forgotten, Spencer was perhaps the most famous British intellectual in America in the latter half of the nineteenth century.

8. For Peirce's criticism of the Spencerian type of evolution, see e.g. "Herbert Spencer's Philosophy" (W6, sel. 45; esp. 397–99); see also "Evolutionary Love" (W9, sel. 30).

9. Wright's discussion of cosmical weather is found in his 1864 "A Physical Theory of the Universe," which is included in *Philosophical Discussions*, edited by Charles Eliot Norton (New York: Henry Holt and Co., 1877), 1–34.

10. The paper "On the Movements and Habits of Climbing Plants" appeared in the *Journal of the Linnean Society of London* (Botany) 9 (1865): 1–118, but received little attention until 1875, when it was published as a book (London: John Murray). *The Descent of Man, and Selection in Relation to Sex* (London: John Murray) appeared in 1871.

---

[*]The text breaks off mind-sentence at the end of the sheet, suggesting Peirce may have continued on at least one subsequent sheet. No further sheets, however, have been identified.

11. John Stuart Mill, *Examination of Hamilton and of The Principal Philosophical Questions Discussed in his Writings* (Longman, Green, Longman, Roberts & Green, 1865).

12. Peirce's began reading Kant's *Critique of Pure Reason* at the age of sixteen in 1855.

13. On Peirce's (extreme) scholastic realism, see Mateusz W. Oleksy, *Realism and Individualism: Charles S. Peirce and the Threat of Modern Nominalism* (Lodz: Lodz University Press, 2008); John F. Boler *Charles Peirce and Scholastic Realism: A Study of Peirce's Relation to John Duns Scotus* (Seattle: University of Washington Press, 1963); and Rosa M. Mayorga, *From Realism to "Realicism": The Metaphysics of Charles Sanders Peirce* (Lanham: Lexington Books, 2007).

14. For a more extensive account by Peirce on Spinoza's *Ethics*, see his *Nation* review thereof (P 582; *Nation* 59 [8 November 1894]: 344–45).

15. Chauncey Wright, *Philosophical Discussions: With a Biographical Sketch of the Author by Charles Eliot Norton* (New York: H. Holt, 1877). Charles Eliot Norton (1827–1908) was an American social critic and progressive social reformer, and Professor of Art History at Harvard (1875–1898).

16. Alexander Bain (1818–1903), Scottish empiricist philosopher and a lifelong friend of John Stuart Mill.

17. Nicholas St. John Green (1830–1876) is most well known for his "Proximate and Remote Cause," which appeared in *American Law Review* in 1870 and which landed him a position as lecturer at Harvard's law school. In 1873 he accepted a law professorship at the new Boston University Law School. Green rejected, as did Oliver Wendell Holmes, the logical formalism current at Harvard, arguing that law is a social institution that evolves from the struggles of human experience. A collection of Green's essays appeared posthumously as *Essays and Notes on the Law of Tort and Crime* (Menasha, Wis.: George Banta Pub. Co., 1933).

18. Also in 1870 Green published "Contributory Negligence on the Part of an Infant," and in 1874 he published "The Three Degrees of Negligence." Both appeared in *American Law Review*.

19. Benjamin Franklin Butler (1818–1893), a lawyer, politician, and Civil War general ran unsuccessfully for Governor of Massachusetts in 1871–72, though he got elected for that position a decade later.

20. Peirce developed a view very similar to this in "Dmesis," which appeared in *The Open Court* in 1893 (W9, sel. 4). See also "Why do We Punish Criminals," which Peirce wrote earlier that year for the *Independent*, but was rejected for publication (W8, sel. 52 and W8:464–65).

21. John Fiske (1842–1901), American historian and author of *Outlines of Cosmic Philosophy, Based on the Doctrine of Evolution: with Criticisms on the Positive Philosophy* (London: MacMillan, 1874).

22. Francis Ellingwood Abbot (1836–1903), a classmate of Peirce at Harvard, became a Unitarian minister in 1864, and in 1867 joined the Free Religion Association. In 1870, he founded the independent weekly the *Index*, and ran it for a decade—the weekly was considered the unofficial mouthpiece of the Free Religion

Association. Abbot's 1890 *The Way Out of Agnosticism* received a scathing review by Josiah Royce, which caused a public controversy in which Peirce sided with Abbot (W8, sel. 40; see also W8:433–38). In his 1903 *Nation* obituary of Abbot (P 1031), Peirce praised Abbot's philosophical work and his love of truth, explicitly citing *Scientific Theism* (see below).

23. Francis Ellingwood Abbot, *Scientific Theism* (Boston: Little, Brown, & Co., 1885).

24. See "Some Questions of Four Incapacities" (W2, sel. 22), and Peirce's 1871 review of Fraser's *The Works of George Berkeley* (W2, sel. 48).

25. William James (1842–1910) An extensive discussion of James's views, especially as related to the Metaphysical Club, is found in Part II of Louis Menand's *The Metaphysical Club* (New York: Farrar, Straus and Giroux, 2001).

26. Oliver Wendell Holmes Jr. (1841–1935); an extensive discussion of Homes is found in Part I of *The Metaphysical Club.*

27. For a discussion of this, see the introduction.

28. James Mark Baldwin, *Dictionary of Psychology and Philosophy,* 3 vols. (New York: Macmillan, 1902–5), 2:322. Peirce there writes: "The doctrine appears to assume that the end of man is action—a stoical axiom which, to the present writer at the age of sixty, does not recommend itself so forcibly as it did at thirty."

29. Peirce first made this move publicly in "What Pragmatism Is," *Monist* 15 (April 1905): 161–81; reproduced in EP2 as sel. 24, see esp. 334–35. References are to the British philosopher Ferdinand Schiller (1864–1937), the Italian mathematician-philosopher Giovanni Vailati (1863–1909), and the Italian magazine *Leonardo*, which was edited by Giovanni Papini and Guiseppe Prezzolini. For a more detailed description, see e.g. Cornelis de Waal, *On Pragmatism* (Belmont: Wadsworth, 2005), chapters 4–5.

30. In an aborted attempt written earlier that year under the title "Studies in Meaning," Peirce writes of the first two essays that they (R 630:02):

> attracted no great attention, though a French version by the author was printed in the *Revue Philosophique*. But some years later Professor William James, brought the matter before the philosophic world, (pressing it, indeed, further than Mr. Peirce, who continues to acknowledge, not the *existence,* but yet the *reality,* of the Absolute, as set forth, for example, by Royce); and in this form Pragmatism has taken a prominent place in the philosophy of today.

31. Gerbert d'Aurillac (ca.946–1003), or Pope Sylvester II (999–1003) has several mathematics textbooks ascribed to him, including *Regulae de numerorum abaci rationibus*. In vulgar arithmethic, as opposed to decimal arithmetic, fractions are represented by two numbers separated by a line, as in ⅗.

32. Francis C. Russell, "A Modern Zeno," *Monist* (1909): 289–308; the passage occurs at the bottom of the opening page. Had the passage made it to the publisher, the choice of example would certainly have been an unfortunate one, as Carus made Russell responsible for readying the Illustrations for publication. See introduction.

33. Arthur Cayley (1821–1895), British mathematician. Felix Klein (1849–1925), German mathematician; see also Chapter 9. Nikolai Lobatchewski (1792–1856) Russian mathematician, and author of *Geometrical Researches on the Theory of Parallels*, trans. G.B. Halstead (Austin: Bulletin of the University of Texas, 1891).

34. In a discarded draft on the same sheet, Peirce puts it as follows: "The context, however, when I came to look into it, showed

35. Nicolas Copernicus (1473–1543), *De revolutionibus orbium coelestium* (on the revolutions of the celestial spheres) (Nurnberg: Johannes Petreius, 1543).

36. Francis Bacon (1561–1626), *Novum Organum Scientiarum* (the new instrument of science), London, 1620. The first book of the *Organon* consists largely of an assault on the syllogistic logic which found its first systematic presentation in Aristotle's *Organon.* In its stead Bacon advocated for induction (aphorism 14).

37. See Robert Leslie Ellis's preface in *The Works of Francis Bacon* (London: Longman, 1859), 3:69f.

38. William Whewell (1794–1866), *Novum Organum Renovatum* (London: John W. Parker, 1858), as the title suggests a renovation of Bacon's work.

39. William Whewell, *History of the Inductive Sciences: From the Earliest to the Present Times,* 3 vols. (London: John W. Parker, 1837).

40. John Stuart Mill (1806–1873), *A System of Logic, Ratiocinative and Inductive: Being a Connected View of the Principles of Evidence, and the Methods of Scientific Investigation* (London: John W. Parker, 1843).

41. Thomas Harriot (1560–1621), British astronomer and mathematician.

42. It is safe to say that one of the forces driving Peirce was precisely to provide such a logic. His completed though unpublished 1894 *How to Reason* and his *Minute Logic* (1901–02) are testimony of how close he came to, or of how far he remained from, writing such a treatise.

43. In the 1870s the Scientific Club of Cambridge, to which Peirce's father belonged, discussed political economy, concentrating on the *Recherches sur les Principes Mathematiques de la Théorie des Richesses* (1838) by Antoine Augustine Cournot (1801–1877). For Peirce's work in mathematical economy in the 1870, see NEM 3:547–54. Alfred Marshall (1842–1924) was a British mathematical economist.

44. Rudolf Julius Emanuel Clausius (1822–1888), German physicist and mathematician; William John Macquorn Rankine (1820–1872), Scottish engineer and physicist; William Thomson, 1st Baron Kelvin (1824–1907), British engineer and physicist.

45. The official Latin name of the Franciscan order is *Ordo Fratrum Minorum,* hence a Franciscan brother is sometimes referred to as minorite.

46. Ludwig Schütz, *Thomas-Lexikon: Sammlung Übersetzung und Erklärung der in sämtlichen Werken des h. Thomas von Aquin vorkommenden Kunstausdrücke und wissenschaftlichen Aussprüche,* 2nd enlarged edition (Paderborn: F. Schöningh, 1895).

47. Francis C. Russell, "A Modern Zeno," *Monist* (1909): 389. See also footnote to this same quotation in Version I.

48. Though the first name Francis could suggest a woman (although in that case it would rather be spelled Frances), Peirce knew very well that Russell, with whom he was personally acquainted, was not.

49. Though an attorney by profession, Russell did write on logic and extensively corresponded with Peirce on the subject. Some of Russell's publications include: "Logic as Relation Lore" (*Monist* 3.2 [1893]: 272–85), "Substitution in Logic" (*Monist* 15.2 [1905]: 294–95), and "Hints for the Elucidation of Mr. *Peirce's* Logical Work" (*Monist* 18.3 [1908]: 406–15. His correspondence with Peirce is found in RL 387.

50. On the role of the doctrine of chances in the Illustrations, and how this differs from traditional treatises on logic, see the introduction; for Peirce's account of the term probability, see also his 1910 letter to Paul Carus (chapter 9).

51. William Whewell (1794–1866), *History of the Inductive Sciences* (London: John W. Parker, 1837).

52. See note 5 above.

53. Matt. 7:16–20: "Ye shall know them by their fruits. Do men gather grapes of thorns, or figs of thistles? Even so every good tree bringeth forth good fruit; but a corrupt tree bringeth forth evil fruit. A good tree cannot bring forth evil fruit, neither can a corrupt tree bring forth good fruit. . . . Wherefore by their fruits ye shall know them."

54. Most of these worshippers, like Richard Congreve, were attracted to Comte's notion of a secular Religion of Humanity, which presented itself as a substitute for Christianity, complete with its own priesthood, prayer, and rituals.

55. In the 1860s, less than three decades after publication of Comte's *Cours de Philosophie Positive*, William and Margaret Huggins showed that the stars were composed of the same elements as found on earth. They showed this using a spectroscope, an instrument that had been invented earlier that decade by Gustav Kirchoff and Robert Bunsen.

56. See note 41 above.

57. Mill, *System of Logic,* Preface, first paragraph and part of one sentence from the second; the former is worth citing whole:

> This book makes no pretence of giving to the world a new theory of intellectual operations. Its claim to attention, if it possess any, is grounded on the fact that it is an attempt not to supersede, but to embody and systematize, the best ideas which have been either promulgated on its subject by speculative writers, or conformed to by accurate thinkers in their scientific inquiries.

58. It might be helpful to reproduce the remainder of the sentence, and the paragraph, as Mill has it:

> or added any fundamentally new process to the practice of it. The improvement which remains to be effected in the methods of philosophizing (and the author believes that they have much need of improvement) can only consist in performing, more systematically and accurately, operations with which, at least in their elementary form, the human intellect in some one or other of its employments is already familiar.

59. In this book, which appears in 1829, James Mill, following Hume and Hartley, develops a psychological theory where mental phenomena are reduced to their simplest components. To Mill the mind is passive and the association of these components into groups and successions can be explained in terms of a single law, the law of contiguity.

60. Immanuel Kant, *Kritik der reinen Vernunft*, 2nd ed., *Akademie Ausgabe*, vol. 3 (Berlin: Georg Reimer, 1911), 165; A186/B230. Translated by Norman Kemp Smith respectively as "subsistence" and "existence of substance."

61. Henrich Hertz (1857–1894), *Principles of Mechanics* (New York: Macmillan, 1899), 14–24; see esp. 18.

62. On Peirce's distinction between generality and indefiniteness, or vagueness, see esp. EP2:394 (1906) where he defines *generality* as referring to the interpreter's "right to complete the determination as he pleases," and *indefiniteness* as "the *sign's* not sufficiently expressing itself to allow of an indubitable determinate interpretation" (emphasis added).

63. Mill, *System of Logic,* Bk. I, Ch. ii "Of the Things Denoted by Names"; see esp. "Part IV. Relations."

64. Mill, *System of Logic,* Bk. I, Ch. ii, §13.

65. Mill, *System of Logic,* Bk. III, Ch. xiv, §1–3.

66. Mill, *System of Logic,* Bk. III ("Of Induction"), Ch. iii ("Of the Ground of Induction,"), §3.

67. Mill writes (*System of Logic,* III.iii.3):

> When a chemist announces the existence and properties of a newly-discovered substance, if we confide in his accuracy, we feel assured that the conclusions he has arrived at will hold universally, although the induction be founded but on a single instance. We do not withhold our assent, waiting for a repetition of the experiment; or if we do, it is from a doubt whether the one experiment was properly made, not whether if properly made it would be conclusive. Here then, is a general law of nature, inferred without hesitation from a single instance; a universal proposition from a singular one.

68. Mill, *System of Logic,* III.iii.3.

69. John Gay (1699–1745), British philosopher and clergyman, was the main inspiration for David Hartley's 1749 *Observations on Man,* which is typically seen as the work that opened the path for the associationalist psychology as it developed in Britain.

70. Durandus of St. Pourçain (1275–1334), French Dominican philosopher and theologian and Petrus Aureolus (1280–1322), French Franciscan philosopher and theologian. Both are generally regarded as transitionary figures between Scotus and Ockham.

71. William of Ockham (ca.1287–1347), Franciscan philosopher and theologian. In 1320 Ockham was still in England, either at Oxford or at London Greyfriars, an intellectual center that at the time was rivaled only by Oxford.

72. Johann Maier von Eck (1486–1543), German theologian and defender of Catholicism during the Protestant Reformation. He was known for his commentaries on Aristotle's *De caelo* and *De anima,* and on the *Summulae* of Peter of Spain.

73. The reference is to Chauncey Wright; see Peirce's discussion of Wright above, with notes.

74. Simon Newcomb (1835–1909), American astronomer and a student of Peirce's father at Harvard.

75. Thomas Bayes (1701–1761), English mathematician and Presbyterian minister, and author of *An Introduction to the Doctrine of Fluxions* (1736), wherein he defended Newton's theory of calculus against Berkeley's objections in *The Analyst*. Bayes is best known for having formulated a specific case of the theorem that bears his name: Bayes's Theorem.

76. On March 5, 1909, which is when he began this manuscript, Peirce wrote to Carus in response to the latter's claim that Peirce got his notion of absolute chance, or tychism, from his father: "my father worshipped Laplace and particularly admired his *Théorie des Probabilités*, and therefore I thought it by far the best to devote my time with him to things more profitable to myself and always avoided talking [to him] about chance" (Open Court Collection, SIU Carbondale, 27.91.25).

77. George Boole (1815–1864), British mathematician and logician who greatly influenced Peirce's own work in mathematical logic. For Boole's discussion of probability theory, see esp. the second half of *An Investigation of the Laws of Thought: on which are Founded the Mathematical Theories of Logic and Probabilities* (London: Walton & Maberly, 1854).

78. August Weismann (1834–1914), German evolutionary biologist who maintained that multicellular organisms consist of germ cells and somatic cells. Inheritance takes place solely through the former—the gametes (egg cells and sperm cells). Somatic cells, which carry out bodily functions, do not function as agents of heredity. Consequently, genetic information cannot pass from somatic cells to germ cells and so to the next generation, thereby ruling out the Lamarckian notion of evolution through aquired characteristics.

79. Grover Cleveland (1837–1908), 22nd and 24th President of the United States, said in his Third Annual Message to Congress (December 6, 1887): "Our progress toward a wise conclusion will not be improved by dwelling upon the theories of protection and free trade. . . . It is a condition which confronts us—not a theory."

80. Bernhard Riemann (1826–1866), German mathematician; see his 1854 lecture Über die Hypothesen welche der Geometrie zu Grunde liegen (*Abhandlungen der Königlichen Gesellschaft der Wissenschaften zu Göttingen* 13 [1867]: 133–52), to which Peirce refers with relative frequency.

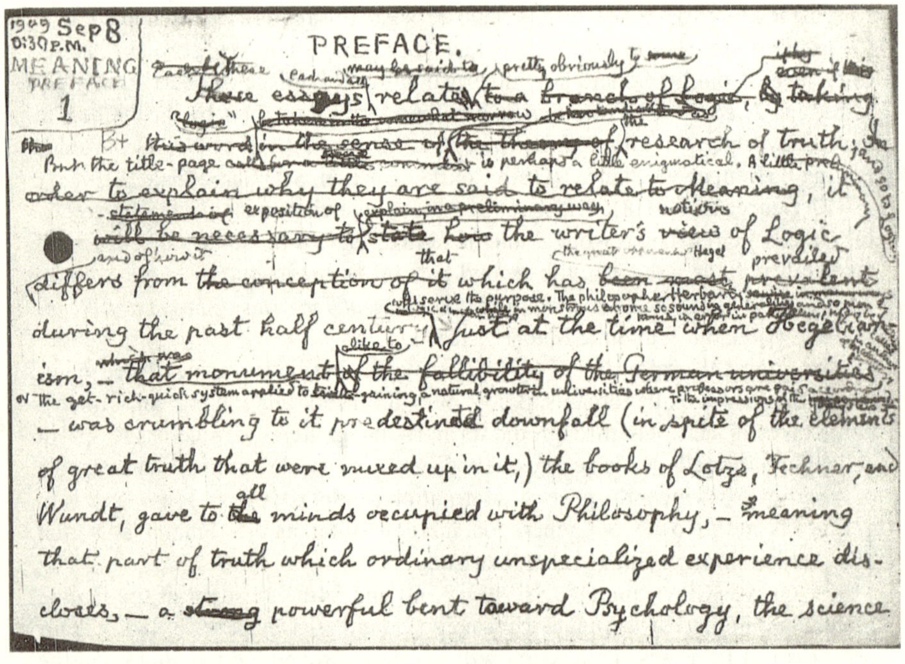
1909 Sep 8
0:30 P.M.
MEANING
PREFACE
1

PREFACE.

These essays relate to a branch of Logic,

research of truth;

But the title-page ... is perhaps a little enigmatical.

order to explain why they are said to relate to Meaning, it

will be necessary to state how the writer's view of Logic

differs from the conception of it which has been most prevalent

during the past half century, just at the time when Hegelianism, — that monument of the fallibility of the German universities

— was crumbling to it predestined downfall (in spite of the elements

of great truth that were mixed up in it,) the books of Lotze, Fechner, and

Wundt, gave to all minds occupied with Philosophy, — meaning

that part of truth which ordinary unspecialized experience discloses, — a powerful bent toward Psychology, the science

Draft of the opening of Chapter 8, Peirce's unfinished Preface to the *Efsays on Meaning* (R 634:2). Most of the sheets related to Peirce's 1909 reworking of the Illustration give the date and time he began working on the sheet in the upper left corner. This draft was not included in the *Collected Papers*, and the editors, in their attempt to determine what to publish and what not, rated it B+.

Chapter 8

E*f*SAYS ON MEANING

By a Half-Century's Student
Of the game[1]

The E*f*says on Meaning *was written in September and October of 1909. Though the text included below contains no direct reference to the Illustrations papers, an earlier draft (R 633), written September 4–6, does. In that draft Peirce clearly stated that the first essay recombines "The Fixation of Belief" and "How to Make Our Ideas Clear," which most likely refers to the still uncompleted work Peirce began that spring, some of which is found in the previous chapter. If this is correct, then the two uncompleted prefaces below are meant to replace the much shorter preface found there. From what Peirce further writes it becomes increasingly clear that revised forms of the remaining four papers of the Illustrations are no longer being planned. As Peirce explained, "The remainder of the volume sets forth some of the matured ideas of a man who from boyhood to old age has longed and laboured to learn all he could of the right methods of the conduct of thought in the research of truth" (R 633:2–3). Instead of reworking those four papers, Peirce is aiming to develop the groundwork for a system of logic, considered as semeiotic. In fact, Peirce envisioned the text below as the preface to a series of essays leading up to a new book on logic, where logic is conceived as semeiotic (p. 268). The emphasis, as* *with the original series, is on the types of reasoning most useful for science, i.e., inductive reasoning and retroductive reasoning (the latter also being called by Peirce hypothesis and abduction).*

Of the various attempts of writing the preface two versions were chosen: the version he wrote last, in which he does not get all that far, and a much more extensive intermediate version. Where material from other versions and drafts significantly enhance the text reproduced, it is included as footnotes. Version I was composed between September 8 and October 18, and its material is found in three different Robin folders: R 634:30, R 635:2–19, and R 637:2–43. Every sheet is numbered by Peirce and given a date and time stamp. The much shorter Version II was written October 22–27, and is found in R 640:2–18. For a further discussion of the history surrounding this text see the introduction.

PREFACE

*[*VERSION I*]*

These essays relate pretty obviously to the research after truth, and so to Logic. The title-page may be a little enigmatic; but setting forth in this preface the writer's point of view of logic and say as briefly as his small skill can, how his view of the subject differs from others which have been almost exclusively taken throughout the fifty years last passed it is hoped to make the purpose of the volume tolerably clear from the onset. Two different sciences study reasoning. They are as sharply distinguished from each other as are anatomy and physiology. The one describes, as minutely as it well can, the different events of consciousness that take place in the course of a reasoning and shows how this series of events differs in different kinds of reasoning. This requires very sly and adroit spying indeed; because nobody—no ordinary person, at any rate,—can genuinely and in good faith *reason,* so as to find out a truth he did not know before, without bestowing his undivided attention upon the subject of his reasoning, be it the habitability of Mars, or the exact reason why four colors are sufficient for distinguishing all the adjacent pairs of regions of any map, or how to reconcile what we appear to know about organic chemistry with what we appear to know about electrons; or any other such matter; and this leaves the reasoner without any available attention to pay to what is going on in the meantime within the arena of his consciousness. The observation of what one is doing and suffering in his inner consciousness,—assuming that it can be done at all, which some shrewdly doubt,—is certainly far more difficult than is, for example, the management of a high-power microscope so as to resolve the most difficult test-objects; and to do simultaneously the two things that it is requisite to do at the same time in order—while concentrating all one's being upon a Reasoning, as one needs to do in order to discover new hypotheses at the same time—directly to observe what immediate consciousness (i.e. oneself, or that part of oneself that *appears*) does and suffers during that reasoning, is harder yet, by far.*

Perhaps the trickster can only trick himself, and observation of one's very self is far more impossible than smelling out one's own nose, or seeing the difference between the states of one's own retina, when it is seeing and when it is in pitch darkness.

> The old Sphinx bit her thick lip,—
> Said, "Who taught thee me to name?
> I am thy spirit, yoke-fellow;
> Of thine eye I am eyebeam.
>
> "Thou art the unanswered question;
> Couldst see thy proper eye,
> Alway it asketh, asketh;
> And each answer is a lie."[2]

It is odd that it should be so excessively difficult to observe what passes in consciousness, considering that consciousness is nothing but immediate knowing that, in short, it is oneself as far as feeling is concerned. Yet, be the fact ever so odd, it certainly is a fact that nobody can accurately observe the events of consciousness, especially while he is busy with reasoning about something else, unless he can catch himself unawares and play a trick on his unsuspecting self.

But if we can really observe ourselves the fact is to some extent explicable, by means of an observation that the writer, who is a highly practiced observer of sensations, had made long before he found it needed for any philosophical argument.[3] At any rate self seems utterly† inexplicable otherwise. This observed fact that perhaps explains the paradoxes is that, in the writer's case at any rate, and he is no doubt much like others in so fundamental a respect,—in spite of the great differences between individuals that Galton /thought to exist/,[4] there is, in reasoning an alternation of two quasi-personalities, one of them proposing arguments, and appealing to the other cooler personality to confirm his judgment of their soundness.

**M-dashes added for clarity; the text in between was added interline.*

†At any rate self seems utterly] *above deleted:* ; and the facts thus explicable seem to be quite

If then, we suppose that these two consciousnesses are so connected that the one can observe the other while the latter's whole attention is occupied with something else, we have a hypothetical explanation of the perplexing circumstances. The two personalities in one person may in that case be likened to an old married couple who know each other so thoroughly that each knows what the other is thinking. It is needless to work this hypothesis out in detail, here; yet it seems to throw light on more than one phenomenon, and will suggest some truths that seem to* be confirmed otherwise.

The second science of reasoning pays no attention whatever to what passes in consciousness, as contradistinguished from the facts reasoned about, except when the need of guarding against the notorious mendacity of self-consciousness affords it indispensible so to say, because its main purpose is to ascertain under what conditions an argument can safely be trusted and how far. Now an argument is a body of ascertained events or other real facts whose reality is believed to show that another specified proposition is true in reality. Although this science finds it requisite in order to generalize its results to study the natural forms of assertions, yet it interests itself solely in ascertaining what sort of relation between the facts asserted in the premises to be real and the fact concluded would be at once needful and sufficient in order that the argument should lead a user of it to the truth in the sense and measure in which it professes or is supposed so to do.† Such a relation is of the same hard stuff as are the facts to which it relates, and has nothing in the world to do with the evolutions of the mind in apprehending those facts, unless, as above intimated, to avoid being deceived by self-consciousness.

This last statement has certain apparent limitations in some cases; but a careful thinker will be able to satisfy himself that they are no more than apparent. They may be examined under the following three heads. *Case N°* 1. The question reasoned about may itself be how a given sort of person thinks under circumstances of a given kind. Here, obviously, modes of thinking being the very facts of the case, must be considered,‡ the very statement we are no examining, *requires* it, and only pronounces

* seem to] *under deleted:* can

†At this point Peirce inserts interline the following subsequently deleted sentence: "To do this it is requisite to study the nature and forms of assertion itself."

‡R 635:21–23 and R 636:2–27 branch off at this point, though not seamlessly. The date and time stamps match, but the text does not flow smoothly from one sheet to the other. The former continues an aborted explication of the doctrine of chances in terms of a betting game with a pair of dice; the latter, hereafter called Branch B, traverses much of the same ground as the text here reproduced and parts of it are here reproduced as footnotes.

the mode of thinking *about these modes of thinking* to be irrelevant. *Case N°* 2. Under this head may be comprised all reasonings from the characters of samples to those of the lots sampled;—these are the only inferences furnishing new and practically reliable information concerning matters of fact, as is here proved. The sample will, in all cases consist of a finite number of objects to be separately examined or investigated; but the lot sampled may either be, theoretically at least, exhaustible by a finite number of such samples, or it may be utterly inexhaustible.* Considering, first, the former alternative. Suppose that from a vein of ore a sample of some ten pounds weight be made up, a part of it from the edges and a part from the centre of the vein, with the utmost care, so as to make it as perfect an image in miniature of the composition of the whole as may be. This sample shall consist of a considerable number of bits, each apparently of one homogeneous kind, and the proportion of each kind in it having been ascertained, and separate analyses of all the kinds having been made, an inference may be drawn as to the value of the part of the vein in sight. For the sake of simplicity (this being a mere example in logic,) it shall be assumed that the whole treatment of the ten-pound sample once it has been made up, will be perfectly free from error. But the operation of forming that sample is an extremely difficult one, involving much reasoning as well as many delicate estimations.† It is best performed without much hesitation or introspection; and yet all the while a watch must be maintained over oneself in order to exclude the influence of desires, of prepossessions, of fatigue, etc.‡ All this self-observation, however, has nothing to do with the inferential procedure. It only serves to assure the sampling chemist that one of his premises is true that asserts that the sample is a "fair" one,

*In a draft of this sheet, Peirce writes instead (R 635:20): "whether this lot be finite, or denumeral, or a general class."

†In Branch B Peirce puts it thus (R 636:2; Sept. 22):

> Now sampling is an exceedingly delicate operation, involving much reasoning. The ore will seldom run anywhere near uniformly; and the sample should be made up of many small pieces so as to perfect an image in miniature of the whole lot sampled as may be.

‡In Branch B Peirce puts it thus (R 636:2–4; Sept. 22–23):

> The man who is doing this has to guard against his own prepossessions. *[…]* the truth of the sample, as an image of the whole lot, and the validity of the sampling,—an operation of reasoning,—requires him to keep sharp watch upon his own mind, lest his decisions to include or exclude this or that bit of stone be influenced by his desires. The best way of making up a sample is from unhesitating, undoubting sense-impressions stimulating the action of a habit formed under the influence of criticisms of many former samplings self delivered as soon as the degree of accuracy of each has come to light, each criticism

which implies that the aggregate of a sufficiently great number of samples taken under the same determination of the sampler's will would (1st) nearly exhaust the vein, while (2nd) the single samples would not much differ from one another in any important respect. If the sample drawn really be "fair" in these two implications of the word, it is plain that the fact of its fairness involves the fact that the proportionate composition of the sample will approximate to that of the part of the vein exposed. Undoubtedly, every inference that the general character of a whole is similar to that of a sample does so far depend upon how the sampler thinks, that its trustworthiness depends upon those separate acts of volition each of which determines that a certain part of the whole shall or shall not be a part of the sample, really being, one and all, or very nearly all, thoroughly obedient to the general determination of the will that the sample shall be a fair one. To make them so, in spite of the wiles of that imp, the subconscious mind, Mr. Jastrow and the writer in our investigation of the question of an *"Unterschiedsschwelle,"* were aided by a well-shuffled pack of cards.[5] In other cases, the writer, when in doubt whether an object should or should not be included in a sample, has let the decision depend upon a throw of dice. To hum and haw over such a question will be either a waste of time or worse; but whoever thinks that it is an easy thing always to be fair commits an error into which a man really desirous of being so seldom does commit. It needs a strong intellect and an upright heart which the best of men cannot always be sure of having. Here, then, is a case in which *how* we feel or think has very much to do with the validity of an inference. Only the importance of the connection does not consist in its being true that our feeling of confidence in the argument either constitutes or is a sure sign of that argument's being valid. It is, on the contrary, much nearer to the truth to say that that importance is due to the fact that no person's confidence in an argument is any sure sign of that argument's validity. For if circumstances permit human choice to be eliminated by allowing the spin of a teetotum[6] to decide whether or not each given portion taken out from the rock shall or shall not go into the sample, in place of human judgment, then if the number of such portions in the sample is sufficiently great the argument will be much improved.

---

being immediately followed by an energetic resolution to make its lesson efficient. Nothing so much interferes with the success of the analyst's endeavour to attain this ideal of a good sampling as an attitude of introspection on his part, although introspection must be made if he has any interest in one kind of conclusion rather than another. But this is only because the sample's being uninfluenced by any desire is a material fact influencing the conclusion, just as something passing in the mind was in the former instance.

When the sample is no finite proportion to the lot sampled, the reasoning from the sample to that of which it is a sample becomes a true Induction.* Suppose that a person has a teetotum of which one edge only is marked, and that he spins it a great many times, while two friends keep tallies of the results of the spins, one tally recording the spins that leave the teetotum resting on its marked edge, and the other those that leave it on other edges; and suppose that a fourth person as soon as the record of each spin has been made, takes out the natural logarithm of M, the number up to that time of spins leaving the teetotum on the *Marked* edge and subtracts from it the natural logarithm of N, the number of spins that have, so far, left the teetotum on edges *Not marked.* Now since at each throw the fourth person's result will change from log M – log N either to log (M+1) – log N or to log M – log (1+N), either it will be increased by $\frac{1}{M}-\frac{1}{2M^2}+\frac{1}{3M^3}-\frac{1}{4M}$etc., i.e. nearly by $\frac{1}{M}$ or else it will be diminished by $\frac{1}{N}-\frac{1}{2N^2}+\frac{1}{3N^3}-\frac{1}{4N^4}$ etc. i.e. nearly by $\frac{1}{N}$. Accordingly the changes will be less and less the longer the experiment is continued. That is to say, the fourth person's latest result will oscillate in an irregular manner, but generally speaking through a smaller and smaller amplitude of oscillation.† To render this evident to the non-mathematical

*The example that follows supersedes the following example (R 635:18–19; Oct. 2):

> When we reason from a sample which is not a finite part of the whole lot, and which could never be exhausted by the sampling, the case is decidedly different, but yet not utterly so. Thus, it is known that in habitable parts of the earth, the sun has for several thousand years risen either 365 or 366 times each year. If nothing more were known, it would be justifiable to believe, i.e. to accept for all purposes, the proposition that the sun always had done so and always would. But since we know that another star larger than the sun might approach so near as to alter the length of the year, and since moreover we have considerable reason to think that suns like ours may cool down so as no longer to shine, and since whatever *may* happen would *be sure* to happen in sufficient time, we reasonably conclude that there will be an end to the present series of such years, but that it will probably continue for many thousands of years yet; and this conclusion has nothing to do with how we think. If one were to provide himself with a thousand quarter-dollar pieces, and pitching each of them up once were to find that about 500 of them turned up heads and all the rest tails, the proportion as more and more of them were pitched oscillating in an irregular manner between a majority of heads and a majority of tails, and the extreme proportions usually approaching equilibrium, he ought to infer that no matter how long the experiment were to continue, this irregular oscillation never would cease, so long as the general conditions remained the same; or, more accurately speaking, he ought to infer *that that inference is to be provisionally accepted, as being the provisional result of a method of inquiry that, carried indefinitely far, would gradually lead toward a result free from error.*

† amplitude of oscillation] *before partially deleted:* ; and the consequence will be that before a hundred spins have been made, the fourth person will exclaim "Oh, it is plain to be seen that my value will be close upon [such and such an amount] *in the long run.*" | *the latter above non-deleted:* In order to make the matter comprehensible to non-mathematical readers, an actual series of experiments has been made and the series of results of the fourth person have been calculated

reader, here is the veritable record of such a series; though it contains but a hundred experiments, a number far too small for much accuracy.

TABLE[*]

S	R	M	N	D	C	S	R	M	N	D	C
1	N	0	1	-∞		31	M	7	24	-1.232	+0.154
2	N	0	2	-∞		32	N	7	25	-1.273	-0.041
3	N	0	3	-∞		33	N	7	26	-1.312	-0.039
4	N	0	4	-∞		34	N	7	27	-1.350	-0.038
5	M	1	4	-1.386		35	N	7	28	-1.386	-0.036
6	N	1	5	-1.609	-0.223	36	N	7	29	-1.421	-0.035
7	M	2	5	-0.916	+0.693	37	N	7	30	-1.455	-0.034
8	N	2	6	-1.099	-0.182	38	N	7	31	-1.488	-0.033
9	N	2	7	-1.253	-0.154	39	N	7	32	-1.520	-0.032
10	N	2	8	-1.386	-0.134	40	M	8	32	-1.386	+0.134
11	N	2	9	-1.504	-0.118	41	N	8	33	-1.417	-0.031
12	N	2	10	-1.609	-0.105	42	N	8	34	-1.447	-0.030
13	N	2	11	-1.705	-0.095	43	N	8	35	-1.476	-0.029
14	N	2	12	-1.792	-0.087	44	N	8	36	-1.504	-0.028
15	N	2	13	-1.872	-0.080	45	N	8	37	-1.531	-0.027
16	M	3	13	-1.466	+0.405	46	N	8	38	-1.558	-0.027
17	N	3	14	-1.540	-0.074	47	N	8	39	-1.584	-0.026
18	M	4	14	-1.253	+0.288	48	N	8	40	-1.609	-0.025
19	N	4	15	-1.322	-0.069	49	N	8	41	-1.634	-0.025
20	M	5	15	-1.099	+0.223	50	N	8	42	-1.658	-0.024
21	N	5	16	-1.163	-0.065	51	M	9	42	-1.540	+0.118
22	N	5	17	-1.224	-0.061	52	N	9	43	-1.564	-0.024
23	N	5	18	-1.281	-0.057	53	N	9	44	-1.587	-0.023
24	N	5	19	-1.335	-0.054	54	N	9	45	-1.609	-0.022
25	N	5	20	-1.386	-0.051	55	N	9	46	-1.631	-0.022
26	M	6	20	-1.204	+0.182	56	N	9	47	-1.653	-0.022
27	N	6	21	-1.253	-0.049	57	M	10	47	-1.548	+0.105
28	N	6	22	-1.299	-0.047	58	N	10	48	-1.569	-0.021
29	N	6	23	-1.344	-0.044	59	N	10	49	-1.589	-0.021
30	N	6	24	-1.386	-0.043	60	N	10	50	-1.609	-0.020

S stands for the number of spins, **R** for the result, **M** for the total of Ms obtained so far, **N** for the total number of Ns obtained so far, **L** for (Log M – Log N), and **C** for $(L_s - L_{s-1})$.[†]

[*]This table, which Peirce left unfinished, is reconstructed using the data Peirce provided. Only the R values of sixty spins could be retrieved, possibly because at least one sheet following the table has not been recovered.

[†]The editor provided the column designations and the legend. Peirce indicates that a sheet numbered 7.4¾ follows this sheet. No such sheet could be identified, nor any possible subsequent inserted sheets.

*The discussion of this argument and of Induction in its general features, is intimately connected with, if not strictly a part of that all-important branch of Logic called the Doctrine of Chances or, somewhat less appropriately, the "Calculus of Probabilities." The study of this branch of logic has been principally left to the mathematicians, who have certainly been, within their proper province, the most accurate of all social groups of reasoners, from the days of Euclid to those of Weierstrass.[7] But they have never paid any particular attention to logical analysis, which is what is most needed throughout logic; for the ordinary subjects of their study have, from their point of view, been singularly free from difficulties of this kind that call for such analysis; and the consequence has been that chance, probability, etc. have been very superficially defined and have been represented to have much more to do with the ways in which men think and believe and feel than in their true essence they have. The second essay in this volume will show that a reasoning such as that here instanced about the teetotum involves a distinction between what may be called the *Main Circumstances* of the series of spins and other circumstances attending that series. The Main Circumstances are those circumstances which, without determining on which edge the teetotum shall rest after any one spin, are such that no matter at what spin the two tallies considered above shall begin, so long as the beginning of taking the tally is settled upon before the first spin whose result is to begin the record is made, and no matter whether the results of all the spins are recorded or only a part of them provided the rule which determines what spins are to be recorded, does not depend upon this result of the recorded spin, and is settled upon before the second recorded spin is over, if they (the Main Circumstances,) were to remain entirely unchanged throughout an endless series of spins, and the quantity log $\frac{M}{N}$ were to be recalculated after each recorded spin, then, as the spinnings went on the value last recorded of log $\frac{M}{N}$ would oscillate in an irregular manner, yet so that there would be one single value, V, that would never cease to be passed over, so that if after any one spin log $\frac{M}{N}$ should be greater than V then after some later one it would be found that log $\frac{M}{N} < V$, while if after any spin log $\frac{M}{N} < V$ were true, there would be a spin at log $\frac{M}{N} > V$ would become true; while if any other value, say U, were considered, there would be either some last time when log $\frac{M}{N} \geqq U$, or else some list time when log $\frac{M}{N} \leqq U$.

[Case] N° 3. It remains to consider a kind of reasoning that, at first blush, seems plainly to falsify the position that the science of logic needs

---

*This sheet begins completing a sentence begun at the previous, now missing sheet. The editor has omitted the following: any time by the former tally, and let N (standing for "Not marked" denote the number recorder by the latter tally.

no support from the science of psychology. This is the kind of inference by which we pass from belief in a surprising fact to giving some degree of credence to a pure hypothesis whose truth would explain that fact and do away with its astonishing character.* The great majority of logicians refuse to class this passage of belief among reasonings. They admit that the passage from "Some priest is a warrior" to "Some warrior is a priest" is an inference. But the passage from "Manetho, the list of Egyptian kings, and a few recently found inscriptions assert that there was a king of the two Egypts named Menes, or MN," to "There really lived such a king," they refuse so to class. If they were right, of course such a mental operation could not constitute an exception to the rule that inference never turns upon the manner in which its copulate premiss is thought. But they are clearly wrong if inference is the passage from acceptance of one or more assertions to a consequent acceptance of another assertion; and it is wiser rather to run the risk of making the definition too broad than too narrow. The genius of inference by which more or less credence is given to an elsewise unsupported proposition because it would explain a fact that without it would be surprising, i.e. contrary to what there was ground to presume, extends with no natural break to cases where no thoroughly sane man can resist the impulse to believe† as, for example, where a man believes that something happened before his eyes five minutes ago because he seems to remember it. His confidence will fade by imperceptible degrees, until if it were a trivial circumstance that he seems to remember nearly seventy years ago, he will be very doubtful about the accuracy of his memory. The logical rules of inference from a surprising fact to a hypothetical state of things that would explain it involves a complexity of conditions,

---

*In Branch B Peirce writes the following (R 636:4; Sept. 23):

> An instance in strong contrast to both the preceding /this text immediately follows the discussion of sampling/ is that of the provisional adoption of a hypothesis, taking this word, not in the sense of a mere "working hypothesis," which only serves to provide a vacant gaze with a definite question to answer, and thus to convert it from seeing into looking, but in the sense of a conjecture that one is more or less inclined to credit because it seems reasonable in itself and will serve to explain some otherwise surprising circumstances. Such a hypothesis is absolutely indispensible in every quite new, every novel, inquiry, for whose conduct no rules have been settled upon.

†In Branch B Peirce continues thus (R 636:5; Sept. 23):

> Every reasonable hypothesis that would explain facts not otherwise explained merits so much credence as to be put to the test. For it is reasonable to believe that an hypothesis which should seem reasonable to an ideally intelligent mind sufficiently well informed would be true; and in default of such a mind one must exert the only one that is available, one's own, and trust to that in such measure as experience seems to warrant or to permit.

one of which is how much time can be allowed for coming to one's conclusion. A general who during a battle must instantly risk the existence of a nation either upon the truth of a certain hypothesis or else upon its falsity, must perforce go upon his judgment at the moment; and his doing so is in so far logical that all reasoning is based upon a tacit assumption that Nature, in the sense of the aggregate of truth, is conformed, more or less, to something similar to the reasoner's Reason. This kind of inference may be called Practical Retroduction.[8] It is ordinarily called a Presumption. If on the contrary, indefinite time can be allowed for judging of the truth of an explanatory hypothesis, the most favorable conclusion is that the consequences that hypotheses that are susceptible of verification should be subjected to systematic and thorough tests. This, which may be called Scientific Retroduction, will be subject to considerations of economy. If, for example, a supposition far from reasonable, yet still possible, would explain a physical phenomenon, and if, in case it be incorrect, there is a way of disproving it at little expenditure of either time, energy, or any other valuable, then it may be worthwhile to clear the ground by taking up the investigation of it forthwith. It will be remarked that the result of both Practical and Scientific Retroduction is to recommend a course of action. It is further to be remarked that the Scientific Recommendation is to collect a Sample of the consequences that would hold true in case the Hypothesis were true, and to find out what ones of them actually are true. For to say that all the knowable consequences of a proposition are true is, on Pragmaticistic principles, the very same thing as to say that that proposition is itself true, while to say that all its consequences excepting those of a specified kind are true, is to say that its whole falsity lies in a certain part of its implication. The logical justification of Retroduction of either kind is as follows.* In the first place, we certainly do thoroughly believe and cannot help so believing, do what we may, that some reasonings are

*The logical justification … is as follows.] In a draft (R 638:2 9; Oct. 4 6), Peirce writes:

> [Although a] Retroductive Recommendation may be almost utterly bad, yet it can be said, that if Retroduction be constantly employed as a method, whenever Deduction and Induction are impossible, it must, on the whole lead us closer to the truth than we can otherwise come. For there is no other rational method. This is the verdict of pure logic, taking no account of the discoveries of the special sciences. But since all modern science depends ultimately on this method, its history furnishes such a sample of intelligent hypotheses, that a student of that history must be blind not to see that man's mind has a certain power of divining the truth, not, indeed, enough of that power often to light upon the truth at the first guess, yet enough, when the guessing is properly conducted, to enable him in a moderate number of guesses to come wonderfully near the truth. For if there be any limit to the number of explanatory hypotheses that might be imagined

sound. For we can free ourselves of a belief only by reasoning ourselves out of it, and to do this is to believe that some reasonings are sound. Now although it is, of course, one thing to believe a proposition, no matter how thoroughly and firmly, and quite another for the proposition to be true, yet practically for the believer they are one and the same. For if his belief is perfect he thinks he is sure it is true and between that and his thinking it *is* true there is no practical difference. We must and do admit, therefore

---

for any surprising phenomenon, that number certainly cannot fall far short of a trillion; and therefore if there were no tendency to guess right, since the examination of a physical hypothesis will busy dozens of investigators for decades, in the three of four centuries in which physical science has been pursued the world would have hardly attained to the power of foreseeing a single physical effect in the most general way. The very first guess that a physicist or chemist makes when he observes a wholly unexpected and surprising phenomenon in the course of his work is that the special cause of it lies somewhere within his laboratory, and that neither the configuration of the planets, nor anything that an antipode fish may have done at the moment occurs to him as likely to have anything to do with the case: and he will in more than ninety-nine cases out of a hundred be right in his conjecture. It is true that al old alchemist might have thought the aspects of the planets not unlikely to be concerned in the matter; but that was because the Egyptians or Babylonians whom the alchemists held in exaggerated respect, and who introduced that kind of hypothesis into chemistry, had fallen into certain erroneous beliefs whose substance, had they been true, would have rendered their hypotheses really quite reasonable. Nobody who takes due account of those circumstances will regard the alchemists' follies as in the least degree tending to refute the opinion that man has a certain imperfect power of divining the truth, although those follies ought to warn us all not to confide in that divining power any further than is necessary to avoid closing a possible avenue to truth. Pure logic rather encourages than forbids inquiry into the truth of any hypothesis which we have an impulse to accept. It only forbids unreasonable belief. The business of Logic (in its department of Critic) is to decide whether, accepting the premises of any such inference of a hypothesis, and accepting whatever no man either actually doubts or is able reasonably to bring himself really and truly to doubt, the acceptance of the conclusion, in a given sense of acceptance, is warranted or not. It is that decision, precisely, which it is maintained, need and should have nothing to do with the manner of the reasoner's thinking, or with any truth that it needs the special science of psychology to discover.

The main question of logic is how it comes about that certain kinds of knowledge expressed or "thought to ourselves" in premises renders it safe for us to give a certain degree or kind of credence to something else that we can say to others or to ourselves in a conclusion inferred from the states of things represented in those premises. Certain knowledge about human nature may serve to put us on our guard against sources of error, and should accordingly be studied after we once clearly comprehend precisely what truth it is which can guarantee other truth. For example, the knowledge that the mind never has complete control at one time of many unrelated thoughts should warn us in reasoning to pay exclusive attention to the pertinent premised facts and not to allow our minds to wander to the operations of consciousness in thinking those facts. It is true that we cannot think of those facts except through some representation of them. But the representation which is the plainest, most easily observed in itself, whose non-representative characters are least obstructive, which brings out the pertinent features of the fact represented with least admixture of non-pertinent features, and even whose

that some reasonings are sound. But to say this is to say that some instinct or natural impulse to believe is in conformity with the real nature of things; and the only question is how far that conformity extends. This can only be ascertained by sampling; and the process of sampling will consist in taking Retroduction after Retroduction and testing the truth of each by as large a sample of its consequences as can conveniently be obtained. This justifies Scientific Retroduction, which simply puts that process of testing in practice for single retroductions; and there is nothing in the justification that cannot be learned from indubitable external observation and equally indubitable reasoning. Any doubt of the result is either insincere, unthinking, or idiotic. Such a general work of testing Retroduction has been carried on extensively throughout the three or four centuries last past; and the general results of the texts are registered, as it were, in the habitual impulses to regard certain hypotheses as reasonable and certain others as fanciful, firstly, of a certain class of students of the history of the inductive science, of whom William Whewell[9] furnishes an example, and who may well find in him their exemplary men alive to the logical side of their subject, and the best of them inspired themselves with high scientific genius (on the reasoning side, at least,) who study that history (and indeed whatever they study), not as an object of curiosity so much as for the in-

---

pertinent features are so glassy and transparent as hardly to be noticed in themselves but only to be seen *[tangentially]*, will, other things being equal, be the one to prefer. That which passes within consciousness is, perhaps, of all modes of representation the worst. If it is capable of being observed at all, it is only with extreme difficulty and by an exclusive attention to it that does not leave any to be bestowed upon its significance, and pretty much all that can any way be observed in it is the feeling of it. It is as if a person slightly deaf, instead of listening, were by means of mirrors try to put his ear under a microscope. It is far better to observe some enduring visible representation of his thought, such as a grammatical sentence of simple structure, an algebraical equation, or a diagram. The best modes of representation for the purpose of reasoning are those kinds that the common herd of humans most hate, and hate precisely because these modes oblige them to think. The Copulate Premiss, i.e. the premises taken as one statement, is a *sign* of the conclusion, and is the very same sign whether in English, German, Italian, Spanish, or French (to one who knows the language,) and whether it be printed, written, spoken, or thought. In consequence of this the most essential part of logic is the study of the relation of Signs in general to that which each represents, or as logicians say, to their "Objects." The word "Symbol" will in this book be used as the common name for that class of Signs which represent, to those that can interpret them, the objects they do quite regardless of any resemblances to them (although such may have influenced the original choice of the signs,) and equally so of any actual connexions therewith, (however close such connexions there may be,) but solely because those interpreters have habits of mind, whether inherited or acquired, that lead them whenever they perceive the signs straightway to think of those Objects. For instance, a diagram will not be called a symbol, although it has some conventional features, because the analogy between the relations of its parts to those of the Object is largely concerned in its representation of the latter; *[MS breaks off here]*

struction in methodeutic[10] that it yields, and who are imbued with a modest, but not diffident, sense of their responsibilities as teachers of their fellow-men in the difficult art of reasoning; secondly, but far more imperfectly, of those who while holding the scales of their judgment justly balanced and equal, have been attentively interested through life in watching the growth and decadence of current hypotheses under the guidance of the writings of deeper students; and thirdly, of those who, by imitative contagion, a few directly, the many by many indirect clews, have contracted the main habits of thought of the first two classes, though mostly with exaggerated prejudices one way or another. The last are the vehicle of the general body of educated opinion of their times, with all its stupidities, but with the outer husk, at least, of its ripe fruit. The vague general assertion that a cautious presumption may be credited if there is no evidence to the contrary is supported by pure logic, which is by no means restricted to the alleged *"a priori* truth," for it is only forbidden to make use of the results of metaphysics and of the special sciences. But as for more definite judgments, preferring one kind of presumption to another, such as that it is presumable that any ordinary book as written by the author whose name it bears, if this be not too famous, and if there is no particular reason to doubt the authorship, but that in the case of an alchemical work the presumption is the other way,—the discussion of such presumptions must be relegated to methodeutic, a branch of logic whose investigations suppose the general principles of logical Critic concerning the degree and quality of validity of the main genera of reasonings to have been already established,[11] and which accordingly is free to make use of all pertinent results of science including, for example, Fechner's law[12] and such principles as that associations by contiguity are often inconvertible, or irreversible, so that an ordinary Christian cannot say his *Pater noster* backwards, etc. etc.

What has now been said ought to show the Reader that the manner of thinking, as contradistinguished from the Real facts thought, has no bearing upon the Reality of the fact inferred. But in order that this truth may not float in the air of his cognition but may attach itself to some experience, it will be well to give one or two examples of the modern ways of discussing logical questions which seem to the writer to be condemned by the reflexions he has endeavoured to suggest.

Many modern students who write concerning what they call "logic" do not use this word in the sense here given to it; and the manner of thinking, although it has nothing to do with the reality of a believed state of things, no doubt is perfectly relevant to the very different department of truth with which they are busied; so that they can be accused of nothing worse than a misuse of the term logic; though even that is a much more serious offence than German philosophers suppose it to be. But this does not apply to Kant, who always speaks of *allgemeine Logik* as having been perfected by Aristotle, which proves that by *allgemeine Logik* he means the subject of Aristotle's Analytics and Hermeneutic; for he certainly manifests no admiration for the book of Categories or for the Topics and Sophistical Elenchi, which last two he pretty certainly never read.*

Now in Kant's much belauded brochure *Über die falsche Spitzfindigkeit der vier syllogistische Figuren,*[13] a work of astounding ignorance, in which Aristotle's doctrine of the relation of the indirect to the direct moods is put forward as Kant's own discovery, he attacks the mood *Baralipton,* or *Bramantip,*[14] saying of it, "It is quite obvious that the conclusion as it stands cannot be derived from the premises at all."†[15] Now it happens that this mood *is* faulty; but it only is so for a reason that Kant entirely overlooked. The objection he makes is, as he would have admitted, at least equally cogent against *Frisesomorum,* of which the following is an example,[16]

> Some luminosity of the interior of a rock lasts long without chemical action;
> No luminosity due to previous exposure to sunlight comes from interior parts of a rock;
> *Hence,* some luminosity that lasts long without chemical change is not due to previous exposure to sunlight.

Kant admits that this conclusion follows from the premises, but he says that this is "only because I can, after each premises, think its immediate consequence," in which its subject and predicate are transposed.[17] This transposition from

---

[Original Peirce footnote] His judgment of the completeness of Aristotle's work, or rather of the logic of the Leibnizians, seems to be somewhat general in Germany, although it is most erroneous and superficial.

†*[Original Peirce footnote]* This sentence is copied from T. K. Abbott's translation, the original not being at hand.

In draft sheet of Branch B Peirce writes that this brochure (R 636:28; Sept. 23):

> was, in substance, little more than Aristotle's reduction of the moods of syllogism to the four distinct moods of the first figure, which reduction he claimed as his own discovery, in spite of its having been made much of in every medieval treatise on the subject. If he ever read the first book of Aristotle's *Analytics* or any of the medieval logics it is difficult to imagine what meaning he attached to the reduction of the figures that differed in any essential particular from his own. In short it was the work of an able but grossly ignorant logician; unless, indeed, we are to say that it was a saltimbanc's hullabaloo to attract and amaze who saw how this ancient doctrine could be made to produce that effect in the existing ignorance both of Aristotle and the middle ages. Nevertheless, this production has been lauded to the skies by sundry German writers.

> "Some luminosity from the interior etc. lasts long without chemical action,"

to

> "Some luminosity lasting long without chemical action is luminosity of interior parts,"

and the corresponding transposition of the terms of the other premises do not cause either to express any different fact from that it expressed in the beginning; but Kant makes the fact that the conclusion follows from the premises due to a possibility of thinking the latter in a particular form. The truth is that what follows at all follows from the facts and not from any particular form of statement.*

No other German treatise on logic has been so much admired in this country of late years as that of Chr. Sigwart, (who is not to be confounded with the elder writer *[Heinrich]* Sigwart);†[18] and §§1 and 3 of his "General Introduction" to that treatise have been most particularly approved.[19] In §1 he describes Logic‡ as "a technical science of Thought, directing us how to

---

*In Branch B Peirce puts it thus (R 636:12–13; Sept. 24):

> whether the conclusion is a sound one or not depends, according to Kant, upon which premise is thought first, or upon which is regarded as the major. He is utterly oblivious of the fact that the conclusion follows, if at all, from the real facts, and not from the thought or expression of them. "The so-called ratiocination in the fourth figure," he says, "contains the matter, it is true, but not the form of our reasoning." Note the last words. He does not see that whatever follows logically follows whether he or anybody thinks it does so, or not.
>
> Another example of the same false doctrine may be drawn from *[Chr. Sigwart's]* highly esteemed treatise on Logic.

†In Branch B Peirce continues as follows (R 636:13–14; Sept. 24):

> The fundamental doctrine from which this author's theory *[i.e., Sigwart's]* is developed is that the only ultimate security in reasoning is a certain FEELING, which he calls the "Gefühl der Evidenz",—and be it remembered that in philosophical German *Gefühl* is used much more strictly than "feeling" is used in literary English: the monstrous misnomer of calling anything a *science* that has no better basis than feeling, becomes even more monstrous still when this word is strictly applied. Every circle-squarer has that feeling to the full. Necessary reasoning,—mathematical reasoning,—which alone can be entirely trusted, and even it must not, until it has been reviewed, scrutinized, scanned, and criticized by the sharpest logic,—deserves its trust-worthiness, not from that circle-squarer's *feeling*, but man's power of recognizing his own meaning. For a reasoning of this necessary kind is logically sound only if the conditionally concluded fact was *already expressed* in the statement of the hypothesis from which it set out. *[Peirce next gives two examples.]*

‡*[Original Peirce footnote]* I am forced to quote from miss Helen Dendy's translation, the accuracy of which, however, is guaranteed by Sigwart himself.

arrive at certain and universally valid propositions,"[20] showing that he uses the word "logic" substantially as the present writer does. Furthermore, like the great majority of German logicians of the present day, he professes to regard Logic as sharply distinguished from Psychology.[21] In order that the reader may have some preliminary, but not altogether vague, notion of how the position taken in this volume (and to be more thoroughly defended in another,) differs from that of the majority of contemporary logicians, a part of a paragraph from Sigwart's §1 with some other passages that throw light upon that one, shall here be quoted, and without any attempt to refute it in this preface, shall be made the subject of some comments with a general expression of dissent.

Sigwart states what he means by "logic," in §1.9, thus:

> We know from the fact that error and dispute exist that our Thought frequently misses its aim ... Since then actual Thought can and does miss its aim, we have need of a discipline which shall teach us to avoid error and dispute, and to conduct Thought in such a manner that the judgments may be *true*—that is, necessary and certain—that is, accompanied by a consciousness of their necessity, and therefore universally valid.
>
> Reference to this aim distinguishes the logical from the psychological treatment of Thought. The latter is concerned with the knowledge of Thought as it actually is.... The logical treatment, on the contrary, presupposes the desire to think the truth; it has no meaning except for those who are conscious of this desire and [except (?)] in that region of Thought that is governed by it. Starting with this aim before it, and investigating the conditions of its attainment, Logic proposes, on the one hand, to set forth the Criteria of true Thought which are due to the demand for necessity and universal validity; on the other hand, to direct us how to conduct the mental operations in such a way that the end may be attained. In one aspect then Logic is a critical discipline [as the present writer would say, is a Critic] having reference to Thought which has already taken place; in the other, it is a technical discipline. But a criticism is of value only as a means to the end, the principal task of Logic and that in which it properly consists, is to be a technical discipline, or Art.[22]

This extract sufficiently shows that the word Logic with Sigwart stands for pretty nearly the same studies that the present writer uses it to name. At the same time, it furnishes the writer with about his thousand-and-first instance to convince him that of all scientific groups the logicians are, in the main, the most blundering of reasoners. If this be the fact, as he believes it is, the reader must be notified of it in order that he may be put upon his guard against the same obtuseness on the part of the present writer. The fact might be explained by supposing that with most of them it is only because they have had so much difficulty in reasoning straight that the men who have become logicians have more than other men been drawn into the study of logic, while because the basis of their logic has been mistaken, it has not done them much good. When Sigwart says that logical studies "presuppose" a desire to know the truth, if he means they prove the existence of such a desire, he is mistaken, since it was known long before logic was studied that all men possess more or less curiosity. It is the first assertion of an ancient book.[23] If he means that a desire to know the truth of a conclusion has been the chief motive for logical criticism of the argument of which it is the conclusion, this is flatly contradicted by the almost uniformly trivial character of the arguments selected for that purpose. Why should it not be an object of natural curiosity to ascertain precisely what it is that renders it possible for us on learning one fact of pair of facts, straightway to recognize that another proposition is true? As for his making the most valuable part of logic to be the *art* of reasoning, or Methodeutic, though this latter is a practical science rather than an art, a parallel argument would apply to several branches of physical theory. Thermodynamics was originally largely studied and greatly stimulated owing to a hope of improving marine engines; but what physicist or what chemist would not today regard its theoretical side as outweighing its practical side, as a kilo to a grain? For though the latter may make life easier, what does life amount to if we cannot enjoy the highest and greatest of all satisfactions, that of adoring the Author of all truth in tracing out his ways of governing his universe. Glancing to a previous sentence, that for a proposition to be true is the same as for it to be necessary and certain, though it may be true without anybody's suspecting it of being so, that a proposition cannot be accompanied by a seeming consciousness of necessity without being certain or even true, and that from such a consciousness universal validity always follows are three more slips of logic.

By "universally valid," Sigwart apparently means holding good for all men, in contradistinction, for example, to judgments of taste; and what he means by the very ambiguous word "necessary" is explained in the following passage from his §1.6:

> In speaking of the *necessity* of our Thought we must guard against a confusion of meaning. From a psychological point of view everything which the individual thinks may be looked upon as necessary, *i.e.* as an activity which results in accordance with general laws from whatever conditions may be present. … But besides this necessity of psychological causality there is another which springs entirely from the *contents* and *object* of Thought itself; which is, therefore, grounded, not upon the variable subjective states of the individual, but upon the nature of the object thought of, and which may so far be called *objective*.[24]

What Sigwart calls the "content" of thought is its predicative value, nearly what the present writer calls its Substance, and one variety of it Meaning. It is also nearly the same as its Definition. When the objective necessity of thought springs from its object, why not speak of it as the necessity of that Object, instead of attributing it to the Thought? And why not study the relations of Objects that constitute that necessity instead of the relations of Thoughts by which it is *thought [to]* be real?

The way is now cleared for bringing forward the passage illustrating the contrast between the two views.* Here it is, and assuredly, there is nothing in it that is extreme or peculiar:

---

*Peirce here originally continues as follows (R 637:20, 32–35; Oct. 10):

> Only it will be desirable first to mention that Dr. Sigwart holds, in common with almost all modern thinkers, what the present writer denies, that existence is the only form of reality, and that, in consequence of that opinion, in this passage he uses the word "existent" when *real* is what he plainly means. That is Real which has characters independently of what contrary or other characters any person or collection of persons may in any sort of idea attribute to it. This is not to say the characters are possessed independently of all Thought about it, nor of whatever might be the final opinion to which sufficient research would lead. Thus the substance of a dream, that is, what is dreamed is *not* Real, since the object dreamed has only such characters as the mind of the dreamer gives it. But the dream itself is a Real fact, if it has occurred, no matter who opines that it has not. A Real object may be External or Internal, i.e., mental, as a dream is. For only that is external whose possession of some character is independent of individuals' opinions, but whose possession of any such Real character is independent of any individuals' thought about *any* subject and of all ideation of any kind. Thus a dream though it Really happens depends for its character on the ideas in the mind of the sleeper, and is therefore an Internal, or mental, Reality. As for existence, it may be used in a wider or a narrower sense. In its wider sense it is synonym of Actual, and is such Being as is both Definite and in all respects Determinate; while in its narrower sense it is a Substance, or substratum which acts on, and is acted on by, the Universe of Substance. To say that an object is in all respects Determinate, is to say that every non-relative predicate is either false of it or free from that falsity; or in other words the Principle of Excluded Middle applies to it. A *disposition,* or kind of "behaviour" (using this word as the chemists use it,) is not In All Respects Determinate. Thus, having a vapor-pressure at 100°C of about one atmosphere is a Real Disposition of water; but it neither involves nor excludes being altered by electrification. It is, on the contrary, of a General nature; and what is General is not subject to the Principle of Excluded Middle, since both "Any S is P" and "Any S is not P" may be false. Neither is it, apparently, true that a Real object should be Indefinite, though our habits of language prevent us from readily comprehending this. But consider a permission. A permission may certainly be real, if we extend the meaning to include permission according to a natural law; say to the law of gravitation. It may include a permission to be positively electrified and a permission not to be so. It thus has two contrary characters. It would seem to be merely a usage of speech that prevents us from saying,
>
> This permission is to be electrified;
> and This permission is *not* to be electrified.

> It may be counted amongst the surest results of the analysis of our knowledge that every assumption of an external world is mediated by Thought, and in some way or another derived by unconscious mental processes from the subjective facts of sensation. Thus, except by Thought, we have no means of ascertaining whether we have really achieved our purpose of knowing the Existent; the possibility of comparing our knowledge with things as they exist apart from our knowledge is forever closed to us.[25]

His use of the words "Existent" and "exist" here is an inaccuracy. For "existent" we should read "Real," and for "exist," "really are." For Real is the proper contrary of Illusion, Delusion, or Figment, while to *exist* means, by virtue of the *ex* in *existere,* to act upon, to react against, the other things that exist in the psycho-physical universe. His view is that we know immediately only our thought and that no reasoning from our Thought to real external objects has any logical validity. The present writer* finds himself forced to dissent from all views which suppose our knowledge by introspection is more immediate than that by external perception. The only Immediate Thought is in judgments of immediate perception. There is, no doubt, such a thing as a vacant gaze that we occasionally experience before fully awakening from sleep; and in this state we see without looking; but that is not Thought. In Thought we are wholly intent upon whatever object

---

> But although the permission in the sense of a promise made is a Real fact, yet in the sense in which it has contradictory attributes, it has, at most, but a germinal being. Of course a fiction, such as the substance of one of Dickens's novels, has a Being, but is not Real. It is a Figment. Nothing prevents a Figment from involving some self contradiction *[MS breaks off at this point]*

*Peirce here originally continues as follows (R 637:36–37; Oct. 11):

> wholly dissents from this reasoning. According to him Thought reveals its Object and nothing else, and of our Thought itself we have only mediate knowledge. Knowledge begins in Immediate judgments concerning the Objects perceived, and these judgments cannot be doubted, at the time, nor until memory becomes doubtful. What the first impressions of Sense may be is a subject for hypothesis. By the time knowledge comes, they are buried beneath subsequent syntheses. We know in Immediate judgments of Perception that our Percepts are involuntary; the shock of an overborne resistance to them is often very perceptible. Nevertheless, though thus exterior to our Will, they might be thought not exterior to our organism, were it not for their perfect consistency, which resists all tests to which dreams quickly yield. Although Berkeley's idealistic reasoning is, in one particular, very bad and inconsistent with his own method as illustrated by the best parts of his *Essay on Vision,* yet his doctrine *affirms* little or nothing that is false. It is only at fault in what he fails to say and in what he denies; in his failure to apply his *esse est percipi* to his own consciousness and in his denial of Matter in the modern sense of the word. So, likewise, the doctrine of the solipseists Ostwald, Heim, etc. (who, however, are at odds as to who the sole being is,) may be considered as simply understanding oneself in a broad sense.

we are thinking about, and that object never is or can be the very Thought we are then in the act of thinking. True Thought takes the form, as the writer believes, of a dialogue. For true thought is deliberate, and deliberation implies the asking of questions, appeals to another personality for confirmation, objections, and other dialogic forms. At any rate, Thought is continually questioning and denying; and consequently if a Thought could have itself for its object, it could deny its own truth and the thinker would land in an Insolubilium of the worst kind, which he could not fail to take note of, which in fact he never does. This knocks the bottom out of the pseudo-psychological opinion that what we immediately know is our own Thought itself. The unity of Thought does not consist in that *Ich denke* which Kant, in his first edition, called the "=*x*," and not *"Ich denke,"* or "I opine."* The unity of thought, if we could view our own consciousness, would probably consist in the *continuity of life of a growing idea;* but as well as we can observe it, it lies in the logical coherency and consistency of argument. For to Think is to reason, as we could not fail to see if the logicians had not mispersuaded us that the only reasoning is Aristotelian syllogism. Now the unity of reasoning is simply the identity of the object reasoned about. An argument is simply a constructure of premises which constitutes a *sign* of the truth *of the truth* of its conclusion, no matter what kind of reasoning it uses; and the question of what its logic is, is nothing but the question in what mode of representation those premises make up a sign of the substance of the conclusion. Now any equivalent signs will be

*In Branch B Peirce writes the following (R 636:25–27; Sept. 28–30):

> There is a celebrated passage in the second edition of the C.d.r.V *[Critik der reinen Vernunft]*, and a really very notable one, in which Kant says that "the *I Think*,"—"DAS: ICH DENKE"—must *be able* to accompany all his ideas, "since otherwise they would not thoroughly belong to me." A man less given to discoursing might remark on reading this, "For my part I don't hold my ideas as my ownty-downty; I had rather they were Nature's, and belonged to Nature's author." However, that would be to misinterpret Kant. In his first edition, he does not call the act "the *I* think," but "the object = X." That which the act has to effect is the consecution of ideas: now the need of consecution of ideas is a *logical* need, and is due not, as Kant thinks, to their taking the form of the *Urtheil*, the assertion, but to their making an *argument*; and it is not "I think" that always virtually accompanies an argument, but it is "Don't you think so?" Nor is this a mere empirical observation, although, even if it were, it would not be ruled out of logic as belonging to the science of psychology, since it is one of the things that the "naïve" man knows, that everybody knows who knows what argument is. Surely it is proper for the logician to assert that there may be Arguments,—or it would be, if it were not well known before the study of Logic began. Now what is *meant* by an Argument and how does it differ from an assertion? It involves assertions; it is assorted. But something more is needed to constitute it an Argument. It is an appeal to its interpreter's own reason to assent to its soundness when once he has assented to the truth of its copulate premiss.

in the same relation to that conclusion; and the preference among different forms of signs should be given that one which is most easily examined, manipulated, preserved, and anatomized. If this be admitted, logic will rarely consider judgments as they are in the mind but will make choice of some form of external sign. This has been the writer's method; and he has found so great a variety of kinds of signs to be indispensable in reasoning of different kinds that it seems to him that in the present state of science in which the study of the nature of the broader classes of signs has been almost completely neglected, it would be well for scientific logicians to recognize the whole field of general semeiotic as consigned to their industry. Another book in which if the author's* powers hold out to complete it, the results of those studies of logic considered as semeiotic to which the only useful part of his days have been consecrated, results which he believes will furnish any person of intelligence who will examine them with clear ideas of how to find the truth, as well as a sound basis for a truly religious life, will be submitted in systematic form to the judgment of thoughtful men. Meantime, this preliminary and unsystematic volume is put forth in the hopes that it may, in some way or through some pen lead to thinking men generally being made acquainted with the point of view in question, which the pragmatists have not quite taken up.

To attain this view-point, it is requisite to know something of the nature of Signs; and that feature of Signs which is particularly to be considered in this volume is their *meaning*. In order that the reader may get a little clearer idea of what this is than the word by itself is likely to convey, it will be well to begin by explaining that† the word Sign will be used throughout the volume to denote an Object perceptible, or only imaginable, or even unimaginable in one sense,—for the word *"fast"*‡ which is a Sign, is not imaginable, since it is not *this word itself* that can be set down on paper or pronounced, but only *an instance* of it and since it is the very same word when it is written as it is when it is pronounced, but is one word when it means "rapidly" and quite another when it means "immovable," and a third when it refers to abstinence. But in order that anything should be a Sign, it must "represent," as we say, something else, called its *Object,* although the condition that a Sign must be other than its Object is perhaps arbitrary, since, if we insist upon it we must at least make an exception in the case

* in which ... author's] in which the author is recording

†The text that follows from here is reproduced, with omissions, under the heading "Signs and Their Objects" as CP 2.230–32. See later note on where the inclusion in the *Collected Papers* ends.

‡ *"fast"*] *"tree"*

of a Sign that is a part of a Sign. Thus, nothing prevents the actor who acts a character in a Historical drama from carrying as a theatrical "property" the very relic that that article is supposed merely to represent, such as the crucifix that Bulwer's Richelieu[26] holds up with such effect in his defiance. On a map of an island laid down upon the soil of that island there must, under all ordinary circumstances, be some position, some point, marked or not, that represents *quâ* place on the map, the very same point *quâ* place on the island.* As to what is meant by saying the Object is *represented* by the Sign, perhaps the best account that can be given of the matter is the following, which may be preceded by the remark, that the mode of representation may be by likeness or analogy, in which case, the sign may be called an *Icon;* or it may be by a real connexion as a certain kind of rapid pulse is symptom of a fever, in which case the sign may be called an indication or *Index;* or finally the only connexion may lie in the fact that the Sign, a word for example, is sure to be interpreted as standing for the Object, in which case the Sign may be called a *Symbol;* and these terms, *Icon, Index, and Symbol,* will so be used throughout the volume.†

A sign may have more than one Object. Thus, the sentence "Cain killed Abel," which is a Sign, refers at least as much to Abel as to Cain, even if it be not regarded as it should, as having "*a killing*" as a third Object. But the set of objects may be regarded as making up one complex Object. In what follows and often elsewhere Signs will be treated as having but one object each for the sake of dividing the difficulties of the study. If a Sign is other than its Object, there must exist, either in thought or in expression, some explanation or argument or other context, showing how, upon what system or for what reason the Sign represents the Object or set of Objects that it does. Now the Sign and the Explanation together make up another Sign, and since the explanation will be a Sign, it will probably require an additional explanation, which taken together with the already enlarged Sign will make up a still larger Sign; and proceeding in the same way, we shall, or *should,* ultimately reach a Sign of itself, containing its own explanation and those of all its significant parts; and according to this explanation each such part has some other part as its Object. According to this every Sign has, actually or virtually, what we may call a *Precept* of explanation according

*[Original footnote by Peirce:] If map and island were both ring-shaped it might not be so. This example is introduced in order to correct an error, (not the sole, nor most important one,) in a certain book that is very well worth reading.

†[Original footnote by Peirce:] This terminology was first introduced by the writer in the *Proceedings of the American Academy of Arts and Sciences,* 1867 May 14. Vol. VII p. 294 [W2:56], only Icons were at first called "likenesses."

to which it is to be understood as a sort of emanation, so to speak, of its Object. If the Sign be an Icon, a scholastic might say that the *"species"* of the Object emanating from it found its matter in the Icon. If the Sign be an Index, we may think of it as a fragment torn away from the Object, the two in their Existence being one whole or a part of such whole. If the Sign is a Symbol, we may think of it as embodying the *"ratio,"* or reason, of the Object that has emanated from it. These, of course, are mere figures of speech; but that does not render them useless.

It is not only essential to a Sign that it should *represent,* i.e. stand in place of or for, an Object, but, if possible, still more so that it should be capable of *Interpretation* by or through a mind, into which it implants a germ which, on development, will affect the conduct of the person to whom that mind appertains; and not until this effect, which throughout the volume will be called the *Interpretant* of the Sign, is brought about will the Sign function as the Sign τὸ τί ἦν εἶναι. That is, the Interpreter must be affected by the Sign, so far as it is the function of the Sign to affect him, as if it were the Object itself that affected him; and while he may not notice the Sign at all, he must recognize the Object, or the Sign will not have accomplished its perfection of being the Sign that it is.* At the same time, the Sign can only represent the Object and tell about it. It cannot furnish acquaintance with or recognition of, that Object; for that is what is meant in this volume by the Object of a Sign; namely, that with which it presupposes an acquaintance in order to convey some further information concerning it. No doubt there will be readers who will say they cannot comprehend this. They think a Sign need not relate to anything otherwise known, and can make neither head nor tail of the statement that every Sign must relate to such an object. But if there be anything that conveys information, and yet has absolutely no relation nor reference to anything with which the person to whom it conveys the information has, when he comprehends that information, the slightest acquaintance, direct or indirect,—and a very strange sort of information that would be,— the vehicle of that sort of information is not, in this volume, called a Sign. Two men are standing on the sea-shore looking out to sea. One of them says to the other, "That vessel there carries no freight at all, but only passengers." Now if the other, himself, sees no vessel, the first information he derives from the remark has for its Object the part of

*On a draft sheet this sentence runs as follows (R 637:31):

> That is to say, the Sign must act upon the mind of the Interpreter in such a way that the latter shall be affected substantially as if by the Object; (for so far as the Sign is deceptive it is not a Sign of its Object;) though the Interpreter ~~must~~ will perceive that it is the Sign and not the Object itself that directly affects him.

the sea that he does see, and informs him that a person with sharper eyes than his, or more trained in looking for such things, can see a vessel there; and then, that vessel having been thus introduced to his acquaintance, he is prepared to receive the information about it that it carries passengers exclusively. But the sentence as a whole has, for the person supposed, no other Object than that with which it finds him already acquainted. The Objects,—for a Sign may have any number of them,—may each be a single known existing thing or thing believed formerly to have existed or expected to exist, or a collection of such things, or a known quality or relation or fact, which single Object may be a collection, or whole of parts, or it may have some other mode of being, such as some act permitted whose being does not prevent its negation from being equally permitted, or something of a general nature desired, required, or invariably found under certain general circumstances.[*] In this volume, something will be said about the relations of signs to their Objects; but what is called on the title page "Meaning" is that which a sign communicates. This may be nothing but a Feeling or emotion, which is all that a performance of instrumental music, for example, commonly expresses. Or the Sign being a command, such as the order "ground Arms," its meaning may be the impulse to obey, which the sign excites. A question is a sort of command. Or the Sign may be an appeal to reason by an argument consisting of known premises, the synthesis of which, which synthesis will be its meaning, may be a new thought. Or the Sign may be an assertion, or "Proposition," to use the logical term, where the Meaning is the substance of an assent to it. Or it may be a mere suggestion to imagination or memory, such as single word may convey. Many "Utterances," as all acts of using Signs will here be called, are purposeless. But a serious Utterance is usually intended to influence either a single act or the reasoned[†] conduct of the Interpreter or Interpreters, and its meaning is that general kind of Conduct that it virtually recommends. Such Signs are mostly Arguments; and the Conclusion of an Argument, considered as matter for belief, is its Meaning; while a secret purpose of an Utterance is the Meaning of some thinking, or "saying to himself" on the part of the Utterer. It will thus be perceived that the subject of this book is no trivial one. In the writer's opinion, it, or rather Semeiotic, the science of Signs in general, ought to be regarded as the foundation of a liberal education, whether for young men or young women.

There is no attempt in these pages to treat this vast subject systematically: a few topics only which press upon the author are considered. He has

[*]The *Collected Papers* rendition ends at this point.

[†] reasoned] *above deleted:* rational

done his best to make the book clear and agreeable; but it would hardly be possible for the most accomplished pen to produce the latter effect; since the subject, of its very nature, requires the reader *to think,* without which he cannot even know what it is all about; and there is no occupation that the general herd of mankind find so intolerably irksome as thinking. "I hate books," will often be heard in tones that leave no doubt of the dept of the feeling; and one who hates books must loathe thinking. So it must be acknowledged that it is only those who enjoy the exercize of rational exertion who will have any patience with the individual who has perpetrated these Essays, especially as he is far from a graceful writer; and because he is accustomed to take certain points of view that happen to be much out of fashion, his modes of expressing himself are likely to be thought even more *gauche* than they really are; and they are bad enough at best.

Pretty much all he can do toward rendering his writing perspicuous, beyond giving concrete examples whenever he can discover the need of them, is as far as possible always to use each word in a single sense *[...]*[*]

*[*Version II*]*

These essays will be found to relate to one department of the research after truth, that is, to Logic. The intention has not, however, been to treat that subject systematically, but rather to give such examples of its discussions as shall persuade the intelligent reader of the serious,—nay, frankly, the tremendous, importance of the subject as it is here conceived, and shall gain hearing for such an exposition of it in a future publication as would only be taken up by a reader already convinced that it was worth reflective study. In that work that idea of the subject that has little by little and very slowly by reason of the extreme caution and laborious self-criticism that was required, come to crystallize in the writer's mind during a half century in which he has thought of nothing else except for its bearing on this sole purpose of his life, and sole field in which he found himself not to be a fool, shall be set forth.

The present reader is likely to be pretty sceptical of the "tremendous importance" of Logic: the author, at any rate, would be so, were he in the

[*]The manuscript breaks off at what may be considered the end of a sentence (there is no period), but not at the bottom of the sheet. Possibly, this means we have reached the point where Peirce stopped working on this branch. Peirce marked the sheet: October 13, 1909, 6:15*[*PM*]*. The version reproduced next, Version II, was begun on October 20.

reader's shoes, and judged of it by such treatises of it as should have come his way. For truly there is but one other branch of the whole tree of heuretic science that is not laden with luscious spiritual peaches compared with the poor little crude persimmon on the branch of modern logic. That other is metaphysics; and even that might not have been in quite so lamentable a condition if it had been grafted on a sound logical stock, and had not been infested with every species of theology. God be thanked that these pasts seem now to be dying out, and without having killed either science or religion. The logic which the reader will find specimens of in this volume is so widely different from any other that has ever been published, especially during the writer's time, that there is no arguing from the qualities of the others to those of his.

The reader ought, too, to be made acquainted with the nature of the importance claimed for this system. He must not suppose that there is any pretension to improve those modes of reasoning which he is employing in his daily business, whose conclusions have hundreds of times been checked and corrected by comparison with facts; nor that the superiority of practice over theory in producing good habits of thinking or any other sort of skill is to be contested here, nor that any maxims of common sense whatever are here to be flatly contradicted, since logic itself must be based upon them. But many of the beliefs that affect a man's cheerfulness and consequent* efficiency, such as his faith in Divine Providence, and his idea of dying, lack, and must lack, the comparison of their conclusions with any positive experience of the facts, if not always entirely, yet often in any sufficient number of cases; so that a man is naturally led to some doubt as to whether the reasons for such beliefs be sound or not, and so to criticize them within himself, and to approve of such criticism, if it be well conducted. Now logic helps him to render that criticism broad, sound, impartial, secure, by tracing the reasonings back to their fundamental principles. Every young man has to decide between arguments *pro* and *con* his adopting a given profession, wooing an individual woman, selling in a certain locality, investing his patrimony in a proposed way, and many other things, without having more than the slightest previous personal experience to guide him; and throughout his life, perplexing questions of evidence are always liable to arise. None of these questions can safely be left to the impulses of the moment. The reasons on both sides have to be weighed and criticized; and to be well-grounded in a sound method of discussing them will be an important advantage.

---

* beliefs ... consequent] *above deleted:* reasonings upon which a man's happiness and

Until long after the writer began his studies, down-right fallacies were accepted as sound reasonings in all the sciences. Faulty arguments are still occasionally met with in words of natural science; while historical criticism is about as much as ever it was a field where errors of logic, both in claiming as "demonstrations" what are at best no better than probabilities, and in not admitting conclusions which ought to be accepted, are much too common. If, then, any scientific truth is important, logic certainly is so.

The writer regards the science of Logic as made up of three studies. Its centre and its heart is the study of logical *Critic*, whose business is to ascertain what descriptions of arguments are sound, and in what the soundness of each consists. Upon this Critic chiefly, should be founded the study of *Methodeutic*, whose purpose is to ascertain the proper order of procedure in any inquiry; while the third component study is required as a preliminary to Critic, its purpose being to learn the different natures of the parts of arguments. One is at a loss by what name to call this analytic division of Logic. 'Analytic' would not do, because this name was pre-empted by Aristotle for his four books on Critic. An argument is a complex fact, or the expression of such a fact, which is a *sign* of the reality of another fact expressed in the argument's conclusion. Every judgment from a Sign is a reasoning, every reasoning is an interpretation of a Sign, and every interpretation of a Sign is, at least, allied to reasoning. Almost any kind of sign may enter into an argument as an important part of it, leading to the conclusion, and enacting a rôle in it similar to Aristotle's "middle." It is accordingly necessary that the introductory part of Logic should examine the different sorts of Meanings of Signs. Aristotle's book *de Interpretatione*, or Περὶ ἑρμηνείας, which treats of the parts of arguments, suggests that this part of Logic might be called *hermeneutic*, the science of interpretations or Meanings. Or it might be called *Universal Grammar*, the grammar of signs in general.

A *Sign*, in the sense in which the word will be used throughout this volume, is anything which represents something else,* its *Object*, to any mind that can *Interpret* it so. More explicitly, the Sign is something that appears, in place of its Object, which does not appear for itself, (at least, not in the respect in which the Sign appears;) so that the Sign, is, as it were, the *species*, or appearance, virtually, or figuratively speaking, emanating from the Object, and capable of producing upon an intelligent being an effect that will throughout this book be called the *Interpretant* of the Sign, an effect which is recognized as due, in some sense to the Object; and it is

* something else] *replaces:* something,—generally something else,—

in producing the Interpretant so that it is referred to the[*] Object, that the sign fulfills the function its fitness for which constitutes it a "*Sign*." This statement is merely a preliminary exposition of the sense of the word Sign, in its fullness; and it will hereafter appear that some of the conditions here specified can, in a Sign that is a part of another Sign,[†] miss fulfillment in some measure without the fragment's[‡] entirely losing its character of being a Sign. In another place, the possibility of a more accurate definition shall be considered.[§]

The facts, 1st, that an argument is a Sign; 2nd, that it is worse than needless, in logical Critic, to consider the argument from any other point of view, or to consider anything but this sign itself, and the Signs of which it is composed, together with the relations of all these to their Objects and Interpretants; 3rd, the fact that the majority of the leading species of Signs may enter into an Argument in such a way that their essential characters have to be considered in Logic; 4th, that the interpretation of any Sign, without which it would not function as a Sign, is somewhat closely analogous to reasoning, even when it is not strictly so; and 5th, the circumstance that, at present, no other social group than the logicians occupy themselves with

---

[*] the] *above deleted:* its

[†] a Sign … Sign] *above deleted:* exceptional case

[‡] fragment's] *above deleted:* Sign's

[§]Peirce inserts at this point a longish, and only partially surviving footnote (R 640:11–15). Some parts are reproduced below, whereas the remainder (how Mill's use of connotation and denotation violates the ethics of terminology, and Peirce's claim that the newness of his view merely consists in thinking things through more thoroughly than others, and the beginning of an example) is omitted.

> For the benefit of readers who have some acquaintance with Logic, it may be pointed out that the relation of a Sign to its Object and its relation to its Interpretant are substantially Mill's *denotation* and *connotation*, which latter term is an offence against the ethics of terminology *[…]* By the *Interpretant* of a Sign is meant all that the Sign can signify, mean, or itself convey of new, in contradistinction to what it may stimulate the observer to find out otherwise, as, for example, by new experiences or by recollecting former experiences. By the *Object* of a Sign, as the term is used by the present writer, is meant that to which the Sign applies but which it does not express otherwise than through some other Sign, or through collateral experience, or through an indication of how the interpreter of it may proceed in order to identify it. Some teachers of Logic have pronounced these two definitions unintelligible; and therefore several examples shall presently be given to aid the understanding of them. But the writer fully admits that 'Interpretant' and 'Object' both are left by the definitions somewhat indefinite, and were intended to be so; since for most purposes it would be idle and only confusing to introduce the fine distinctions requisite to making them more definite; and when greater definiteness is useful, adjectives can and shall be provided to secure that greater definiteness. *[What remains is the very beginning of a paragraph that presents the first of the announced examples.]*

the nature of Signs in general,—all these facts, considered together, have led the writer to think that logicians should, for the present, consider it as a part of their duties to study the essential characters of all the principal classes* of signs, so far as their relations to their Objects and their Interpretants are radically different in those different classes; and therefore he will name the treatise to which these essays are introductory, "A System of Logic, considered as Semeiotic."

Views of Logic more or less similar to this have been held in the past by authors of some of the most esteemed treatises, although none of them have quite reached the author's whole conception. But the views of all the leading schools of Logic of the present day, of which there are three or four, are all decidedly opposed to those of the present writer. That common tendency of them which he most of all opposes is that toward regarding human consciousness as the author of rationality, instead of as more or less conforming to rationality. Even if we can find no better definition of rationality than that it is that character of arguments to which experience and reflexion would tend indefinitely to make human approval conform, there still remains a world-wide difference between that idea and the opinion just mentioned. But the thinkers of our day seem to regard the distinction between being the product of the human mind and being that to which the human mind *would* approximate to thinking if sufficiently influenced by experience and reflection, as a distinction of altogether secondary importance, and hardly worth notice; while to the writer, no distinction appears more momentous than that between "is" and "would be."†

NOTES

1. That this was written in 1909, suggests Peirce began his "study of the game" in 1879—that is, just after completing the Illustrations. In a short draft written between the two versions, Peirce identifies by name a short "Introductory Lecture on the Study of Logic," which he published in 1882 in the *Johns Hopkins University Circular* (W4:378–82), writing that this is where he first put his opinion forward, and that he has held on to it ever since.

2. Ralph Waldo Emerson, "The Sphinx" (*Dial,* 1841).

---

* classes] *above deleted:* kinds

†The text ends at the very top of the sheet in a complete sentence, suggesting no material is missing. This sheet is dated October 23, 1909, 2:45 PM. The abrupt ending suggests that Peirce abandoned the project at this point. See introduction for further details about the compositional history of this document.

3. This is most likely a reference to Peirce's work on the relative brightness of stars, which he began in the second half of the 1860s (see the introduction). Peirce later capitalized on this experience when in 1885 he published, with Joseph Jastrow, "On Small Differences of Sensation," *Memoirs of the National Academy of Sciences* 3 (1885):75–82.

4. Francis Galton (1822–1911), British polymath who founded psychometrics and differential psychology.

5. *Unterschiedsschwelle* is a term of psychophysics introduced by Ernst Heinrich Weber (1795–1878), who defined it as the smallest difference between two stimuli that can still be detected by the perceiver; in English: difference threshold. See also Peirce and Jastrow, "On Small Differences of Sensation."

6. A teetotum is a small spinning top with four sides, each with a letter that tells the player what to do. Conventionally, the letters were TADN, with the T standing for *totum,* take everything from the pot.

7. Karl Weierstrass (1815–1897), German mathematician who is oft considered the father of modern analysis.

8. Peirce also calls retroduction abduction and hypothesis. See the discussion of hypothesis in chapter 6, with endnotes.

9. William Whewell (1794–1866); for a discussion of his contributions, by Peirce, see Chapter 7, with endnotes.

10. Methodeutic, which Peirce also calls Speculative Rhetoric, refers to the part of logic that is concerned with "the general conditions requisite for the attainment of truth" (CP 2.207, 1901); it studies "the methods that ought to be pursued in the investigation, in the exposition, and in the application of truth" (EP 2:260, 1903).

11. In his division of the sciences Peirce typically divides logic into Speculative Grammar, Critic, and Speculative Rhetoric, which each subsequent branch relying on those that precede it while informing the latter of its findings. Speculative Grammar studies the general nature of signs and Critic "classifies arguments and determines the validity and degree of force of each kind" (EP 2:260, 1903). For Speculative Rhetoric see previous note.

12. Fechner's Law states that subjective sensation is proportional to the logarithm of the stimulus intensity.

13. Immanuel Kant, Über die falsche Spitzfindigkeit der vier syllogistische Figuren (Königsberg: J.J. Kanter, 1762). The edition Peirce had at hand when writing this text is *Kant's Introduction to Logic and His Essay on the Mistaken Subtilty of the Four Figures*, translated by Thomas Kingsmill Abbott (London: Longmans and Green, 1885).

14. To help students identify which syllogisms are valid, medieval textbooks developed a doggerel verse of mnemonic names where the vowels of each word indicate the nature of the three propositions in the syllogism. *Baralipton* refers to the following syllogistic form: if every M is L, and every S is M, then some L is S. for example: Every spirit is simple; Everything simple is incorruptible; hence, something incorruptible is a spirit.

15. *Kant's Introduction to Logic*, 87. The text continues as follows, using the example above: "This will be seen at once by comparing it with the middle term. In fact, I cannot say something incorruptible is a spirit, because it is simple; for from the fact that something is simple it does not follow that it is a spirit."

16. The term is defined by Peirce in the *Century Dictionary*, CD:2383.

17. *Kant's Introduction to Logic*, 86.

18. Peirce, apparently not recalling the first name, leaves a blank space in the manuscript. Peirce is most likely thinking of Heinrich Christoph Wilhelm von Sigwart (1789–1844), also a German philosopher and logician, and the father of the aforementioned Christoph Eberhard Philipp von Sigwart (1830–1904).

19. Christoph Sigwart, *Logic,* 2 vols. (New York: Macmillan, 1895). Sigwart's endorsement of Helen Dendy's English translation is found on p. vii. Sigwart presumably also endorsed the capitalization in the English edition.

20. Sigwart, *Logic,* §1, 1.

21. Sigwart, *Logic,* §1.1, 1–2.

22. Sigwart, *Logic,* §1.9, 9–10; text in square brackets is Peirce's.

23. E.g., Aristotle's *Metaphysics* opens this way.

24. Sigwart, *Logic,* §1.6, 6.

25. Sigwart, *Logic,* §1.6, 7.

26. Sir Edward George Earle Bulwer-Lytton (1803–1873). His play, *Richelieu; or, The Conspiracy,* was first staged in 1839.

Chapter 9

LETTER TO PAUL CARUS

Peirce wrote this undated letter in the second half of August 1910. It is most likely a response to a letter Carus had sent him in which he practically threatened Peirce that he would publish the Illustrations whether Peirce sent him corrections or not. The letter is preceded by an aborted July 19, 1910 draft in which Peirce aimed to explain the corrections he deemed "indispensable in the reprint of my Popular Science Monthly *papers" (RL 77:208–18). In that letter, Peirce, having tried hard to rewrite the first two Illustrations articles in the previous year, agreed to republish the Illustrations in its original form, prefaced by a note explaining a major error that needed to be corrected. Peirce wrote this letter such that it could act as the printer's copy for that preface, writing "only on one side of the paper, so that, if I would suddenly die in one of the fits to which I am almost daily subject, what I write could be set up and printed." Both in the July draft and in the letter printed below, Peirce took the main error of the Illustrations to lie in its nominalism, even though in his own estimation he had already moved away from that at the time. As he explains in the draft:*

> *The greatest and most decisive of all questions of metaphysics, is, according to the way of thinking which has gradually ripened in me since 1873 (which is the date of my formulating the opinion expressed in the two articles that are the two parts of that Essay on Pragmatistic Clearness),*[1] *is the question whether*

> *generals are real; which question I would like to phrase as Whether anything* Indeterminate *is Real (since we define an Individual, meaning what is the contrary of a General, or that which is* omni modo determinatum*),*[2] *and which question I would supplement by asking Whether anything* Indefinite *is Real.*

Unfortunately, the July draft breaks off before the nominalism-realism discussion really gets underway. In the August letter, reprinted below, Peirce reiterates the plan of publishing the Illustrations virtually unaltered and adding a preface "in which I state in general terms how what I then say ought to be altered." This time, however, the letter does not contain the preface, even if in draft form, but is much more preliminary, nor is it limited to the main error of the series, its nominalism, as it also addresses issues with the later articles, especially his account of the doctrine of chances, distinguishing probability from both plausibility and likelihood. In addition Peirce is suggesting adding a discussion of the existential graphs, an idea that is very skeptically received by Open Court.

Paul Carus received the letter on August 29, 1910 and immediately forwarded it to Francis C. Russell whom he charged with readying the Illustrations for publication. Russell had three copies made at the Open Court office in Chicago: one for himself, one for Carus, and one for Peirce. The last was sent to Peirce on September 17, 1910 together with the original holograph letter. The holograph letter is preserved in Peirce's papers at Harvard's Houghton Library (RL 77: 220–41), with the exception of the first sheet, which seems to be lost. Of the typescript sent to Peirce, only the first sheet was found (RL 77:242). The remaining two typescripts survive in the Open Court Collection at the Morris Library of Southern Illinois University in Carbondale (Coll. 27, Box 91), one containing corrections made by Russell (comparing the typescript with the original letter), the other containing only a few corrections by Carus. The letter is included in Volume 8 of the Collected Papers *(CP 8.214–38), albeit with significant omissions. A more detailed discussion is found in the introduction.*

My Dear Doctor Carus:[*]

Ever since I was paid that money by you and Mrs. Carus, I have been engaged with all my energy, allowing only for such as I had to expend

---

*The first part of this letter is transcribed from the typescript that was made at the Open Court Publishing Company, taking into account Francis Russell's corrections to the typescript. For this part the original holograph is lost, with the exception of one page (see next footnote). A later footnote indicates where transcription from the holograph begins.

upon my wife's health and upon getting this house habitable and in salable condition, in trying to write an article or articles for you upon the second grade of clearness,[3] i.e., that which results from analytic definition and upon corrections to the errors and other faults of the articles of mine that appeared in the *Popular Science Monthly* in 1877 and 1878[4] to which I should be glad if you would add a reprint of the article of January, 1901 which requires no correction.[5]

I have written a great deal but am satisfied with but the smaller part of it, and the result is that I worked, until,—what with worries, too,—I was in a state of downright nervous exhaustion from which I have now been for a good number of months recovering, but owing to the decay of my powers and being so sick of writing about the same things, I cannot even write yet but a few equivalents of pages of this letter very slowly and for not over three hours and with many days when I can do nothing. Since I got your letter I have—often trying in vain to accelerate my speed,—the slowness of which is largely due to the labor of writing with the extreme precision required, gradually been forced to the conclusion that since you are very reasonably impatient, my best course is simply to write a preface in which I state in general terms how what I then say ought to be altered, and I will here (if my strength holds out) try to indicate the points I should make.

In regard to the first Essay consisting of the first two articles, the principal positive error is its nominalism, especially as illustrated by what I said about Gray's stanza, "Full many a gem" etc., pp. 3 to 5 et seq.[6] I must show that the *will be's,* the actually *is's,* and the *have beens* are not the sum of the reals. They only cover actuality. There are besides *would be's* and *can be's* that are real. The distinction is that the *actual* is subject both to the principles of contradiction and of excluded middle; and in *one* way so are the *would be's* and *can be's.* In *that* way a *would be* is but the negation of a *can be* and conversely.* But in another way a *would-be* is not subject to the principle of excluded middle; both *would be X* and *would be not-X* may be false. And in this latter way a *can be* may be defined as that which is not subject to the principle of contradiction. On the contrary, if of anything it is *only* true that it *can be X* it *can be not-X* as well.

It certainly can be proved very clearly that the Universe does contain both *would be's* and *can be's.*

Then in regard to the second article, I ought to say that my three grades of clearness are *not,* as I seemed then to think, such that either the first or the second are superseded by the third although we may say that they are acquired,—*mostly,*—in the order of those numbers. I ought to describe,

* error is its nominalism ... and conversely.] Transcribed from RL 77:220.

if only in a paragraph, how to train oneself and one's children in the first grade of clearness, so that, for example, one will recognize a millimetre length when one meets with it; and so with colors.[7] I have done a great deal of work in training myself to this kind of clearness.[8] It would if put together amount to two or three years of industry; and I should recommend systematic* exercises of the sort to everybody.[9] Useful as that is, however, I don't hesitate to say that the second grade of clearness is far more important, and all my writings of late years illustrate that.[10] Still, I continue to admit that the third grade is the most important of all and a good example of it is Wm. James who is so phenomenally weak in the second grade, yet ever so high above most men in the third. But there is no reason why all three should not be symmetrically developed.

The bulk of these *Popular Science* articles, after the first two, are occupied with a criticism of the underlying principles of Laplace's *Théorie analytique des probabilités* and Mill's *System of Logic,*—two writers of a high order which have had and still have a great and deplorable influence.[11]

Before the third article on Probability I should like to insert a short and easy account of my Existential Graphs; because when that system is well in hand, it becomes so much easier to show great faults of Laplace and Mill; and that shorter account, *I could now easily write*.[12]

It would also be well to show how all numbers involve *essentially* nothing but ideas of *succession*.[13] Then I should like to point out how utterly Laplace fails to define what he means by *probabilité,* his account of it resting upon what he calls the également *possibles,* which I maintain has none but the vaguest meaning.[14] I ought on my side to define *probability.* For that purpose, I should have to begin by distinguishing three ways—three quite different *directions* so to speak, as different as the *X, Y, Z* of a system of orthogonal coördinates—in which cognitions can fall short of absolute certainty,—or rather of *mathematical* certainty, which is not absolute, because blunders may have been committed in reaching it.

The names which I would propose for general adoption for the three different kinds of *acceptability* of propositions are

plausibility
verisimilitude
probability†

*From this point onward the transcription is made from Peirce's original holograph letter as it is preserved at the Houghton Library.

†*Marginal note by Peirce:* I am not satisfied with the third. I would rather suggest the phrase "so it would appear."

The last alone seems to be capable of a certain degree of exactitude of measurement.

By *plausibility,* I mean the degree to which a theory ought to recommend itself to our belief independently of any kind of evidence other than our instinct urging us to regard it favorably. All the other races of animals certainly have such instincts: why refuse them to mankind? Have not all men some notions of right and wrong as well as purely theoretical instincts? For example, if any man finds that an object of no great size in his chamber behaves in any surprising manner, he wonders what makes it do so; and his instinct suggests that the cause, most plausibly, is also in his chamber or in the neighbourhood.

It is true that the alchemists used to think it might be some configuration of the planets; but in my opinion this was due to a special derangement of natural instinct. Physicists certainly today continue largely to be influenced by such plausibilities in selecting which of several hypotheses they will first put to the test.

By *verisimilitude* I mean that kind of recommendation of a proposition which consists in evidence which is insufficient because there is not enough of it, but which will amount to proof if that evidence which is not yet examined continues to be of the same complexion as that already examined, or if the evidence not at hand and that never will be complete, should be like that which is at hand. All determinations of probability ultimately rest on such verisimilitudes. I mean that if we throw a die 216* times in order to ascertain whether the probability of its turning up a six at any one throw differs decidedly from $^1/_6$ or not, our conclusion is an affair not of *probability* as Laplace would have it, by assuming that the antecedent probabilities of the different values of the probability are equal, but is a verisimilitude or as we say a *"likelihood."* That Laplace is wrong can be demonstrated since his theory leads to contradictory results. But perhaps the easiest way to show it is wrong is to point out that there is no more reason for assuming that all the values of the probability are equally probable than for assuming that all the values of the *odds* are so; or that all the values of the logarithms of the odds are so, since this is our instinctive way of judging of probabilities, as is shown by our "balancing the probabilities."

Having thus defined *plausibility* and *verisimilitude,* I come to define *probability.*[15] None of the books contain a definition of mathematical probability (which is what I mean by "probability," however measured) which will hold water. For the sake of simplicity, I will define it in a particular example. If, then, I say that the probability that if a certain die be thrown

---

* 216] *above deleted:* 3600

in the usual way it will turn up a number divisible by 3 (i.e. either 3 or 6) is $^1/_3$, what do I mean? I mean, of course, to state that that die has a certain habit or disposition of behaviour, in its present state of wear. It is a *would be* and does not consist in actualities or single events in any multitude finite or infinite. Nevertheless a habit does consist in what *would* happen under certain circumstances if it should remain unchanged throughout an endless series of actual occurrences. I must therefore define that habit of the die in question which we express by saying that there is a probability of $^1/_3$ (or odds of 1 to 2) that if it be thrown it will turn up a number divisible by 3 by saying how it *would* behave if, while remaining with its shape, etc. just as they are now, it *were to be* thrown an endless succession of times. Now it is very true that it is quite impossible that it should be thrown an infinite succession of times. But this is no objection to my supposing it, since that impossibility is merely a physical, or, if you please, a metaphysical one, and is not due to any logical impossibility to the occurrence in a finite time of an endless succession of events each occupying a finite time. For when Achilles overtook the tortoise he had to go through such an endless series (endless *in the series,* but not endless *in time)* and actually did so; or if he didn't another did.

Very well, I will further suppose that tallies are kept during the throwings, one tally of the throws turning up 6 or 3, and the other tally of the throws turning up 1, 2, 4, 5; and further I will suppose that after each throw the number that the latter tally has reached shall be divided by the number that the former tally has reached. And I will use the expression that this quotient changes its value at every new throw, instead of saying that a new quotient differs from the last. When the quotient changes from being greater than 2 to being* less than or greater than 2 to being either just 2 or on the opposite side of 2 to what it was before, as for example if it passes from being 21:10 to being 21:11 or from 21:11 to 22:11 or from 25:12 to 25:13, etc., I shall say it *touches* 2 (meaning strictly that it either comes to 2† or passes across 2). Then after the first throw it will be either 0 or ∞ and there it may remain for any number of throws. But after it has once moved away it never will return to either of these values, but after it has finally recovered from the effects of the first throws it will oscillate in a very irregular way, and soon it will "touch" (or pass over) some other values *for the last time* although nobody can *know* that it is to prove to have been for the last time; and then values still nearer to 2 will be touched or traversed for the last

* to being] *before deleted:* not greater *[above deleted:* less*]* than 2 (the ratio changing say from 21:10 to 21:11) or changes the other way

† comes to 2] *above deleted:* reaches

time. And in its endless series* there will be no value except only 2 that it would† not touch or traverse for the last time *excepting* only the value 2. And this "would be" is what constitutes the habit which we state in saying that the odds against its turning up a number divisible by 3 are 2:1, or that the probability of its turning up a 6 or a 3 is $^1/_3$.

Aug 23

I could not write before. Now there are two points to be particularly noticed about the definition of probability. The first is that the probability of an event may be an *irrational* number. In that case *every* value of the quotient of one tally divided by the other will get finally left (so that the quotient never again touches or traverses it). Some will be left because they are too small and others because they are too large. But any one you please will get left for one or the other of these two reasons. This must be taken into account in drawing up the definition.

The other point is that our throws and tallies will give the probability in question only on condition that the fact that at certain throws (determined by their ordinal places in the series) had turned up, such and such of them numbers divisible by 3, and the rest of them numbers not of that kind, will make absolutely no difference in the probability that any given throw (determined as above) will turn up a number divisible by three. This statement, of course, refers to what would happen if a series of throws long enough to cover the highest *in order* of throwing of the determinate throws were to be repeated *endlessly*.

In short that the numbers turned up are each absolutely independent of what has been or will be turned up on other occasions. This comes to saying that the tallies need not be of consecutive throws but may be every truth or any other endless series so long as whether a given throw be counted on the tally or not is nowise influenced by whether that throw turns up a number divisible by three or another kind of number. The definition must be made explicit upon this point.

[This latter condition will make quite a puzzle for those who deny absolute chance; yet experience proves (*far* more certainly than it proves determinism), that this condition can be fulfilled *in all cases*.][16]

The definition of probability resulting in this way.

Now from the above definition of probability together with the ordinal definitions of

* series] *before deleted*: it will

† would] *above deleted*: will

1. The whole numbers, as expressed in the secundal system in which,

10	means	2
11	"	3
100	"	4
101	"	5
110	"	6
111	"	7
1000	"	8
	etc.	

But of course this description is not a definition.

2. Of addition.

3. Of multiplication.

4. Of involution.

5. Of ratio.

6. Of the rational positive values as those which result from putting

$$N:1 \equiv N$$

whatever whole cardinal[*] number (including 0) N may be. And such ratios I call the ratios of the zero grade.

And from running through any complete series of all the ratios[†] of any given *rational*[‡] *grade* from any ratio N:1 to N+1:1 inclusive and putting between every successive pair of ratios of grade *m*, (which two we may denote by p:q and r:s) a new ratio

$$(p+r):(q+s)$$

and all the ratios of grade *m* together with all so inserted constitute the ratios of grade *m*+1.

And all the ratios of all finite grades are all the positive[§] rational values.

7. And if (p:q) *[and]* (r:s) are two ratios[**] of the same[††] grade, whether successive or not, then if they have different values and p:q is less than r:s then p:q <

* cardinal] *interline insertion*

† ratios] *above deleted*: ordinals

‡ rational] *interline insertion*

§ positive] *interline insertion*

** ratios] *after deleted*: successive

†† same] *above deleted*: any

$(p+r):(q+s) < r:s$.[*] But if $p:q \nless r:s$ then $p:q \nless (p+r):(q+s) \nless r:s$.[†] And all ratios of real grades[‡] are real values and two real values only differ by what other real values one higher (or lower) than which the other is not. From these definitions I will deduce every valid rule of the doctrine of chances.

I will then show that Laplace's definition of probability, first, has no meaning at all unless it be the expression of a subjective state of mind, and second, that necessary, and quite easily perceived, consequences which can be drawn from his own explanations and expositions of it are revolting to common sense.[17]

In order to get this matter straightened out, I think it would be well to change the place of the Sixth Paper and place it directly after the third.[18] Then I would append a *Correction* in which I should state that the division of the elementary kinds of reasoning into three heads was made by me in my first lectures and was published in 1869 in Harris's *Journal of Speculative Philosophy*.[19] I still consider that it had a sound basis. Only in almost everything I printed before this century (?) I more or less mixed up Hypothesis and Induction. *Hypothesis* is, however, the most expressive term as well as that best supported by the historical usage of logicians, generally, for what I had in mind (rather confused by my unsound reflexions upon it).[20]

The general body of logicians had also at all times come very near recognizing the trichotomy. They only failed to do so by having so narrow and formalistic a conception of inference (as necessarily having formulated judgments for its premises) that they did not recognize Hypothesis (or, as I now term it, *retroduction)*[21] as an *inference*. The thing was always recognized and all logicians *almost* (perhaps all, though the Stoics and others denied the validity of induction) distinguished *induction* from necessary reasoning.

When one contemplates a surprising or otherwise perplexing state of things (often so perplexing that he cannot definitely state what the perplexing character is) he may formulate it into a judgment or many apparently connected judgments, he will often finally strike out a hypothesis, or problematical judgment, as a mere possibility, from which he either fully perceives or more or less suspects that the perplexing phenomenon would be a necessary or quite *probable* consequence.[22]

That is a *retroduction*. Now three lines of reasoning are open to him. He may proceed by mathematical or syllogistic reasoning at once to demonstrate that consequence. That of course will be deduction.

---

[*] $p:q < (p+r):(q+s) < r:s$] *emended from*: $p:q < (p+r):(q:s) < r:s$

[†] $p:q \nless (p+r):(q+s) \nless r:s$] *emended from*: $p:q \nless (p+r):(q:s) \nless r:s$

[‡] of real grades] *interline insertion*

Or he may proceed still further to study the phenomenon in order to find other features that the hypothesis will *explain* (i.e. in the English sense of explain, to deduce the facts from the hypothesis as its necessary or *probable* consequences). That will be to continue reasoning *retroductively,* i.e., by hypothesis.

Or, what is usually* the best way, he may turn to the consideration of the hypothesis, study it thoroughly and deduce miscellaneous observable consequences, and *then* return to the phenomena to find how nearly these agree with the actual facts. This is not essentially different from *induction.* Only it is (most usually)† an induction from instances which are not discrete and numerable. I now call it Qualitative Induction. It is this which I used to confound with the second line of procedure, or at least, not to distinguish it sharply.

A good account of Quantitative Induction is given in my paper in *Studies in Logic, By Members of the Johns Hopkins University,* and its two rules are there well developed.[23] But what I there call Hypothesis is so far from being that, that it is rather Quantitative than Qualitative Induction. At any rate, it is treated *mostly* as Quantitative. Hypothesis proper is in that paper only touched upon in the last section.

There is a third kind of Induction. In order to show this, it is requisite to define Induction.

Now the essential character of induction is that it infers a *would be* from *actual singulars.* These singulars must, in general, be finite in multitude and then, as I show in my Johns Hopkins paper, the inductive conclusion can be (usually) but *indefinite* and can *never* be *certain.* Even if the series of singular instances, as in the case of Achilles and the Tortoise be known to hold *without end,* a special postulate is requisite to make the conclusion hold. [It is a distinguished merit of the strictly ordinal view of fractions that it brings this point out in perfect clearness, in conjunction with F. Klein's doctrine of Measure (in one of the early volumes of the *Mathematische Annalen*).[24] For there may be a discontinuity at that very point, so that e.g. 0.111111 (continued without end) does not amount to $^1/_9$, just as if $^1/_9$ and all following numbers were shoved along, which would not derange the arithmetic at all if the definitions are strictly ordinal.]‡

But in ordinary cases an induction would become both *precise* and *certain,*—though even then it would not be apodictic certainty,—if the

* usually] *interline insertion*

† (most usually)] *interline insertion*

‡ ordinal.]] *closing square bracket added by the editor*

instances were of denumeral (or simply endless) multitude. Therefore, defining* induction as the sort of inference which produces verisimilitude or likelihood, that is, which regards an endless series of *actualities*† as conclusive evidence of a *would-be* since it is the best evidence possible when we are not behind the scenes.

From this point of view any plausible proposition that is supported by instances in every respect is justifiable so long as one keeps on the alert for the first exception. Of course, such an induction has the very minimum of likelihood, yet it has *some;* and we very often find ourselves driven to accept it. The world has always turned on its axis so far as we know about once every 24 hours and therefore we presume‡ (vaguely) that it always will continue to do so. In every case that has been sufficiently inquired into, every human being has been born of a woman not a maiden. So almost everybody feels sure it always will be found so. People are far more confident of it than they have any right to be. All former generations of men have died off. Therefore, people say, they always will. In one sense I suppose this is certain. But that they always *would* even if there were no accidents, seems to me as weak an inference as any that I would not positively condemn as utterly worthless. I call this kind of thing *crude induction.*

I must confess that although my explanation of the validity of induction seems to me to be far superior to any other, I am not altogether satisfied with it, or rather with its results.§ Quantitative Induction depends upon the possibility of making a truly representative sample. That is to say, the examples composing it must be chosen as possessing the conditional character, which is easy enough, but also so that the choice of them shall not be influenced, one way or the other, by whether or not they possess the consequent character. They must be such on the whole in that course of experience to which the induction is to be applied. That, to be sure, cannot but be the case, should the entire class sampled be alike in respect to the consequent character. But the further this ideal state of things is from being realized the more extremely difficult it becomes to get a truly representative sample, and the result, after every precaution has been taken, is that we cannot expect any great precision in inductive conclusions when the class is anywhere near being equally divided between individuals that do and that do not possess the consequent character. However, this is not owing

---

* defining] *before deleted*: *verisimilitude*, or *likelihood*, as the sort of approach to perfect certainty that a sound induction possesses,

† of actualities] *interline insertion*

‡ presume] *above deleted*: suppose

§ , or rather with its results] *interline insertion*

to any *falsity* in my theory but to the essential imperfection of induction itself when applied to these cases.

As for the validity of the hypothesis, the retroduction, there seems at first to be no room at all for the question of what supports it, since from an actual fact it only infers a *may be,—may be* and *may be not.* But there is a decided leaning to the affirmative side and the frequency with which that turns out to be an actual fact is to me quite the most surprising of all the wonders of the universe. I hope there are few men who are so often deceived as I. They surely cannot be many. Yet I could tell you of conjectures pronounced by me with a confidence that I could not comprehend, and that were verified amazingly. I simply should not dare to tell them; I feel my credit would not support such tales: I should be a fool to expect it. So I will not enter upon that. But I will point to some of the great steps of science. Faraday was decidedly a cautious reasoner. What made him so strangely confident that electricity in all the variety of its workings never acts at a distance? What an unaccountable thing was Dalton's atomic theory and its ready reception. What in the world should prevent,—even if there were atoms,—197 atoms of one substance from joining 719 of another, to make a chemical compound, both numbers being prime? That question prevented me, as a boy, from believing in atoms; and I was only convinced when new facts were discovered. How was it that Galileo, after glancing at one pretty evidently absurd hypothesis, at once struck upon the idea that the speed of a falling body increases uniformly with time? I am ashamed at being obliged to confess that this volume contains a very false and foolish remark about Kepler. When I wrote it I had never studied the original book as I have since. It is now my deliberate opinion that it is the most marvelous piece of inductive reasoning I have been able to find. J.S. Mill says it was not reasoning at all, but only description.[25] But that only shows that he had not the slightest idea of the problem. There may be nineteen hypotheses in the book, though I cannot now say how I made out that number. But there are certainly nothing like that number to explain any one the numerous circumstances of the motion of Mars which Kepler[*] succeeded in reducing to order. The facts he had before him were simply observations of the place of Mars[†] in the heavens at various dates. His problem, as Mill seems to suppose was not simply to draw a line on the celestial globe, through all these points, and to see that that line was an ellipse with the sun at one

[*]Though throughout this text Peirce mostly wrote Kepler at this instance he wrote Keppler, a spelling he preferred because it is how Kepler signed his name in his correspondence. See W8, sel. 49 and W8:452; see also R 339d:581 (1908).

[†]Mars] *before deleted*: relatively to the different stars at many di

of the foci. For, to begin with, Mill seems to forget that Kepler told where Mars was at every time *in space* with its three dimensions, and not merely, what the *projection of its path* upon the celestial sphere was.*

A pretty good example of a retroduction is the case of stopping an unknown person in the streets of a large city, asking some bit of information and crediting it. What makes this a more characteristic case is that one will have selected the man to interrogate, and will have made the selection almost without being conscious of having done so.

The justification of retroduction is that one *must* trust one's instincts through life or be content with a passivity that cannot content him.

All positive knowledge must be reached if at all by an operation that begins with a conjecture. I do not mean that one reflects upon this at the outset, I mean that afterward when one subjects one's behaviour for the day or for any other marked period to self-criticism this is how one ought to vindicate good retroductions.

As for deduction, Kant gave the principle of that. In the premises the conclusion is, in substance, asserted.

The practical lesson that experience with retroductions ought to teach men,—especially *male* men,—is that there is much to be learned from one's instincts, and that consequently pains should be taken to let them develope in directions in which they prove to be sound, and to correct them in the directions in which they have proved treacherous.

The universe which man inhabits, the Real Universe, is the Universe of his conclusions, such as they *would be,* if properly corrected and if duly enlarged by new experiences. This is the true Kantian doctrine.[26]

Often we ought to put much faith,—or, at the very least, high hope,—in ratiocinations wholly unsupported by inductions—i.e. by positive evidence, (*positive* meaning pragmatistic, but with emphasis on the restraining tendency of experience). If one feels a strong inward tendency to hope for a future life of activity, work, further enlightenment, it is bad logic not to yield to it, while remembering that it is but a justifiable hope, and while on the watch to note whether it is really doing us good or not. Nothing has been a greater curse to a large part of the human race than religions that have been outlived. At the same time they may be keeping the ruder

*Peirce inserted, centered in the white space following the end of the paragraph "Aug", which suggests that he did not complete the letter on August 23. Though Peirce neglected to mention the day, it can be conjectured that he continued the letter either the next day (i.e., the 24th) or shortly thereafter, as in an unsent cover letter dated August 26, 1910 Peirce mentions "a long letter that should have gone two days ago" (RL 77:252–53).

part of the community from crime and ruin, they may be undermining the characters of a higher class, who ought to remember that they are of the stuff that dreams are made of.[27] That they have been worked over with a view to squeezing money out of the laity to support a despicable clergy. Fortunately no protestant congregation is yet rich enough to answer to this description. Whether or not the Mormons, the "Christian Scientists," the Romanists, the Spiritualists do or do not, it is not my business to say. They are all loaded with gold or keenly in quest of it. Religions are wholesome while they are *poor and persecuted.*

I want to put into the volume a sufficient refutation of Mill's system of Logic. To begin with it falls into the fatal error of Nominalism. Now how hard it is to form an exact and true definition of Nominalism, is shown by my own treatment of Gray's stanza ten years after I had declared myself a Realist (in the sense in which we call most of the scholastics realists, or say "a scholastic realist.")[28]

Nominalism might from the etymology of the name be supposed to consist in holding "would-bes" to be as the general nature of names.

But it ought not to remain long in doubt that a *would-be* is something that can be appropriately clothed in words or thoughts, that it is, or would be, *applicable* to singular actuals, and that it has a meaning, that is, is capable of definition, or logical analysis. Now these are the essential characteristics of a particular kind of *sign.*[29]

A sign in general is 1st something Real, that is 2nd applicable to an object different from itself and already known to the person to whom it is a sign, and 3rd is capable of interpretation in the mind of that person, so that it will (or would if accepted as veracious sign)* have some effect upon him of a kind that it was calculated or fit to have.[30]

Signs in general I divide in ten ways.[31] One of these divisions is according to the substance, or mode of being, of the sign itself. For a sign may be, 1st an existent thing or actual occurrence, 2nd it may be a *would-be,* or habit, 3rd, it may be a mere *can-be.*[32]

Then there are 3 divisions that relate to the Object. One according to the form under which the Sign presents its Object.[33] This is of course the object *as the sign represents it,* i.e. the Immediate Object. There are two divisions that concern the Real Object.[34] One according to the Mode of Being of it, according to which the Sign will be Abstract, Concrete, or the Sign of a Habit, or would be. The other according to the relation, or connexion between the Sign and the Real Object, making the sign either an Icon, an Index, or a Symbol.

*Closing parenthesis added.

There are then 6 divisions of signs that refer to the Interpretant. Now we can regard the Interpretant in 3 ways, or rather there are 3 distinct things which may properly be regarded as the Interpretant.[35] For any thing that the sign, *as such,* effects may be considered as the Interpretant. And this may be 1st something merely subjective, the vague determination of consciousness effected by the sign, 2nd the *actual event* that some signs by virtue of really acting as such bring about. For instance, let the sign be a military word or command. Then the instant action of the whole rank of men addressed will be the Dynamical Interpretant, as I call it. The third sense in which we may properly speak of the Interpretant is that in which I speak of the Final Interpretant meaning that Habit in the production of which the function of the Sign, as such, is exhausted.

I only make use of one division according to the 1st of the above, the Initial Interpretant, as I call it.[36] This relates to the number of independent respects in which the sign is indefinite, that is respects which the sign points out but leaves indefinite. For example, the main difference between the verbs "is beneficial," "benefits," "gives," and "sells" is of this nature.

I recognize 2 distinctions referring to the Dynamical, or Actual, Interpretant. First as to its Nature as Feeling, Action, or Affective habit, and then as to the manner in which the sign makes its appeal, as Imperative, or calling for Assent, or submitting to consideration. Then I take account of 3 divisions referring to the Final Interpretant. 1st according to the nature of its Purpose, 2nd according to the nature of the influence the sign is intended to exert, and 3rd according to the nature of the assurance the sign affords to an interpreter that has rightly apprehended it. The last 3 divisions are especially logical in the ordinary idea attached to that word.

Of course there are any number of other distinctions which a student of signs may occasionally have to attend to, but I think that these ten are as many as an ordinary Intellect can carry ready for.[*37]

NOTES

1. In this letter, as in chapters 7 and 8, Peirce seeks to recombine the first two Illustrations articles into a single essay. As he puts it in the July letter: "The first two constitute one Essay upon what I call *Pragmatism* or the doctrine of *pragmatic clearness* which consists in a clear conception of that *Habit of Conduct* in which any given concept would work out its actualization" (RL 77:229).

---

[*]With its period lacking this sentence concludes the letter, which is unsigned. The typescript made at the Open Court confirms that this is the end of the letter that Peirce had sent them.

2. Entirely determined being.

3. The reference is to manuscripts R 643–50, composed between December 12, 1909 and August 5, 1910.

4. See chapters 7 and 8 for a sampling of this. For a more detailed discussion of this period, see the introduction.

5. "Pearson's *Grammar of Science.* Annotations on the First Three Chapters." *Popular Science Monthly* 58 (January): 296–306. Alternatively, Peirce suggests that this paper could be published "either in whole or omitting the first four and a half or five pages" (RL 77:247).

6. Peirce initially wrote the first two articles as a single essay. In chapters 7 and 8 Peirce seeks to recombine them. Peirce's original discussion of Gray's stanza is found above in chapter 2, sect. IV.

7. See also chapter 2, sect. I, with endnotes. The latter refer to R 254 (ca.1895) where Peirce separates sensuous from logical clearness and gives an extensive analysis of what the first grade of (sensuous) clearness for a color amounts to (R 254:4–6).

8. Peirce's early scientific work in astronomy, which culminated in his 1878 *Photometric Researches* (Leipzig: Wilhelm Engelmann), required him to make very precise color differentiations. See also W5, item 122, "On Small Differences of Sensation."

9. See the fifth lecture of Peirce's 1898 Cambridge Conference Lectures; see esp. RLT:182–87.

10. See esp. R 643–50. Note, though, that Peirce's earlier work for the *Century Dictionary* too has a strong focus on the second grade of clearness.

11. Pierre Simon, marquis de Laplace, *Théorie analytique des probabilités* (Paris: Ve. Courcier, 1812) and John Stuart Mill *A System of Logic, Ratiocinative and Inductive* (London: J.W. Parker, 1843). For an extensive discussion of Mill and Laplace, see esp. chapter 7, version II.

12. The existential graphs refer to a way of doing mathematical logic that is geometric rather than algebraic. For Peirce's discussion of the graphs, see esp. CP 4.347–584 (1902–1908) and RLT:146–64 (1898). For what Peirce might have had in mind with this "short and easy account," see his discussion of the graphs in his June 22. 1911 letter to J.H. Kehler (NEM 3:159–210, esp. 3:162ff). For an accessible introduction to the graphs, see Don Roberts, *The Existential Graphs of Charles S. Peirce* (The Hague: Mouton, 1973).

13. Peirce spent considerable energy on the axiomatization of numbers. See esp. "On the Logic of Number," which appeared in the *American Journal of Mathematics* in 1881 (W4, sel. 38). For other, unpublished, texts see e.g. "The Axioms of Number" (W4, sel. 24; 1880), "Fundamental Properties of Number" (W5, sel. 45; 1886), "Logic of Number" (W6, sel. 21; c.1887), and CP 4.153–58 (from "Recreations in Reasoning"; R205, ca.1897).

14. Laplace, *Théorie analytique des probabilité*, iv–viii.

15. An extensive traditional definition for probability, authored by Peirce, is

found in CD:4741, esp. sub 2, which concerns what Peirce here calls "mathematical probability" (Peirce authored 1–3).

16. The doctrine that there is real chance in the world Peirce named *tychism* (W8:135). See especially his *Monist* article "The Doctrine of Necessity Examined" (W8, sel. 24), and his "Reply to the Necessitarians. Rejoinder to Dr. Carus," which also appeared in the *Monist.* The latter, together with assorted documents and the two papers by Carus that Peirce is responding to will be included in W9 as sel. 42–49.

17. Some of those consequences are discussed in chapter 7.

18. Given that Peirce also suggests combining the first two articles into a single essay this suggestion is somewhat ambiguous. It can mean moving "Deduction, Induction, and Hypothesis" between "The Doctrine of Chances" and "The Probability of Induction," which is what the editors of the *Collected Papers* suggest (CP 8.227n), or it can mean that Peirce wants to switch "Deduction, Induction, and Hypothesis" with "The Order of Nature." Note that "The Doctrine of Chances" and "The Probability of Induction" also originated as a single paper (see the headnote to "The Doctrine of Chances," as well as the introduction to the present volume).

19. The "first lectures" are the Harvard lectures of 1865. See esp. "Lecture VIII: Forms of Induction and Hypothesis" (W1, sel. 33). Though Peirce refers here to his 1869 "Grounds of the Validity of the Laws of Logic: Further Consequences of Four Incapacities," which is the third paper in a series of three, the reference should have been to the second paper, "Some Consequences of Four Incapacities" *Journal of Speculative Philosophy* 2 (1868): 140–57; also W2, sel. 22, see esp. W2:217–20. William T. Harris (1835–1909) was the journal's editor.

20. On the historical usage of the term hypothesis, see Peirce's longish footnote at W2:218f.

21. Meaning literally "to draw back" (the Latin *retro* meaning "back" and *ducere* meaning "to lead") retroduction reasons from consequent to antecedent (EP2:441, 1908). Following Giulio Pacio's translation of Aristotle's ápagogè in Bk 25 of the *Prior Analytics,* Peirce also calls this type of inference "abduction" ("leading away"; from the Latin *ab,* for "away," and *ducere,* for "to lead"). See CD:9; the description sub 2 is Peirce's.

22. Compare with the following from the 1903 Harvard Lectures on Pragmatism, which is the passage most often quoted by contemporary philosophers of science (CP5.189):

> Long before I first classed abduction as an inference it was recognized by logicians that the operation of adopting an explanatory hypothesis—which is just what abduction is—was subject to certain conditions. Namely, the hypothesis cannot be admitted, even as a hypothesis, unless it be supposed that it would account for the facts or some of them. The form of inference, therefore, is this:
>
> The surprising fact, C, is observed;
> But if A were true, C would be a matter of course,
> Hence, there is reason to suspect that A is true.

> Thus, A cannot be abductively inferred, or if you prefer the expression, cannot be abductively conjectured until its entire content is already present in the premiss, "If A were true, C would be a matter of course."

23. *Studies in Logic* (Boston: Little, Brown, and Co., 1883) was edited by Peirce. The paper referred to, "A Theory of Probable Inference," is found at pp. 126–81, and is accompanied with two appendices, "Note A" and "Note B" (pp. 182–203). All three are reproduced in W4, sel. 64–66.

24. In "Über die sogenannte Nicht-Euklidische Geometrie" (*Mathematische Annalen* 4 [1873]: 112–45) Felix Klein argues that with non-Euclidean geometry we need not alter our notions of straight lines and planes if we extend our notion of measurement.

25. John Stuart Mill *A System of Logic,* 8th ed. (London: Longmans, Green & Co., 1872), Bk. III, ch. ii, sect. 3.

26. In a February 5, 1911 letter to Lawrence Lowell Peirce phrases it thus: "So far, we must admit that Kant's notion of the limits of human understanding has a truth in it. It is not the absolute universe that man has a power of knowing, but only the special universe of human science. Within that universe what seems to man reasonable has a certain likelihood of being true for that universe" (RL 256).

27. Shakespeare, *The Tempest* Act 4, Scene 1, 156–57.

28. See "How to Make Our Ideas Clear," sect. IV. Earlier confessions of to a Scotistic or scholastic realism are found at the conclusion of his 1868 "Some Consequences of Four Incapacities" (W2:238–42) and in his 1871 review of the Fraser edition of Berkeley's *Works* (W2:468–69).

29. What follows is a short account of Peirce's semiotics, or his "doctrine of signs." For an accessible introduction to this doctrine, see James Liszka, *A General Introduction to the Semeiotic of Charles Sanders Peirce* (Bloomington: Indiana University Press, 1996). A far briefer and more rudimentary account can be found in Cornelis de Waal, *Peirce: A Guide for the Perplexed* (London: Bloomsbury, 2013), Chapter 5.

30. Over the years Peirce gives various definitions for sign, all invariably triadic. To give one example, in a 1909 letter to William James, Peirce writes: "A Sign is a Cognizable that, on the one hand, is so determined (i.e., specialized, *bestimmt*) by something *other than itself,* called its Object . . . while, on the other hand, it so determines some actual or potential Mind, the determination whereof I term the Interpretant created by the Sign, that that Interpreting Mind is therein determined mediately by the Object" (EP 2:493).

31. For other discussions of the tenfold division see CP 2.254–63 and Peirce's correspondence with Victoria Lady Welby (esp. EP2:483f). The three division allow for only ten classes of signs as some combinations are not possible. For instance, a qualisign can only be iconic, as it cannot point at anything beyond itself; it can only display some quality of the sign. The following classification is found in CP 2.254–63:

Peirce's Tenfold Classification of Signs				
	The sign's phenomenological quality	How that sign could have been determined by its object	How the result can determine an interpretant	Example (given by Peirce)
1	Qualisign	As an icon	As a rheme	A feeling of "red"
2	Sinsign	As an icon	As a rheme	An individual diagram
3		An an index	As a rheme	A spontaneous cry
4			As a dicent	A weathercock
5	Legisign	As an icon	As a rheme	A diagram, apart from its factual individuality
6		As an index	As a rheme	A demonstrative pronoun
7			As a dicent	A street cry (identifying the individual by tone, theme)
8		As a symbol	As a rheme	A common noun
9			As a dicent	A proposition (in the conventional sense)
10			As an argument	A syllogism

32. Compare with EP2:484.

33. As Peirce puts it two years earlier in a letter to Victoria Lady Welby: "It is usual and proper to distinguish two Objects of a Sign, the Mediate without, and the Immediate within the Sign. . . . The Mediate Object is the Object outside of the Sign . . . The Sign must indicate it by a hint; and this hint, or its substance, is the *Immediate* Object" (SS:83, 1908).

34. Peirce also calls this the dynamical object, preferring the latter term because there are situations where this so-called real object "is altogether fictive" (EP2:498, 1909). It is the object "which, from the nature of things, the Sign *cannot* express, which it can only *indicate* and leave the interpreter to find out by *collateral experience*" (ibid).

35. The distinction drawn is between the immediate, the dynamical, and the final interpretant.

36. Peirce most often calls this the immediate interpretant; other designations are emotional interpretant and impressional interpretant.

37. In addition to his ten-fold classification, which is a product of three trichotomies, Peirce also devises a table of twenty-eight classes of signs, based on six trichotomies, and a table of sixty-six classes of signs, based on no less than ten trichotomies (EP2:481). The reduction to sixty-six classes of signs is considerable, as ten trichotomies can be combined in no less than 59,049 ways (id). See also Irwin Lieb's account of Peirce's classification of signs (SS:160–66).

Index